S0-BHV-837

# LIGHT CONSTRUCTION TECHNIQUES

# JOHN E. BALL
**Northeast Louisiana University**

# LIGHT

# CONSTRUCTION

# TECHNIQUES

## From Foundation To Finish

RESTON PUBLISHING COMPANY

*A Prentice-Hall Company*

Reston, Virginia 22090

*Library of Congress Cataloging in Publication Data*

BALL, JOHN E

    Light construction techniques.

    Includes index.
    1. Building.  I. Title.
TH146.B27       690       78–24429
ISBN 0–8359–4035–7

© 1980 by Reston Publishing Company, Inc.
*A Prentice-Hall Company*
Reston, Virginia 22090

All rights reserved.
No part of this book may be reproduced
in any way, or by any means,
without permission in writing from the publisher.

10  9  8  7  6  5

PRINTED IN THE UNITED STATES OF AMERICA

# Contents

# Preface

This text presents the primary practices of light construction. Those practices that are common to carpentry are covered as well as practices of such other crafts as masonry, flooring, ceiling systems, roofing, ceramic tile, and plastering.

Many illustrations are included to help the reader to understand specific methods. These illustrations are not meant to re-place the printed word but to augment and clarify each topic.

**LIGHT CONSTRUCTION TECHNIQUES** will provide a basic construction education for students in high school, vocational school, apprentice training, and adult-education classes. It should also prove invaluable to students of architecture, to builders, and to "do-it-yourselfers."

## ACKNOWLEDGMENTS

The author wishes to thank the following for contributing illustrations:

American Olean Tile Co.
Division of National Gypsum Co.
Lansdale, Pa. 19446

American Plywood Assoc.
1119 A. St.
Tacoma, Wash. 98401

Andersen Corporation
Bayport, Minn. 55003

Architectural Woodwork Institute
Chesterfield House   Suite "A"
505 S. Chesterfield Rd.
Arlington, Va. 22209

Armstrong
Lancaster, Pa. 17604

Asphalt Roofing Mfg. Assoc.
757 Third Ave.
New York, N.Y. 10017

Automated Building Components
7525 N.W. 37th Ave.
Box 2037
AMF Miami, Fla. 33149

Bird & Sons
East Walpole, Mass. 02032

Boise Cascade
P.O. Box 2885
Portland, Ore. 97208

Brick Institute of America
1750 Old Meadow Rd.
McLean, Va. 22101

Cast Iron Soil Pipe Institute
2020 K St. N.W.
Washington, D.C. 20006

Caterpillar
Peoria, Ill. 61602

Copper Development Assoc., Inc.
405 Lexington Ave.
New York, N.Y. 10017

Conwed Corp.
St. Paul, Minn. 55101

Council of Forest Industries of British
Columbia
1500/1055 W. Hastings St.
Vancouver, B.C., Canada V6E 2H1

Crown Aluminum
P.O. Box 61
Route 501 South
Roxboro, N.C. 27573

Eljer
Wallace Murray Corp.
Three Gateway Center
Pittsburgh, Pa. 15222

Fir & Hemlock Door Assoc.
700 Yeon Building
Portland, Ore. 97204

Frantz Manufacturing Co.
301 W. Third St.
Sterling, Ill. 61081

GAF
140 W. 51st St.
New York, N.Y. 10020

Georgia Pacific
900 S.W. Fifth Ave.
Portland, Ore. 97204

Gypsum Assoc.
1603 Orrington Ave.
Evanston, Ill. 60201

Hammond Valve Corp.
1844 Summer St.
Hammond, Inc. 46320

Hercules (Polymere Department)
Home Furnishing Div.
Wilmington, Del. 19889

H.C. Products
P.O. Box 68
Princeville, Ill. 61559

House of Plans
2216 Justice
Monroe, La. 71201

Johns-Manville
22 East 40th St.
New York, N.Y. 10016

Kaiser Aluminum
Kaiser Center
300 Lakeside Dr.
Oakland, Calif. 94604

Keuffel & Esser Co.
20 Whippang Rd.
Morristown, N.J. 07960

Kinkead Industries
(Subsidiary of U.S. Gypsum)
Chicago, Ill. 60646

Knape & Vogt
2700 Oak Industrial Dr. N.E.
Grand Rapids, Mich. 49505

L. E. Johnson Products, Inc.
2100 Sterling Ave.
P.O. Box 1126
Elkhart, Ind. 46514

Lowes Companies, Inc.
Box 1711
North Wilkesboro, N.C. 28656

Masonite Corp.
29 North Wacker Dr.
Chicago, Ill. 60606

Monroe Publication
1868 Shore Dr.
St. Petersburg, Fla. 33707

*Minimum Property Standards[1]
U.S. Department of Housing and
Urban Development
Washington, D.C. 20402

National Forest Products Assoc.
Forest Industries Building
1619 Massachusetts Ave., N.W.
Washington, D.C. 20036

National Oak Flooring Mfg. Assoc.
804 Sterick Building
Memphis, Tenn. 38103

National Particleboard Assoc.
2306 Perkins Pl.
Silver Spring, Md. 20910

Owens-Corning Fiberglass
Home Building Products Div.
National Bank Building
Toledo, Ohio 43601

Portland Cement Assoc.
Old Orchard Rd.
Skokie, Ill. 60076

Red Cedar Shingle & Handsplit
Shake Bureau
5510 White Building
Seattle, Wash. 98101

Reynolds Aluminum
Richmond, Va. 23201

Sargent
New Haven, Conn. 06510

Shakertown
Box 400
Winlock, Wash. 98596

Stanley
New Britain, Conn. 06050

Teco
5530 Wisconsin Ave.
Washington, D.C. 20015

Tile Council of America, Inc.
P.O. Box 326
Princeton, N.J. 08540

*U.S. Army Engineer School

*U.S. Dept. of Agriculture
(Farmers Bulletin No. 2213)

U.S. Forest Products Lab
Madison, Wis. 53707

U.S. Gypsum
101 S. Wacker Dr.
Chicago, Ill. 60606

U.S. Plywood (Champion International)
777 Third Ave.
New York, N.Y. 10017

U.S. Steel
600 Grant St.
Pittsburg, Pa. 15218

*Voluntary Product Standard P. 532-70
U.S. Government Printing Office

Warner & Swasey Co.
31700 Salon Rd.
P.O. Box 39127
Salon, Ohio 44139

Western Wood Products Assoc.
Yeon Building
Portland, Ore. 97204

Western Wood Moulding & Millwork
P.O. Box 25278
Portland, Ore. 97225

W. R. Grace & Co.
62 Whittemore Ave.
Cambridge, Mass. 02140

[1]All starred entries are government operated organizations.

# CHAPTER 1

# PRELIMINARY CONSTRUCTION PROCEDURES

## CONSTRUCTION PLANS

The first step in the construction process is to acquire a complete set of plans. From the plans a general contractor can estimate the costs, arrange for financing, and begin construction. There are several sources for a set of plans, but in most cases they are provided by the architect or designer. A standard set of plans usually includes floor plans, foundation plan, details, elevations, and plot plan.

### Floor Plan

The floor plan is an extension of a sketch with dimensions and details added. (See Figure 1–1.) Generally, the sketch is worked out between the architect or designer and the client, and reflects the ideas of both parties. The floor plan consists of individual parts, but each part plays a significant role in its development. Doors are added to provide entrances, and windows provide a source of ventilation and light. Individual rooms are planned, to include the kitchen, bathroom, bedrooms, dining room, living room, and utility room. Electrical outlets and fixtures are also indicated on the plan.

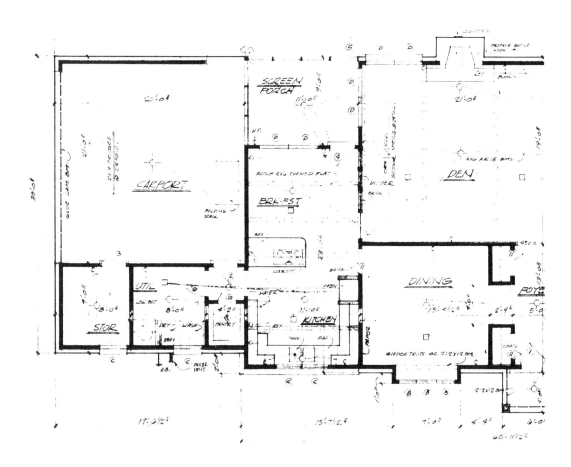

## Foundation Plan

The foundation plan is used in the construction of the foundation. (See Figure 1–2.) It includes: the overall size and shape; the location of footings, beams, and pilasters, and all pertinent dimensions and notes. It also gives the scale to which the foundation plan was drawn. Footing details accompany the plan and show the composition of the individual footing and its relationship to the structure. The footing is an enlarged portion at the lower end of the foundation wall and is used to distribute imposed loads.

FIGURE 1–1.   Floor plan

| NO. | SIZE | DESCRIPTION |
|---|---|---|
| | DOOR & WINDOW SCHEDULE | |
| 1 | 3068 X 13/4" | RAISED PANEL W/ SMALL WINDOW (SEE ELEV.) |
| 2 | 2868 X 13/4" | H.C. FLUSH W/ HALF GLS. TOP |
| 3 | 2868 X 13/8" | H.C. FLUSH EXT. DOOR |
| 4 | 2868 X 13/4" | 9 LT. TOP PANEL BOTTOM |
| 5 | 3068 X 11/8" | WOOD SCREEN DOOR |
| 6 | 2868 X 13/8" | H.C. FLUSH INT. DOOR |
| 7 | 2868 X 13/8" | GRAISED PANEL |
| 8 | 2668 X 13/8" | H.C. FLUSH INT. DOOR |
| 9 | 2068 X 13/8" | " " " " |
| 10 | 2068 X 11/8" | LOUVERED FOLDING DOOR |
| 11 | 2868 X 11/8" | " " " " |
| 12 | 4068 X 11/8" | LOUVERED BI-FOLD |
| 13 | 5068 X 11/8" | " " |
| A | 2860 | 9/6 LT. S/H ALUM. |
| B | 2060 | 1/1 LT. S/H ALUM. W/ DIA. LT. SCREENS |
| C | 2834 | 6/6 LT. S/H ALUM. |
| D | 3044 | 6/6 LT. S/H ALUM. |
| E | 2030 | 4/4 LT. S/H ALUM. OBS. GLASS |

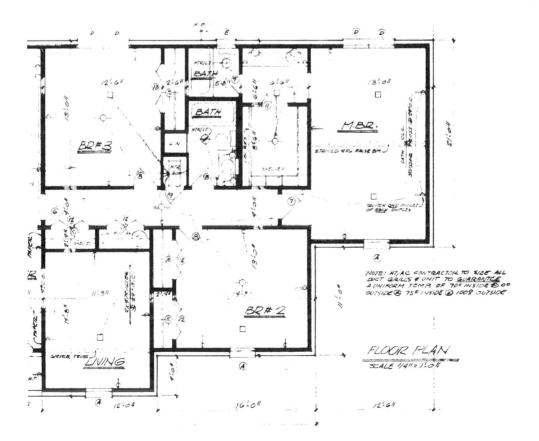

NOTE: HT/AC CONTRACTOR TO SIZE ALL DUCT GRILLS & UNIT TO GUARANTEE A UNIFORM TEMP. OF 70° INSIDE @ 0° OUTSIDE @ 75° INSIDE @ 100° OUTSIDE

FLOOR PLAN
SCALE 1/4" = 1'-0"

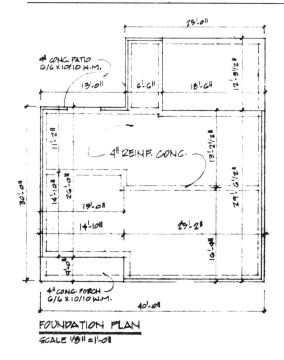

FOUNDATION PLAN
SCALE 1/8"=1'-0"

## Detail

A detail is a particular feature of a building, such as a fireplace. (See Figure 1–3.) The detail is usually drawn to a scale larger than $\frac{1}{4}'' = 1'0''$, fully dimensioned, and it contains the correct symbols. Some of the other features usually detailed are windows, doors, cornices, stairs, and kitchen cabinets.

## Elevation

An elevation is a modified orthographic drawing (a drawing in correct projection) that shows one side of an object. (See Figure 1–4.) Most buildings require four elevations, one of each side. Interior elevations

FIGURES 1–2. (left) Foundation plan, and 1–3 (below) Typical detail drawing

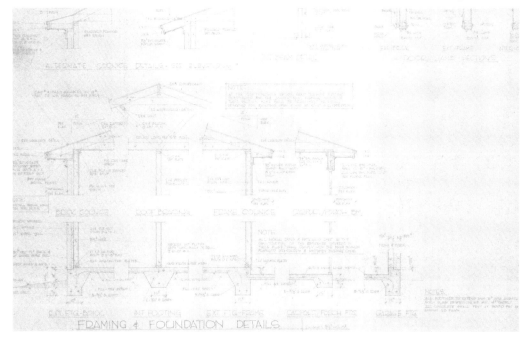

FIGURE 1-4. Elevations

may also be shown, but in most cases an interior elevation is considered a detail.

All surface materials on the elevations are indicated by symbols or notes. Grade lines, floor and ceiling levels, and some of the principal dimensions should also be included.

Some of the features identified on an elevation are windows, doors, roofs, and surface materials.

## Framing Plan

A framing plan describes the shape of the building, shows the location of the individual framing members, and indicates how the members are joined. (See Figure 1-5.)

There are several types of framing plans, but the two most common are for the roof and floor. Although these are not always included in the construction, they should be. The framing plan can be used as a reference if there is any question about how the roof or floor is framed.

FIGURE 1-5. Floor framing plan

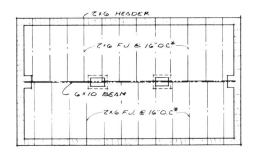

## Plot Plan

A plot plan is a top view of the lot and building. (See Figure 1–6.) A correctly drawn plot plan should locate the building on the lot and show any natural features that exist. It should also indicate the lot size and shape, showing necessary elevations and contours. Roads, setbacks, and the dimensions of front, rear, and side yards are shown on this plan. In addition, the locations of walks, driveways, steps, patios, and porches are indicated. The elevation of the first floor and the finish grade at each primary corner are also indicated, as are any trees that need to be removed, and

the location and identification of utility service lines. The finish grade is the final grade required by the specifications. It is needed so that the earth can be properly placed around the building.

## Plumbing, Heating, and Air Conditioning Plans

The plumbing plan indicates the location of all water lines, plumbing fixtures, drains, and soil and waste pipes. (See Figure 1–7.) The plumbing plan also has isometric details, which precisely describe the important features.

The heating and air conditioning plan is basically a floor plan for piping, heating ducts, furnaces, and other climate control equipment. This plan shows the locations and sizes of the ducts and the locations of thermostats, plenum, and registers.

## Architectural Symbols

In architectural drawings, symbols are used to indicate certain features. The American Standards Association (ASA) and the United States Department of Defense have standardized certain symbols:

FIGURE 1–6.  Plot plan

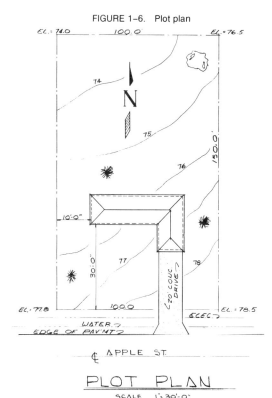

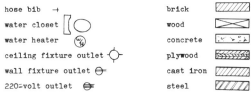

## Dimensions and Notes

A complete drawing must have both dimensions and notes. Dimensions give specific sizes, and notes give related informa-

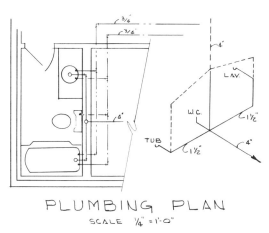

PLUMBING PLAN
SCALE ¼" = 1'-0"

FIGURE 1–7. Plumbing plan

| DOOR & WINDOW SCHEDULE | | |
|---|---|---|
| MK | SIZE | DESCRIPTION |
| 1 | 3068 × 1¾" | H.C. FLUSH EXT. DOOR |
| 2 | 2868 × 1¾" | H.C. FLUSH EXT. HOLLYWOOD |
| 3 | 2868 × 1¾" | H.C. FLUSH EXT |
| 4 | 6068 | ALUM SLD GLS DR. |
| 5 | 2668 × 1⅜" | H.C. FLUSH INT. |
| 6 | 2068 × 1⅜" | " " " |
| 7 | 4068 | BI-FOLD |
| | | |
| A | 2844 | 6/6 LT S/H ALUM. |

FIGURE 1–8. Door and window schedule

tion. The correct dimensions are just as important as the views that depict an object. Without the proper dimensions and notes, the workers would not know what sizes and types of objects to obtain.

## Schedule

A schedule is a collection of organized notes placed in a convenient location on a set of plans. (See Figure 1–8.) The schedule is enclosed by a heavy border and is identified by a tile strip. The three types of schedules usually found in a set of plans are the window schedule, the door schedule, and the room finish schedule.

## Specifications

Specifications are a written set of instructions conveying any information that cannot be easily placed on the working drawings. (See Figure 1–9.) These instructions are a legal document, a basis for bidding, a guide for construction, and also provide technical descriptions. Specifications also indicate quality and kind of materials, workmanship, colors, and finishes.

## CONTRACTS AND BUILDING REQUIREMENTS

After the construction plans have been completed, it is usually necessary to enter into a binding agreement with a contractor. Without a contract an owner has no legal recourse if the contractor should fail to meet his obligations.

There are different types of contract. Some of the more popular ones are:

### The Single Contract

Once the plans and specifications have been completed, the owner can hire a general contractor. It is that person's responsibility to coordinate and direct the various parts of the job. The general contractor will perform certain functions with his own men,

FIGURE 1-9.
Specifications

FHA Form 2005
VA Form VB4-1852
Rev. Sept. 1964

For accurate register of carbon copies, form
may be separated along above fold. Staple
completed sheets together in original order.

Form approved.
Budget Bureau No. 63-R055.17.

☒ Proposed Construction
☐ Under Construction

# DESCRIPTION OF MATERIALS

No. _____
(To be inserted by FHA or VA)

Property address _31 MAGNOLIA_ City _MONROE_ State _CA._

Mortgagor or Sponsor _MONROE BLDG. CO._ _1918 APPLE ST._
(Name) (Address)

Contractor or Builder _PEACH CONST. CO._ _2608 N. 18 ST_
(Name) (Address)

## INSTRUCTIONS

1. For additional information on how this form is to be submitted, number of copies, etc., see the instructions applicable to the FHA Application for Mortgage Insurance or VA Request for Determination of Reasonable Value, as the case may be.
2. Describe all materials and equipment to be used, whether or not shown on the drawings, by marking an X in each appropriate check-box and entering the information called for in each space. If space is inadequate, enter "See misc." and describe under item 27 or on an attached sheet.
3. Work not specifically described or shown will not be considered unless

required, then the minimum acceptable will be assumed. Work exceeding minimum requirements cannot be considered unless specifically described.
4. Include no alternates, "or equal" phrases, or contradictory items. (Consideration of a request for acceptance of substitute materials or equipment is not thereby precluded.)
5. Include signatures required at the end of this form.
6. The construction shall be completed in compliance with the related drawings and specifications, as amended during processing. The specifications include this Description of Materials and the applicable Minimum Construction Requirements.

## 1. EXCAVATION:
Bearing soil, type _CLAY LOAM (Footing shall extend 6" into undisturbed soil)_

## 2. FOUNDATIONS:
Footings: concrete mix _5 SACK_ ; strength psi _2500 PSI_ Reinforcing _3 5/8 Ø RODS_
Foundation wall: material _____ Reinforcing _____
Interior foundation wall: material _____ Party foundation wall _____
Columns: material and sizes _____ Piers: material and reinforcing _____
Girders: material and sizes _____ Sills: material _____
Basement entrance areaway _____ Window areaways _____
Waterproofing _____ Footing drains _____
Termite protection _____
Basementless space: ground cover _____ ; insulation _____ ; foundation vents _____
Special foundations _____
Additional information: _____

## 3. CHIMNEYS:
Material _____ Prefabricated (make and size) _____
Flue lining: material _____ Heater flue size _____ Fireplace flue size _____
Vents (material and size): gas or oil heater _5" METALBESTOS_ ; water heater _3" METALBESTOS_
Additional information: _____

## 4. FIREPLACES:
Type: ☐ solid fuel; ☐ gas-burning; ☐ circulator (make and size) _____ Ash dump and clean-out _____
Fireplace: facing _____ ; lining _____ ; hearth _____ ; mantel _____
Additional information: _____

## 5. EXTERIOR WALLS:
Wood frame: wood grade, and species _#2 Cedar_ ☒ Corner bracing. Building paper or felt _#15 FELT_
Sheathing _Celotex_ ; thickness _½"_ ; width _48"_ ; ☒ solid; ☐ spaced ____ " o. c.; ☐ diagonal; _____
Siding _____ ; grade _____ ; type _____ ; size _____ ; exposure ____ "; fastening _____
Shingles _____ ; grade _____ ; type _____ ; size _____ ; exposure ____ "; fastening _____
Stucco _____ ; thickness ____ "; Lath _____ ; weight ____ lb.
Masonry veneer _BRICK @ 80 PER M_ Sills _BRICK_ Lintels _None_
Masonry: ☐ solid ☐ faced ☐ stuccoed; total wall thickness ____ "; facing thickness ____ "; facing material _____
Backup material _____ ; thickness ____ "; bonding _____
Door sills _Conc_ Window sills _____ Lintels _____
Interior surfaces: dampproofing, ____ coats of _____ ; furring _____
Additional information: _____
Exterior painting: material _Exterior Latex_ ; number of coats _____
Gable wall construction: ☐ same as main walls; ☒ other construction _Hardboard Over 2x4 Framing_

## 6. FLOOR FRAMING:
Joists: wood, grade, and species _____ ; other _____ ; bridging _____ ; anchors _____
Concrete slab: ☐ basement floor; ☒ first floor; ☒ ground supported; ☐ self-supporting; mix _2500 PSI 5 SACK_ thickness _4_ "
reinforcing _____ ; insulation _____ ; membrane _____
Fill under slab: material _4" GRAVEL; 8" EARTH_ thickness _12_ ". Additional information: _FILL NOT TO EXCEED 24" IN DEPTH_

## 7. SUBFLOORING: (Describe underflooring for special floors under item 21.)
Material: grade and species _____ ; size _____ ; type _____
Laid: ☐ first floor; ☐ second floor; ☐ attic _____ sq. ft.; ☐ diagonal; ☐ right angles. Additional information: _____

## 8. FINISH FLOORING: (Wood only. Describe other finish flooring under item 21.)

| LOCATION | ROOMS | GRADE | SPECIES | THICKNESS | WIDTH | BLDG. PAPER | FINISH |
|---|---|---|---|---|---|---|---|
| First floor | | | | | | | |
| Second floor | | | | | | | |
| Attic floor | sq. ft. | | | | | | |
| Additional information: | | | | | | | |

FHA Form 2005
VA Form VB4-1852

8

DESCRIPTION OF MATERIALS

**9. PARTITION FRAMING:**
Studs: wood, grade, and species _#2 Cedar_ size and spacing _2x4 @ 16" O.C._ Other _____
Additional information: _____

**10. CEILING FRAMING:**
Joists: wood, grade, and species _#2 Y PINE_ Other _____ Bridging _____
Additional information: _1X4 STRIPPING @ 16" O.C_

**11. ROOF FRAMING:**
Rafters: wood, grade, and species _#2 Y PINE_ Roof trusses (see detail): grade and species _____
Additional information: _____

**12. ROOFING:**
Sheathing: wood, grade, and species _½" PLYWOOD_
Roofing _Asphalt Compo Shingles_, grade _C_ ; size _____ ; type _____ ; ☐ solid; ☐ spaced _____ " o.c.
Underlay _____ ; weight or thickness _240_ ; size _12 x 36_ ; fastening _G.L.W. Nails_
Built-up roofing _____ ; number of plies _____ ; surfacing material _____
Flashing: material _G.I._ ; gage or weight _26 Ga._ ; ☐ gravel stops; ☐ snow guards
Additional information: _____

**13. GUTTERS AND DOWNSPOUTS:**
Gutters: material _____ ; gage or weight _____ ; size _____ ; shape _____
Downspouts: material _____ ; gage or weight _____ ; size _____ ; shape _____ ; number _____
Downspouts connected to: ☐ Storm sewer; ☐ sanitary sewer; ☐ dry-well. ☐ Splash blocks: material and size _____
Additional information: _____

**14. LATH AND PLASTER**
Lath ☐ walls, ☐ ceilings: material _____ ; weight or thickness _____ Plaster: coats _____ ; finish _____
Dry-wall ☒ walls, ☒ ceilings: material _Gypsum Board_ ; thickness _3/8_ ; finish _Textured_
Joint treatment _PaoFa tape; Float, Sand & Size_

**15. DECORATING:** *(Paint, wallpaper, etc.)*

| Rooms | Wall Finish Material and Application | Ceiling Finish Material and Application |
|---|---|---|
| Kitchen | ¼' Ash Plywood Panelling | Textone; 2 Coats Latex |
| Bath | Textone, 2 Coats Latex | do |
| Other | ¼" Ash Plywood Panelling | do |

Additional information: _____

**16. INTERIOR DOORS AND TRIM:**
Doors: type _Flush, H.C. Interior_ ; material _Mahogany_ ; thickness _1⅜"_
Door trim: type _Sanitary_ ; material _Fir_ Base: type _Sanitary_ ; material _Fir_ ; size _3 ¼"_
Finish: doors _Enamel_ ; trim _Enamel_
Other trim *(item, type and location)* _____
Additional information: _____

**17. WINDOWS:**
Windows: type _Single Hung_ ; make _Bull_ ; material _Aluminum_ ; sash thickness _____
Glass: grade _SSB_ ; ☐ sash weights; ☒ balances, type _Spiral_ ; head flashing _____
Trim: type _Sanitary_ ; material _Fir_ Paint _Enamel_ ; number coats _____
Weatherstripping: type _Fiber_ ; material _Felt_
Screens: ☒ full; ☐ half; type _Exterior_ ; number _All_ ; screen cloth material _Aluminum_
Basement windows: type _____ ; material _____ ; ☐ screens, number _____ ; ☐ Storm sash, number _____
Special windows _____
Additional information: _____

**18. ENTRANCES AND EXTERIOR DETAIL:**
Main entrance door: material _Mahogany_ ; width _36"_ ; thickness _1¾"_ Frame: material _Fir_ ; thickness _5/4_
Other entrance doors: material _Mahogany_ ; width _32"_ ; thickness _1¾"_ Frame: material _Fir_ ; thickness _5/4"_
Head flashing _____ Weatherstripping: type _Neoprene_ ; saddles _____
Screen doors: thickness _1⅛"_; number _1_ ; screen cloth material _Alum_ Storm doors: thickness _____", number _____
Combination storm and screen doors: thickness _____"; number _____ ; screen cloth material _____
Shutters ☐ hinged; ☐ fixed. Railings _____ Louvers _12' GI EACH GABLE_
Exterior millwork: grade and species _C & BETTER 4" #FIR_ Paint _EXTERIOR LATEX_ ; number coats _____
Additional information: _____

**19. CABINETS AND INTERIOR DETAIL:**
Kitchen cabinets, wall units: material _3/4 ASH PLYWOOD_ ; lineal feet of shelves _SEE PLANS_ ; shelf width _12"_
Base units: material _3/4 ASH PLYWOOD_ ; counter top _PLASTIC LAM._ ; edging _PLASTIC LAMINATE_
Back and end splash _____ Finish of cabinets _VARNISH_ ; number coats _5_
Medicine cabinets: make _____ ; model _____
Other cabinets and built-in furniture _SEE PLANS_
Additional information: _VANITY MIRROR 30" x 40" IN BATH_

**20. STAIRS:**

| Stair | Treads | | Risers | | Strings | | Handrail | | Balusters | |
|---|---|---|---|---|---|---|---|---|---|---|
| | Material | Thickness | Material | Thickness | Material | Size | Material | Size | Material | Size |
| Basement | | | | | | | | | | |
| Main | | | | | | | | | | |
| Attic | | | | | | | | | | |

Disappearing: make and model number _____
Additional information: _____

2

9

## 21. SPECIAL FLOORS AND WAINSCOT:

| | LOCATION | MATERIAL, COLOR, BORDER, SIZES, GAGE, ETC. | THRESHOLD MATERIAL | WALL BASE MATERIAL | UNDERFLOOR MATERIAL |
|---|---|---|---|---|---|
| FLOORS | Kitchen | 1/8" Vinyl Asbestos Tile | Tile | Wood | Conc |
| | Bath | do | do | do | do |
| | | do | do | do | do |

| | LOCATION | MATERIAL, COLOR, BORDER, CAP. SIZES, GAGE, ETC. | HEIGHT | HEIGHT OVER TUB | HEIGHT IN SHOWERS (FROM FLOOR) |
|---|---|---|---|---|---|
| WAINSCOT | Bath | Ceramic Tile 72" Around Tub | | 72" | |

Bathroom accessories: ☒ Recessed; material _Chrome_ ; number _2_ ; ☒ Attached; material _Chrome_ ; number _2_
Additional information: _Soap & Grab, Paper Holder, Towel Bars_

## 22. PLUMBING:

| FIXTURE | NUMBER | LOCATION | MAKE | MFR'S FIXTURE IDENTIFICATION NO. | SIZE | COLOR |
|---|---|---|---|---|---|---|
| Sink | 1 | Kitchen | Spats | 26890 | 32x21 | Stainless |
| Lavatory | 1 | Bath | '' | 32680 | 18" Ø | White |
| Water closet | 1 | '' | '' | 39280 | | '' |
| Bathtub | 1 | '' | '' | 63870 | 5'0" x 16" | |
| Shower over tub △ | 1 | '' | Chrome Fittings | | | |
| Stall shower △ | | | | | | |
| Laundry trays | | | | | | |
| | | | | | | |

△ ☒ Curtain rod   △ ☐ Door   ☐ Shower pan: material _____
Water supply: ☒ public; ☐ community system; ☐ individual (private) system. ★
Sewage disposal: ☒ public; ☐ community system; ☐ individual (private) system. ★
★ Show and describe individual system in complete detail in separate drawings and specifications according to requirements.
House drain (inside): ☒ cast iron; ☐ tile; ☐ other _____ House sewer (outside): ☐ cast iron; ☒ tile; ☐ other _____
Water piping: ☐ galvanized steel; ☐ copper tubing; ☐ other _____ Sill cocks, number _____
Domestic water heater: type _Gas_ ; make and model _Spats 30_ ; heating capacity _22.5_
_____ gph. 100° rise. Storage tank: material _G.I._ ; capacity _50_ gallons.
Gas service: ☒ utility company; ☐ liq. pet. gas; ☐ other _____ Gas piping: ☒ cooking; ☒ house heating.
Footing drains connected to: ☐ storm sewer; ☐ sanitary sewer; ☐ dry well. Sump pump; make and model _____
_____ ; capacity _____ ; discharges into _____

## 23. HEATING:
☐ Hot water. ☐ Steam. ☐ Vapor. ☐ One-pipe system. ☐ Two-pipe system.
☐ Radiators. ☐ Convectors. ☐ Baseboard radiation. Make and model _____
Radiant panel: ☐ floor; ☐ wall; ☐ ceiling. Panel coil: material _____
☐ Circulator. ☐ Return pump. Make and model _____ ; capacity _____ gpm.
Boiler: make and model _____ Output _____ Btuh.; net rating _____ Btuh.
Additional information: _____
Warm air: ☐ Gravity. ☒ Forced. Type of system _Closet Installed Central Heating Unit_
Duct material: supply _G.I._ ; return _____ Insulation _Fiberglass_ thickness _6'_ ☐ Outside air intake.
Furnace: make and model _MONROE 1686-8_ Input _80,000_ Btuh.; output _64,000_ Btuh.
Additional information: _____
☐ Space heater; ☐ floor furnace; ☐ wall heater. Input _____ Btuh.; output _____ Btuh.; number units _____
Make, model _____ Additional information: _____
Controls: make and types _MONROE MODEL 1620-AB_
Additional information: _____
Fuel: ☐ Coal; ☐ oil; ☒ gas; ☐ liq. pet. gas; ☐ electric; ☐ other _____ ; storage capacity _____
Additional information: _____
Firing equipment furnished separately: ☐ Gas burner, conversion type. ☐ Stoker: hopper feed ☐; bin feed ☐
Oil burner: ☐ pressure atomizing; ☐ vaporizing _____
Make and model _____ Control _____
Additional information: _____
Electric heating system: type _____ Input _____ watts; @ _____ volts; output _____ Btuh.
Additional information: _____
Ventilating equipment: attic fan, make and model _GILLEY 36"_ ; capacity _____ cfm.
kitchen exhaust fan, make and model _KITCHEN SPECIAL HD 36-A_
Other heating, ventilating, or cooling equipment _____

## 24. ELECTRIC WIRING:
Service: ☒ overhead; ☐ underground. Panel: ☐ fuse box; ☒ circuit-breaker; make _Square D_ Number circuits _10_
Wiring: ☐ conduit; ☐ armored cable; ☒ nonmetallic cable; ☐ knob and tube; ☐ other _____
Special outlets: ☐ range; ☐ water heater; ☐ other _____
☐ Doorbell. ☒ Chimes. Push-button locations _EXTERIOR DOOR_ Additional information: _____

## 25. LIGHTING FIXTURES:
Total number of fixtures _SEE PLAN_ Total allowance for fixtures, typical installation, $ _125.00_
Nontypical installation _____
Additional information: _____

3

DESCRIPTION OF MATERIALS

10

## 26. INSULATION:

| Location | Thickness | Material, Type, and Method of Installation | Vapor Barrier |
|---|---|---|---|
| Roof | | | |
| Ceiling | 4" | BLOWN MINERAL WOOL | None |
| Wall | 3½" | BATT TYPE STAPLED TO STUDS | Yes |
| Floor | | | |

HARDWARE: (make, material, and finish.) _Ruston - Polished Brass & Chrome_

SPECIAL EQUIPMENT: (state material or make and model.)

| | |
|---|---|
| Venetian blinds _____ Number _____ | Automatic washer _Plumbing Only_ |
| Kitchen range _____ | Clothes drier _____ |
| Refrigerator _____ | Other _____ |
| Dishwasher _____ | |
| Garbage disposal unit _____ | |

## 27. MISCELLANEOUS: (Describe any main dwelling materials, equipment, or construction items not shown elsewhere; or use to provide additional information where the space provided was inadequate. Always reference by item number to correspond to numbering used on this form.)

PORCHES: _SEE GARAGES_

TERRACES:

GARAGES: _INTEGRALLY POURED REINF CONC FOOTING AROUND PERIMETER OF 4" REINF. CONC SLAB WITH MONOLITHIC FINISH. ROOF CONST AND ROOFING TO BE SAME AS HOUSE._

### WALKS AND DRIVEWAYS:

Driveway: width _____ ; base material _____ ; thickness _____ "; surfacing material _____ ; thickness _____ "
Front walk: width _____ ; material _____ ; thickness _____ ".  Service walk: width _____ ; material _____ ; thickness _____ "
Steps: material _____ ; treads _____ "; risers _____ ".  Cheek walls _____

### OTHER ONSITE IMPROVEMENTS:

(Specify all exterior onsite improvements not described elsewhere, including items such as unusual grading, drainage structures, retaining walls, fence, railings, and accessory structures.)

_Grade Yard To Provide Adequate Drainage_

### LANDSCAPING, PLANTING, AND FINISH GRADING:

Topsoil _____ " thick: ☐ front yard; ☐ side yards; ☐ rear yard to _____ feet behind main building.
Lawns (seeded, sodded, or sprigged): ☐ front yard _____ ; ☐ side yards _____ ; ☐ rearyard _____
Planting: ☐ as specified and shown on drawings; ☐ as follows:

| | | |
|---|---|---|
| _1_ Shade trees, deciduous, _1½_ " caliper. | Evergreen trees. _____ ' to _____ ', B & B. | |
| _____ Low flowering trees, deciduous, _____ ' to _____ ' | Evergreen shrubs. _____ ' to _____ ', B & B. | |
| _____ High-growing shrubs, deciduous, _____ ' to _____ ' | Vines, 2-year _____ | |
| _____ Medium-growing shrubs, deciduous, _____ ' to _____ ' | | |
| _____ Low-growing shrubs, deciduous, _____ ' to _____ ' | | |

IDENTIFICATION.—This exhibit shall be identified by the signature of the builder, or sponsor, and/or the proposed mortgagor if the latter is known at the time of application.

Date _April 14_              Signature _J. P. Apple_

                             Signature _Arnold S. _____

FHA Form 2005
VA Form VB4–1852

4

U.S. GOVERNMENT PRINTING OFFICE  1964 OF—742-879

but the majority of the work will be carried out by specialty contractors, or subcontractors. Specialty contractors are often referred to as "subs" and perform only one function, such as flooring, plumbing, roofing, cabinet making, or bricklaying. It is the general contractor's responsibility to see that all subcontractors' work is performed satisfactorily.

## Subcontracting Separate Contracts

If the owner wishes, he can award contracts to the "subs" and supervise the construction process himself. But without a thorough knowledge of the construction industry, this is not recommended, because it means the owner must manage and coordinate specialized functions normally carried out by the general contractor. If the owner is able to supervise the "subs," however, he can save the margin of profit usually made by the general contractor.

## Competitive Bids

In competitive bidding the owner has several general contractors bid on a single project. Before a general contractor makes a competitive bid, he must estimate the cost of the project. Once bidding is opened, the owner has the right to accept any bid, whether it is the lowest or the highest. The project can then be awarded either under an all-inclusive contract or by means of separate contracts.

## Speculative Building

Speculative building refers to the construction of a project that will be offered for sale. Frequently a general contractor will develop land and build tract housing for sale to the general public. In this type of construction the contractor does not have to work with an owner. He is concerned with selling the structure once it has been built. A speculative building must be sold quickly after completion to prevent interim financing from decreasing the contractor's profit margin.

## Licensing

The licensing process is established to make certain a contractor meets minimum qualifications, which in turn gives the owner a certain amount of consumer protection. The criteria used for licensing contractors vary with state and local regulations. In many parts of the country a license is required to bid on jobs.

## Building Permit

In most cases, a building permit is required before the contractor can start work on a structure. To obtain a building permit the owner or contractor must file a formal application and submit drawings and specifications to the proper agency. If the plans, specifications, and application meet the necessary standards, the permit is issued.

## Building Codes

Building codes are a collection of building standards that have been created and acted on by the appropriate government agency. They are designed to protect the health, safety, and general welfare of a community.

Communities have the option of developing and writing their own code, or they can adopt a standard code. Some of the standard codes are: the National Building

Code, Uniform Building Code, Southern Standard Building Code, National Electrical Code, and National Plumbing Code.

If a structure is not built according to the adopted code, building officials have the responsibility and the right to order discontinuance of the illegal construction project.

## Restrictions and Legal Aspects of a Lot

Before a lot is purchased, a thorough legal investigation of the property should be made. The investigation should include the location of easements and zoning regulations or ordinances. An easement is the legal right for a party other than the owner to have access to a portion of the property. An example of an easement is the area under utility lines. Because of the easement the placement of a fence might be prohibited or the planned future expansion of an existing structure might be stopped. Therefore, it is imperative to know the legal ramifications and possible involvements of any easements before the lot is purchased and construction is started.

Zoning regulations also must be investigated. These regulations usually govern the size of a building, and they can also stipulate the materials and architectural style. The advantage of a zoning regulation is that it keeps structures consistent in areas of the community, such as prohibiting a factory in a residential area.

## Mortgage

A mortgage is a personal promissory note. In the event that the borrower cannot repay the loan, the mortgage gives the lender the right to have the property sold. Three types of mortgage—conventional, FHA, VA —are available to home buyers who qualify.

A *conventional loan* can be made by commercial banks, savings banks, savings and loan associations, life insurance companies, or individuals. Because the government is not directly involved, the lender can charge any interest rate within the legal limit. If a buyer defaults, the lender usually tries to regain his money by foreclosing the mortgage on the house.

*FHA-insured home loans* are made by private lending institutions and are guaranteed by the Federal Housing Administration. If the buyer defaults, the lender is protected by FHA insurance. The borrower is charged the rate of one-half of one percent a year on the average scheduled mortgage-loan balance. In addition to the down payment, the borrower pays the settlement cost and other initial charges, including the closing cost.

*VA-insured home loans* are made to eligible veterans by private lending institutions and are guaranteed by the Veterans' Administration. A VA loan differs from an FHA loan in that the VA guarantees only part of the loan, but the lending institution is unlikely to experience a loss, because it can foreclose on the mortgage and resell the property.

## Sales Contract

The sales contract obligates the seller to sell and the buyer to buy, and it is usually the first document to be signed. In most cases a cash deposit is given as good faith and binds the contract. If for any reason the buyer changes his mind about the purchase and decides not to buy, he usually loses his deposit. The contract should be clearly stated in terms that the average buyer can understand, and carefully read before it is signed.

## Evidence of Good Title

Once the sales contract has been signed, it is necessary to have a title search made. Evidence of good title assures the applicant that he actually owns or will own the property and that it is free of prior claims. A title certificate or title opinion is usually prepared from public records.

# CHAPTER 2

# FOUNDATION LAYOUT

## CONSTRUCTION EQUIPMENT

Before any grading is done, property lines must be known along with easements, accesses, and encumbrances on the property. In addition, before construction can begin on a building and before the building can be laid out, the job site must be cleared and brought to proper elevation. To clear the land and make the proper cuts and fills, heavy equipment is used. The type of machinery used on a job depends on the site, location, and the size of the job. Of the many different types of machines used in light construction, some of the more common are backhoes, wheel loaders, track-type loaders, and bulldozers. These machines are used to prepare the job site, excavate for the basement, dig the footings, backfill around the foundation, and level the job site when the construction has been completed.

### Backhoes

A backhoe is used to excavate below the surface of the ground. This machine is sometimes also called a hoe, back shovel, or pull shovel. It operates by first digging and then pulling the load toward the power unit. It can be used to dig trenches, and excavate for basements and footings. The width of the bucket varies from 18 inches to 36 inches, and the capacity ranges from 4 cubic feet to 10 cubic feet. The buckets are designed to load easily and dump cleanly. If the backhoe is operated in cohesive soils, the bucket can be equipped with a cleaner bar. (See Figure 2–1.) Fitted to the inside of the bucket, the bar automatically operates when the bucket is opened for dumping.

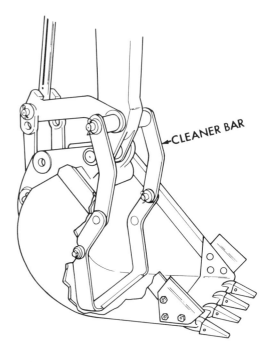

FIGURE 2–1. Backhoe bucket, cut away to show a cleaner bar

The backhoe can also be equipped with a ripper tooth for ripping frozen ground, asphalt, or other hard materials. The tooth is fitted with a reusable drive pin that can be easily removed and replaced.

### Loaders

The two basic types of loaders are the wheel loader and the track-type loader. (See Figure 2–2.) The machine's primary purpose is to load excavated soil into trucks. But it can be used to excavate loose soils, and if a multipurpose bucket is used, a loader can strip topsoil, bulldoze, clean up debris, move logs, and carry pipe.

Some loaders are equipped with auto-

FIGURE 2–2.  Track-type loader

matic bucket controls, which allow the operator to set the bucket at both a preset height and a digging angle. Then the bucket stops automatically at the desired height, empties the load, and returns the bucket to the preset digging angle.

The capacity and reach of a loader vary with the manufacture and type of loader. (Figure 2–3 shows typical dimensions for a track loader.)

## Bulldozers

A bulldozer is a tractor-driven machine that has a broad, blunt blade mounted in front of it. The conventional bulldozer has the blade mounted perpendicular to the line of travel, so it can push the dirt forward. An angledozer has its blade mounted at an angle, so that the dirt is pushed forward and to one side. Some blades can be adjusted to perform as either bulldozers or angledozers.

Bulldozers can be classified as either crawler tractors or wheel tractors. In most cases a crawler tractor is more efficient in soft and muddy soils. The undercarriage of the crawler tractor is designed with wide-track shoes and a long track form. The wheel tractor, designed for articulated

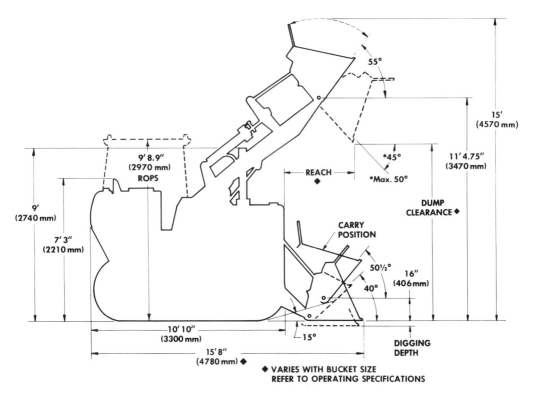

FIGURE 2-3.  Dimensions of a typical loader

steering and fast maneuvering, operates more efficiently on firm ground

The typical dimensions of a bulldozer are shown in Figure 2-4.

FIGURE 2-4.  Dimensions of a typical bulldozer

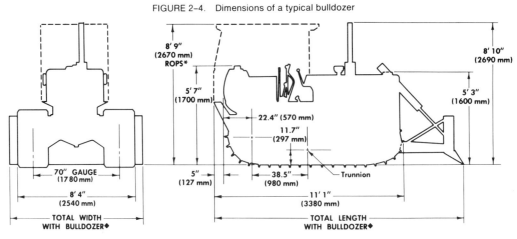

◆VARIES WITH TYPE OF BULLDOZER

## BUILDING LAYOUT AND EXCAVATION

### Staking Out

After the site has been cleared, the foundation is staked out, with the plot plan used as a guide. First the property lines are located. (See Figure 2–5.) Then, measuring from the property lines, a corner of the building location is determined and a stake (1) is placed to mark the spot. The stake is called the hub. Measuring from the hub and property line, a second corner is located and another stake (2) is placed. The other corners can be located either by use of the 6-8-10 method (following) or by use of a level or level-transit. (The use of these instruments is discussed in the second half of this chapter.)

The 6-8-10 method is based on the geometrical fact that a triangle with sides of 3, 4, and 5 or multiples of these numbers will contain a right triangle. The rule for determining the 6-8-10 method is that *the length of the hypotenuse of a right-angle triangle is equal to the square root of the sum of the squares of each side.* The formula is: $A = \sqrt{B^2 + C^2}$.

FIGURE 2–5.    Sequence of steps in staking out the foundation

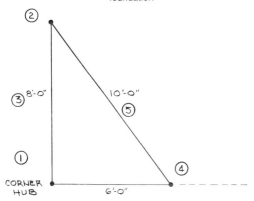

To use the 6-8-10 method, a square corner is laid out: a stake (3) is located at a point 8 feet from one hub. This point should fall on a line between the two previously placed hubs. A second stake (4) is then placed at a right angle and 6 feet from the corner stake. To determine if the building lines are at a right angle, the diagonal (5) between the two stakes is measured; the distance should be 10 feet. It will be possible to locate the third corner by measuring along the previously placed line (1 to 4) to the appropriate location. The fourth corner can be located in the same manner as corner 1.

To use a level or level-transit to stake out the foundation, it is necessary to lay out the first corner by measuring from the front and side property lines. Then the second corner is located by measuring from the first stake and the property line. The transit is then placed over the first corner and leveled up.

When the tripod is placed over the stake a plumb bob should be attached to the level or level-transit; it should fall directly over the hub. The instrument is then focused on the second stake and locked into position. Next, the horizontal circle is turned on the transit until the vernier is on zero. The telescope is then rotated 90 degrees, and the third corner will be located in line with the telescope. The correct distance is measured from the corner stake, and the leveling rod is moved back and forth until it is lined up with the telescope. Rather than moving the leveling rod to determine the point, a straight line can be run with a transit by first locking the horizontal motion clamp screw. The telescope is then rotated in a vertical plane, to locate any number of points on the same line. To locate the fourth corner, the transit must be placed over another corner.

## Batter Boards

Once the building has been staked out, batter boards are set 3 or 4 feet from the hub. These boards are 2 × 4 stakes with horizontal 1 × 4s, called ledgers, nailed to the top. (See Figure 2–6.) In most cases batter boards are L-shaped, but they can be constructed of only two staked and one horizontal board.

The ledgers *must* be properly located, for not only do they hold the strings to mark the building line, but they also indicate the height of the foundation wall, or finished elevation of a concrete slab.

To locate the ledgers, it is necessary to first determine the differences in the elevations of the hubs. (See Figure 2–7.) This is done by first setting up and leveling a builder's level or level-transit at a convenient location. A leveling rod is then placed at each corner, and a reading is taken to locate the highest corner. If the reading at hub A is 5 feet and the reading at hub B is 5½ feet, hub B is of course 6 inches higher than hub A.

Once the highest corner has been found, the height above grade of the foundation

wall or slab elevation is subtracted from the highest elevation reading. Most building codes stipulate that the top of the foundation wall must be a minimum of 8 inches above grade. The leveling rod is then placed next to the stakes and is raised until the desired elevation is reached. Next, the bottom of the rod is marked, to specify the location of the top of the ledger. This procedure is carried out on each stake. With the marks on the stakes used as guides, the ledgers are nailed to the stakes.

Once the batter boards are in place, strings are tightly stretched between them. At the intersection of the two building lines, a plumb bob can be used to determine if the two lines intersect over the hub. The lines are held in place on the ledger by a small saw kerf or nail. After all the building lines have been placed, the diagonals should be checked. If the lines form a square or rectangle, the diagonals should be of equal length.

## Excavation

In basement construction, once the batter boards have been set and the building lines have been stretched, the perimeter of the rough excavation can be set. This excavation is marked with stakes and usually extends beyond the building line approximately 2 feet. The addition of the 2 feet beyond the building line allows for space to place the foundation wall. Once the stakes have been placed, the building lines are removed and excavation is begun.

The depth of the excavation varies, but it can be determined by a study of the wall details from the architectural plans. In most cases the depth is taken from the highest elevation on the perimeter of the excavation.

FIGURE 2–6.   Batter boards

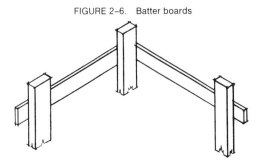

FIGURE 2–7.   Obtaining the difference in elevation between two visible points

## SURVEYING INSTRUMENTS

### Leveling Instruments

In light construction, the level or level-transit is commonly used to lay out foundations. The level is a precision instrument that is used for locating points in a horizontal plane. Most levels consist of a telescope with a focusing knob resting on a leveling base plate. The level is equipped with a spirit level and four leveling screws so that it can be properly adjusted. Some levels are equipped with a horizontal circle graduated from 0 to 360 degrees. The level is most often used to determine elevations and horizontal angles, but it can be used for grading, leveling, and aligning fences and driveways.

The level-transit is very similar to the level, except that it can pivot in a vertical plane. (See Figure 2–8.) This added feature

FIGURE 2–8.   Builder's transit level

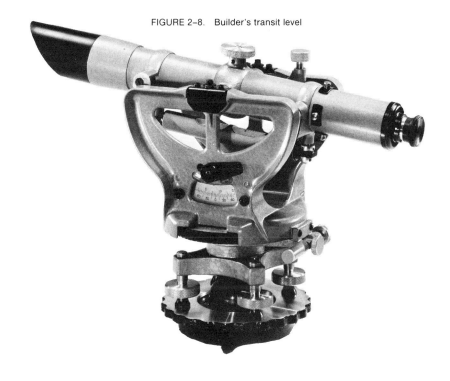

allows the level-transit to measure vertical as well as horizontal angles. The level-transit is also very useful in plumbing a vertical wall.

## Vernier Scales

If a building is squared or a horizontal angle is needed, the vernier scale is used to make sure the angle is correct. The circles and verniers are graduated in numerous ways, but four of the most common are:

- Graduated 30 minutes reading to one minute (See Figure 2–9)
- Graduated 20 minutes reading to 30 seconds (See Figure 2–10)
- Graduated 15 minutes reading to 20 seconds (See Figure 2–11)
- Graduated 20 minutes reading to 20 seconds (See Figure 2–12)

In Figure 2–9 the horizontal circle is graduated to half-degrees, from 0 to 360 degrees. Each division of the vernier represents one minute. To correctly read the measurement, the degrees are read on the horizontal circle and the minutes are read on the vernier. The minute is indicated where the mark on the vernier lines up with a mark on the horizontal circle. From left to right, Figure 2–9 reads 17 degrees, 25 minutes; from right to left, it reads 342 degrees and 35 minutes.

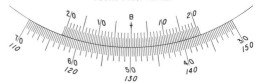

FIGURE 2–10.   Vernier scale: graduated 20 minutes reading to 30 seconds

In Figure 2–10 the horizontal circle is graduated to 20 minutes, from 0 to 360 degrees. Each division of the vernier is 30 seconds. From left to right, the vernier reads 130 degrees, 9 minutes and 30 seconds; from right to left, it reads 229 degrees, 50 minutes and 30 seconds.

In Figure 2–11 the horizontal circle is graduated to 15 minutes, from 0 to 360 degrees. Each division of the vernier is 20 seconds. From left to right, the vernier reads 8 degrees, 24 minutes, and 20 seconds; reading from right to left, it reads 351 degrees, 35 minutes and 40 seconds.

In Figure 2–12 the horizontal circle is

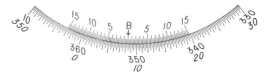

FIGURE 2–11.   Vernier scale: graduated 15 minutes reading to 20 seconds

FIGURE 2–12.   Vernier scale: graduated 20 minutes reading to 20 seconds

FIGURE 2–9.   Vernier scale: graduated 30 minutes reading to 1 minute

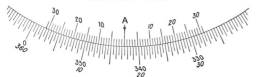

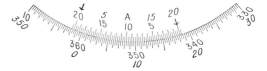

graduated to 20 minutes, from 0 to 360 degrees. Each division on the vernier is 20 seconds. Reading from left to right, the vernier reads 358 degrees, 30 minutes and 40 seconds; reading from right to left, it reads 341 degrees, 49 minutes and 20 seconds.

Both the level and level-transit are mounted on a tripod (See Figure 2-13),

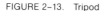

FIGURE 2-13. Tripod

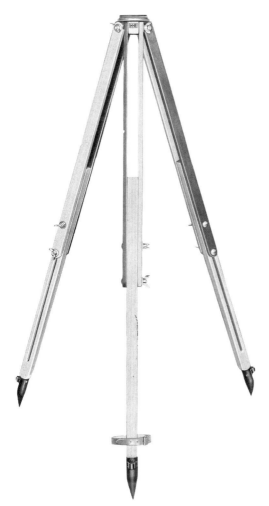

which is usually constructed of specially treated hardwoods that will not bend or warp, even under adverse field conditions. The tripod is available with either fixed or extension legs. Extension legs allow lengthening, or retracking for easier handling.

## Setting Up the Instrument

Before the instrument is removed from the case, the tripod should be placed in the proper location, with its head appearing to be level and the legs spread about 3½ feet. The points of the tripod legs should be firmly placed. If extension legs are used, they should be checked to make sure all the nuts are securely tightened.

The instrument is then carefully removed from the carrying case by means of a firm grip on the base plate. The telescope should *not* be grasped during removal. Then the instrument can be screwed firmly (but not *too* tightly) onto the tripod. If the instrument is placed over a hub, or if a plumb bob is attached to the bottom of the instrument, the plumb bob should be properly located before the instrument is leveled.

To level the instrument, the horizontal clamp screw must be loosened; then the telescope is positioned directly over two leveling screws. To level the telescope, one leveling screw should be turned in a clockwise direction while the other leveling screw is being turned counterclockwise. These leveling screws should be turned uniformly to center the bubble of the spirit level.

After the bubble is centered, the telescope must be rotated 90 degrees and positioned directly over the other two leveling screws. These two leveling screws are then adjusted until the telescope is level.

After the second leveling operation, the telescope can be rotated to any desired angle and still be level. But it is good practice to check the spirit level before each reading.

## Care of the Instrument

Since the level or level-transit is a precision piece of equipment, there are certain rules that should be followed:

- The instrument should be kept clean and dry, and placed in a carrying case when not in use.
- To keep the instrument from getting wet when in use, a plastic cover should be carried and used as needed.
- The leveling screws should be snug, but not overtight. Overtightening could warp the leveling base plate.
- The instrument should always be carefully carried in both hands, and not slung over the shoulder.

# CHAPTER 3

# FOUNDATION SYSTEMS

A FOUNDATION SYSTEM IS THE PORTION OF A building that receives the weight of the superstructure and is the base upon which it is built. Without an adequate foundation system, a building will settle, causing walls, floors, and ceilings to crack. Because of this possibility, a good foundation system should minimize settlement and create a strong and dependable working surface.

## SETTLEMENT

Unless the foundation system rests on bedrock, there is likely to be some differential settlement. If the settlement is uniform and slight, no appreciable amount of damage will occur. But if the settlement is excessive and uneven, problems will result. Settlement can be controlled by designing and proportioning the footings so that they will transmit an equal load to the supporting soil. In most cases, if a structure is placed on a dense base, such as sand or gravel, there will be very little settlement.

## SOIL CLASSIFICATIONS

Soils are usually classified by the size of the particles of which they are composed. When soils are classified in this manner they usually fall into one of the groups listed below:

- *Cobbles and boulders.* Any soil particle that is larger than 3 inches in diameter
- *Gravel.* Soil particles that are larger than ¼ inch, but smaller than 3 inches
- *Sand.* Particles that are smaller than ¼ inch, but are larger than a #200 sieve
- *Silts.* Particles smaller than 0.02 mm and larger than 0.002 mm in diameter
- *Clay.* Soil particles that have a diameter smaller than 0.002 mm

There are two basic types:
- *Spread Foundations:* These distribute the weight of the superstructure over a large area of soil by means of individual footings.
- *Pile foundations:* These direct the weight of the structure through weak soil to a bearing surface that can withstand the weight of the building.

But if the structure is placed on a cohesive soil, such as clay, settlement may be extended for a number of years. Settlement is caused by the reduction of voids between soil particles and the displacement of the soil. In cohesive solids large numbers of voids are created by the minute and scaly grains of soil.

When a structure is built over cohesive soil, its weight slowly compresses, or displaces, the soil and the structure begins to settle.

- *Loam.* A type of soil that contains an even mixture of sand, silt, and clay

Soils can further be classified as cohesionless soils, cohesive soils, miscellaneous soils, and rock. Cohesionless soils are soils such as gravel and sand. These soils do not bind together under pressure and are often used for capillary stops. A capillary stop is an area between two surfaces that is large and porous enough to stop capillary action.

Cohesive soils, however, will stick together. Cohesive soils usually include dense silt, medium-dense silt, hard clay, stiff clay, firm clay, and soft clay. A cohesive soil is sometimes classified as an expansive soil, for soils of this nature tend to change as the moisture content changes.

Miscellaneous soils usually include glacial till and conglomerate. Conglomerate, as a soil classification, is a mixture of sand, gravel, and clay.

Rock, for identification purposes, is subdivided into four basic groups; massive, foliated, sedimentary, and soft. Massive rock is very hard and has nearly vertical or horizontal joints. Foliated rock is also hard, but any joints are usually at an angle. Sedimentary rock includes: hard shells, sandstones, limestones, and silt stones. Soft rock includes those rocks that are especially soft and are not displaced from their natural bed.

The ability of a particular soil to support an existing load varies with the type and weight of load and the type of soil. Most building codes specify the allowable bearing pressure for various types of soils. Table 3–1 can be used to determine the load-carrying capacities of different types of soil.

**TABLE 3–1    BEARING CAPACITY OF SOILS**

| Type of Soil | Allowable Bearing Strength, PSF[1] |
|---|---|
| **COHESIONLESS SOILS** | |
| Dense sand | 6,000 |
| **COHESIVE SOILS** | |
| Dense silt | 3,000 |
| Medium dense silt | 2,000 |
| Hard clay | 6,000 |
| Stiff clay | 4,000 |
| Firm clay | 2,000 |
| Soft clay | 1,000 |
| **ROCK** | |
| Massive | 100,000 |
| Foliated | 80,000 |
| Sedimentary | 40,000 |
| Soft | 20,000 |

[1] Pounds per square foot

Reprinted with permission of National Forest Products Association.

## PILE FOUNDATIONS

Piles are not used extensively in light construction, but they are used to minimize settlement in low-lying coastal areas, and in some other areas. Piles are vertical members that can be made from wood, concrete, steel, or a combination of wood and concrete. A vertical member that transmits loads through soil that has a poor bearing capacity to soil that can support the load is called a *point-bearing pile*. (See Figure 3–1.) A point-bearing pile rests on bedrock, or on soil that is dense enough to withstand the weight of the piles and superstructure, or on soil that has been compacted by the driving of the piles.

A *friction pile* is one that is supported by only the friction between soil and pile. (See Figure 3–2.) In the utilization of skin fric-

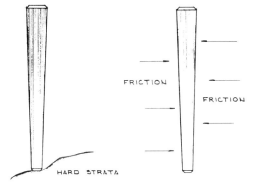

FIGURES 3–1 (left)  Point-bearing pile, and 3–2 (right)  Friction pile

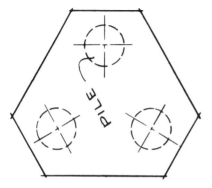

FIGURE 3-3.  Pile cap

tion, the pile incorporates the use of its total bearing surface. A friction pile is usually used in cohesive soils and a point-bearing pile is usually used in granular soils.

The load of the structure is transmitted to the piles by means of pile caps, shown in Figure 3–3. The pile caps are usually constructed of reinforced concrete, and rest on a minimum of three piles. In light construction, the piles are usually placed on at least 8-foot centers.

## Timber Piles

Economic advantages, lightness, and increased friction often make it more feasible to use timber piles rather than piles constructed of other materials. Timber piles may decay, and for this reason the cell structure is often impregnated with creosote or some other preservative. If a preservative is not used, the piles should be made of timber that will resist decay and not deteriorate when exposed to an excessive amount of moisture. A recommended timber for piles is either cedar or cypress. If the piles are placed below ground water level, of course, there is no moisture and no danger of decay. Decay occurs only when the cell structure has a 20-percent moisture

content or more and has an adequate oxygen supply. Consequently, it is wise to use wood piles only below the ground water level unless coated with a preservative.

Almost any timber can withstand the driving that is required in the placement of piles; but to aid in the placement, the ends of the piles are sometimes blunted or fitted with a metal shoe. For maximum strength, the piles should be free of checks, knots, splits, or shakes, and have a uniform taper from head to toe.

## Cast-in-Place Piles

Cast-in-place, precast, composite, and pipe piles are most often used in heavy construction, rather than light construction. Their use is dependent on location, size of building, availability of materials, architect, and engineer.

*Cast-in-place* piles are formed by filling voids in the ground with concrete. These voids are of two types: shell and shell-less. A shell pile is a thin metal shell driven into the ground and then filled with concrete. The shell is driven into the ground by means of a steel mandrel. When the shell reaches the desired depth, the mandrel is removed. If necessary, reinforcement is added and the shell is filled with concrete. This type of pile is used when the lateral action of the soil is so slight that it produces no appreciable amount of skin friction. Consequently, most shell piles rest on solid bearing surfaces.

A shell-less pile is formed when a shell is inserted into the ground, filled with concrete and then removed. Such a pile may also be produced by first driving a steel pipe into the ground. After the pipe is removed, the hole is filled with concrete. A pile of this type is usually used where the soil has enough cohesive ability to support a relatively smooth surface. Another variation of the shell-less pile is one in which the hole is bored rather than punched. Bored holes may vary from 6 to 36 inches in diameter. When a greater bearing surface is required, the bored holes are sometimes flared at the bottom. (A dead load is a stationary load imposed on a structure, a constant weight; a live load is a variable load that may be placed upon a structure.)

## Precast Concrete Piles

Precast concrete piles are usually square or octagonal in shape and are formed by pouring concrete around reinforcement above grade level. The reinforcement usually consists of vertical bars supplemented with hoops or spirals. The size of the individual piles differs according to location, bearing capacity of the soil, and the dead load and live load to be imposed on the piles. When the piles are driven through cohesive soil, the tip is usually blunt, but to facilitate driving in sand and gravel, the tip is tapered. If the pile is expected to penetrate a hard stratum, the tip of the pile may be covered with a metal shoe.

## Composite Piles

Composite piles are made from a combination of timber and concrete. (See Figure 3–4.) The upper portion of the pile, where strength and durability are needed, is made

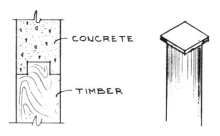

FIGURE 3–4 (left)   Composite pile, and 3–5 (right)
Pipe pile

of concrete, while the lower portion is usually made of timber. The wooden portion of the pile is first driven to the required depth; concrete is then placed in the void left by the driving of the pile. The concrete is connected to the timber by means of either a simple butt joint or a more elaborate tenon joint. The advantage of composite piles is that the cost is lower than for all-concrete piles.

## Pipe Piles

Pipe piles are usually made from heavy steel pipes filled with concrete, as shown in Figure 3–5. The pipes can be driven with their ends open or closed. If they are driven with their ends open, a small pipe is inserted to force compressed air into the larger pipe chamber. This blows the soil from the interior of the pile. Water is sometimes used as well to aid in the removal of the soil. To increase the surface-bearing capacity of the pipe, a steel plate is welded to the top of the pipe pile.

## SPREAD FOUNDATIONS

A spread foundation is used to spread the building load over a sufficient area of soil in order to minimize settlement and to insure an adequate bearing surface. The load of the superstructure in spread foundation systems is transmitted to the footings by

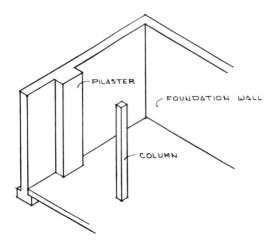

FIGURE 3–6.  Load-transmitting elements in a
spread foundation

means of foundation walls, pilasters, columns, or piers. (See Figure 3–6.)

## Foundation Walls

The foundation wall must receive the load
of the superstructure, transmit the load to
the footings, and retard moisture that
might seep into the basement or crawl
space. Because of concrete's resistance to
lateral pressures and its tremendous compressive strength, it is often used to build
foundation walls.

If the foundation wall is poured separately from the footing, a key should be
used to help the wall resist pressures of any
lateral movement, as shown in Figure 3–7.
(A key is a piece of concrete that protrudes
from the foundation wall and fits into a recessed area of the footing.) If the height of
a foundation wall exceeds ten times its
thickness, the wall should be reinforced
with vertical reinforcement bars, and if the
length of the wall exceeds 20 feet, the walls
should be reinforced with pilasters.

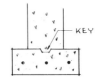

FIGURE 3–7.  Footing key

## Wall Forms

Many different types of wall forms or forming systems are used for foundation walls.
But in most cases the wall forms consist of
form boards, 2 × 4 studs, horizontal walers
(braces used to hold studs in place),
spacers, and tie rods. (See Figure 3–8.) The
form boards are usually constructed of
tightly matched 1-inch sheathing or ¾-inch
plywood. The 2 × 4 studs are placed 2 feet
on center and are used to support the form
boards. If the studs are longer than 4 feet,
two 2 × 4 walers should be placed on 2-foot
centers. Tie rods are used to fasten the two
form boards together. The rods are inserted

FIGURE 3–8.  Wall forms

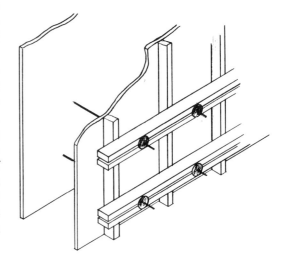

through both walls. Then the nut washers are tightened, and the assembly is toenailed to prevent vibration during the placement of the concrete. Before the rods are completely tied off, metal spreaders are placed between the two form boards. The spreaders, which are available in different widths, are used to maintain a constant width between the form boards. When the concrete is vibrated, the spreaders are easily knocked to the bottom of the forms.

After the wall forms are in position, they should be aligned and braced. To align the forms, a string is stretched along the top and a form aligner brace is used to tie the wall off. In most cases this brace is a 2 × 4 that is nailed to a 2 × 4 stud and a ground stake.

If there is to be an opening in the foundation wall, some provision should be made for it. One method is to construct the opening frame and place it between the form boards. To keep the frame in proper alignment after the forms are removed, a key is placed around the frame. When the frame is put into the forms it should be properly braced to resist the pressures of the poured concrete.

Once the concrete has set, the forms are stripped and stacked for future use. Because forms are used more than once they should be treated with form oil to prevent them from sticking to the concrete, and also to allow them to be easily stripped from the concrete. The tie rods are then broken or extracted from the foundation wall, and the holes are grouted over.

## Concrete-Block Foundation Walls

If a solid concrete foundation wall is not used, a concrete-block foundation wall is usually preferred. In most cases a concrete-block wall can be erected faster and more economically than foundation walls made of other building materials. When the masonry units arrive at the job site they should be stored on planks or pallets and covered with a tarpaulin. The blocks should never be allowed to come into contact with any moisture before their placement.

The first course of blocks should be laid in a full bed of mortar, properly aligned, leveled, and plumbed. The blocks should be laid with the thickest edge of the face shell up to provide a wider mortar bed. As the units are placed, the face shell and ends of all the units should receive a full bed of mortar. Each course must be exactly aligned, leveled, and plumbed.

To strengthen the foundation wall, the blocks should be laid in a lap bond. They should also be placed so that no two vertical joints correspond. The joints should be completely filled with mortar, with metal reinforcing placed on alternate courses. The metal reinforcing prevents lateral forces from dislodging the foundation wall. Additional support of the foundation wall is provided by pilasters, column projections that occur every 25 feet. The pilasters should be at least 2 feet wide and as thick as the foundation wall. For additional strength, the cores of the masonry units can be filled with concrete.

To act as a termite barrier and distribute the load from the floor beams, the foundation wall should be capped with solid units. There are three basic techniques used to top a foundation wall; stretcher blocks with cores full of mortar; solid-top blocks; and 4-inch-thick units. If a stretcher block is used, a piece of metal lath is placed under the top course; then

the cores are filled with mortar and troweled smooth. A solid-top block is the same
size as a regular unit, but it has a 4-inch
solid top. Individual units 4 inches in thickness can also be used to top foundation
walls.

To tie the frame structure to the foundation system, 18-inch-long anchor bolts are
placed in the cores of the masonry units.
The bolts are held in position by a 4-inch
steel washer placed on the end of the
anchor bolt; the cores are also filled with
concrete. The bolts should be placed on 4-
foot centers, and every sill member should
have a minimum of two anchor bolts securing it to the foundation wall.

## Stone Foundation Walls

In locations where there is an abundant
supply of native stone, it is often more economical to use stone for foundation walls.
Such walls are usually referred to as either
rubble or ashlar walls. Rubble walls are
made of irregularly shaped and sized
stones, while ashlar walls are made of
stones that are cut. Rubble work usually
includes stones of granite, hard sandstone,
and conglomerate, while ashlar foundation
walls are usually made from sedimentary
rock formations. A stone foundation wall
should be a minimum of 18 inches thick,
and laid in portland cement mortar.

## Pilasters

Pilasters, an integral part of most foundation walls, are used to support the ends of
beams and to increase the rigidity of the
wall. They are also used to transmit the
load of the superstructure to the footings.
A pilaster should be a minimum of 2 feet in

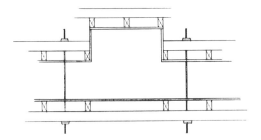

FIGURE 3–9.  Formed pilaster

width and as thick as the foundation wall.
They should be placed on 25-foot centers,
on the inside of the foundation. In a concrete foundation wall, a pilaster is formed
as an integral part of the foundation wall.
(See Figure 3–9.)

## Columns

Columns are usually made of wood or steel,
and are used to help support the superstructure. The majority of columns are
steel, solid timber, or laminated timber.
There are various built-up columns, but
they are only about 75 percent as strong as
one-piece columns of the same dimension.
The three most common types of built-up
columns are solid-core, cover-plated, and
box. (See Figure 3–10.)

To help in preventing decay the columns
should be positioned so they will not come
in direct contact with moisture. One of the
basic techniques used is to elevate the column by placing a base under it. (See Figure
3–11.) For optimum performance from a
column, it should be in a vertical position,
with the weight evenly distributed over the
base.

Columns must be firmly anchored to the
floor so they won't be dislodged. One
anchoring technique is to bolt a metal strap

SOLID-CORE          COVER-PLATED          BOX

FIGURE 3–10.  Built-up columns

to the column and to the floor. (See Figure 3–11.)

## Piers

Piers support the girders and can be divided into two classes; exterior and interior. Exterior piers are subjected to both compressive loads and forces produced by the wind. Interior piers are located within foundation walls and receive only compressive loads. (See Figure 3–12.)

Piers may be constructed of concrete, solid masonry, or hollow masonry filled with concrete. Unless reinforcement is added in the construction of the exterior pier, it should not be built larger than three times its least dimension. If the pier is built of concrete or solid masonry, it may be built to a height of ten times its least dimension. However, if a pier is constructed of hollow masonry the height should not exceed four times its least dimension. In light construction the piers should be spaced 6 to 8 feet on center, and project 18 inches above grade.

## Footings

In many geographical locations, the earth's crust is not strong enough to support a foundation wall and superstructure. In such cases, the weight of the foundation wall and structure is spread, by means of footings, over a greater surface area. The footing is usually a concrete pad, although other materials may be used. Concrete is used as a footing material because of its

FIGURE 3–11.  Column anchorage

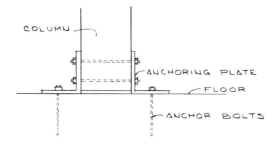

FIGURE 3–12.  Interior pier

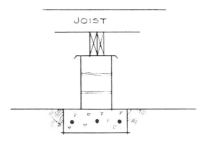

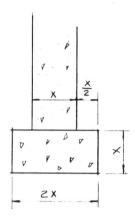

FIGURE 3–13.  Footing size

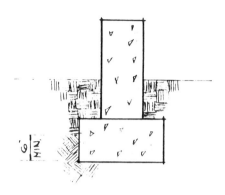

FIGURE 3–14.  Footing placement

ability to maintain relatively high strength qualities, its resistance to decay, and its dependability.

Several methods are employed in the construction of footings. In the most commonly used method, they are stepped at the base of the foundation wall, with the projection equaling one-half the width of the footing, as shown in Figure 3–13. In light construction, very little calculation needs to be done in establishing load limits for the size of the footings. But if there is a need to calculate the footing size, both the dead and live load should be considered. The dead load is the permanent load superimposed on the foundation system, and includes all stationary structural materials such as exterior walls, floor systems, cabinets, fireplaces, roof systems, and interior partitions. The live load includes the weight of objects or forces that are not permanent, such as snow, wind, rain, people, and furniture.

The thickness of a footing should equal the thickness of the foundation wall, or be at least 12 inches thick. The footing should extend into undisturbed soil 6 inches, and should be placed below the frostline. (See Figure 3–14.) If for some reason the footing cannot extend into undisturbed soil, the fill should be well tamped and contain no vegetation.

Footings are formed either by placing form boards in the desired location or excavating a trench to the proper width and depth. If form boards are used, they should be properly braced, as shown in Figure 3–15. In most cases 2 × 4 stakes are placed on 3-foot centers and are braced to a ground stake. If the form boards are not level and are higher than necessary, grade nails can be placed on the inside of the

FIGURE 3–15.  Braced form boards

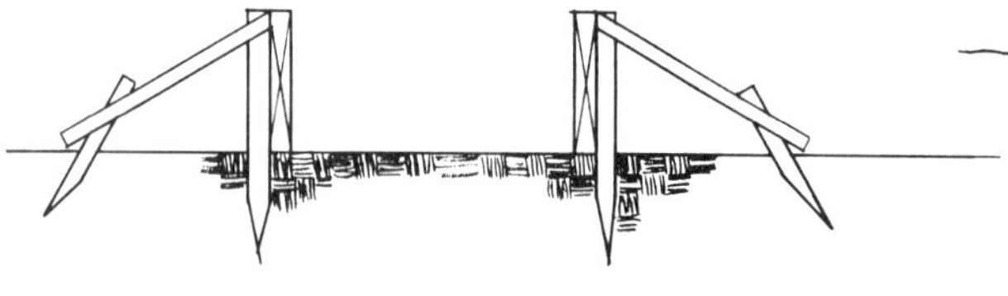

FIGURE 3–16.  Placement of grade stakes

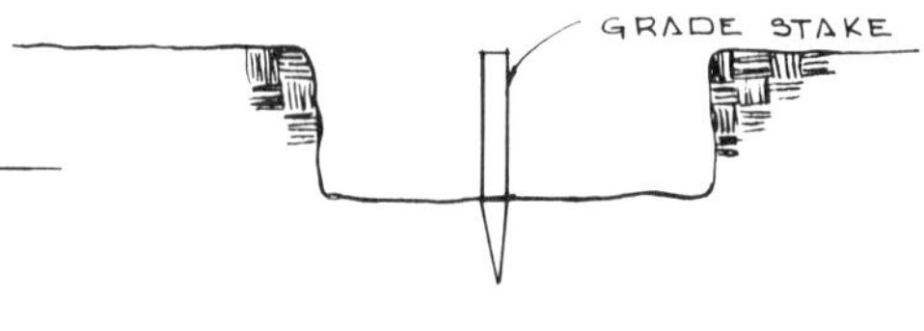

forms. The nails are placed at the correct elevation and the concrete is screeded (leveled) to them.

If a trench is used to form the footings, it is necessary to place grade stakes at the proper elevation. (See Figure 3–16.) When the concrete is poured, it is screeded to the grade stakes. Once the concrete has been struck, the grade stakes can be removed.

In some cases the footing and the foundation wall are formed together. When this method is used, strap ties can be used to attach the form boards, and ground stakes can be used to level the forms.

### Frost Heave

Some soils do not drain as readily as others. Consequently, the moisture retained in the soil may freeze. If the ground is saturated with moisture and the water freezes, the soil expands and begins to move. This reaction of the soil to freezing water is called frost heave. If the footing is not below the frost line and frost heave takes place, the results could mean the tilting of foundations and cracking of brickwork. If the expansion of the soil were uniform under the foundation, no appreciable amount of damage would occur, because the entire structure would fluctuate by the same amount. But, this is not the way it happens. The perimeter of the foundation system is subjected to more frost action than the interior, and therefore a significant difference occurs in the movement of these two areas of the slab.

The frost line varies considerably in the United States. In parts of Maine it may be at 5 feet, while parts of Florida may not have a frost line. (See Figure 3–17.)

FIGURE 3–17. Average frost depth in inches

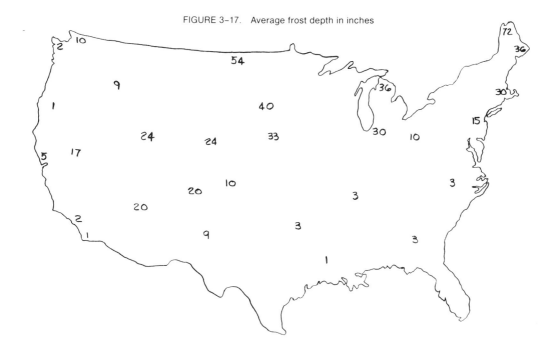

In geographical locations where severe winters are experienced, it is recommended that the footings be placed 1½ feet below the frost line. Thus, the depth to which the footing extends below grade is determined both by drainage conditions and geographical location.

### Stepped Footings

When buildings are located on steep grades, it is often more feasible and economical to use stepped footings than other types. The use of stepped footings eliminates the need for excavation, thus reducing the cost of the structure. The vertical step of the footing should not exceed three-fourths of the distance of the horizontal step, as shown in Figure 3–18. Moreover, to obtain the optimum amount of structural value from the footings, the horizontal steps should be as long as possible.

The reinforcement of the stepped footings should be an integral part of the concrete. The vertical reinforcement bars should be shaped to extend a minimum of 18 inches into the horizontal step. To achieve a structurally sound footing, the horizontal and vertical steps should be poured as an integral unit and should be placed on undisturbed soil, well below the frost line.

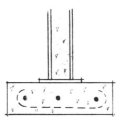

FIGURE 3–19.   Column footing

### Column Footings

A column footing is an independent footing that supports a single column or post, and carries a concentrated load. (See Figure 3–19.) The footings should be at least 8 inches thick and measure 2 feet by 2 feet. If a column footing supports a light load, it could consist of nothing more than a concrete mat, but if a heavy load is placed on the footing, reinforcement bars should be placed longitudinally and transversely.

### Reinforced Footings

To alleviate some of the shear and tensile weaknesses of plain footings, reinforcement bars can be added to footings. The steel, combined with the concrete, furnishes the tensile strength needed to balance the compressive strength. When the footings are spread over a wide area, the reinforcement is placed perpendicular to the foun-

FIGURE 3–18.   Stepped footing

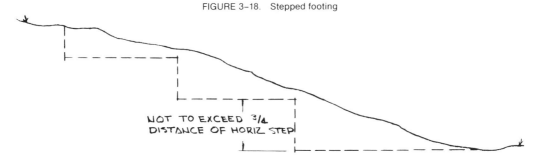

NOT TO EXCEED ³/₄ DISTANCE OF HORIZ STEP

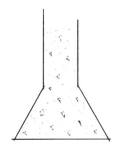

FIGURE 3-20.   Flared footing

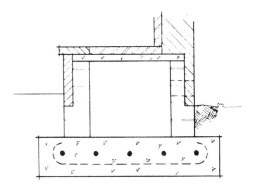

FIGURE 3-21.   Chimney footing

dation wall; but in most cases the reinforcement rods are parallel to it.

### Flared Footings

A flared footing is a continuation of the foundation wall, but with its base projecting 30°, as shown in Figure 3–20. In most cases, the foundation wall and footings are an integral unit and are usually poured at the same time.

### Chimney Footings

Because of the tremendous amount of weight that rests upon chimney footings, special attention should be given to their design. They should be dimensioned 6 inches longer and wider than the base of the chimney. In addition, they should be reinforced with No. 5 reinforcing rods placed both transversely and longitudinally. (See Figure 3–21.)

### Dampproofing

Some foundation systems are built on soil in which surface and ground water are problems. Unless the problem is alleviated, the water may erode the soil from beneath the foundation, or it may seep into the basement. Surface water is a result of rain, melting snow, or water discharged from downspouts. Ground water is a result of

water rising by capillary action. Surface water should be directed away from a structure by sloping grades. The grade should be a minimum of 5 percent for a grassed perimeter and 1 percent when the area adjacent to the structure is paved. The grade provides a natural drain and helps in dispersing surface water from the area of the foundation wall.

One of the better methods of controlling surface water is to discharge it before it reaches the foundation system. If the soil has a high gravel content or is extremely sandy, it will probably drain naturally; but if the soil consists largely of clay deposits, some type of drainage system should be employed. A drainage system of 4-inch tile laid in a base of wash gravel or crushed stone can transport any existing water away from the foundation. (See Figure 3–22.) The tile should be laid with open joints covered with 30-pound felt, or have perforated openings along the length of the pipe. The pipe should have a slight fall so the water can be carried from the foundation and into a storm sewer. To enable the water to flow freely into the drain tile, a backfill of wash gravel should be placed over the drain tile.

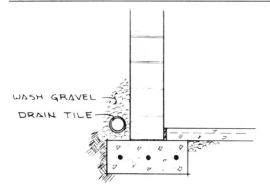

FIGURE 3-22.   Drainage system

At the intersection of the footing and the foundation wall, a fillet should be formed with tar or mortar. The rounded interior corner enables the foundation system to reject the entry of water at the intersection of foundation wall and footing.

To complete the dampproofing, the foundation wall should be either mopped hot with asphalt or tar, covered with polyethylene, or parged* with two applications of cement grout. Any of these three procedures acts as a moisture stop. If the foundation wall is parged, the first coat should be applied about ¾ inch thick and allowed to partially dry. The surface of the plaster should then be roughened with a scratcher and allowed to dry for a minimum of 24 hours. (See Figure 3–23.) To provide a good suction bond, the roughened surface should be slightly dampened before the second coat, also ¾-inch thick, is applied.

*To parge means to coat a surface with plaster; it is usually associated with foundation walls.

FIGURE 3–23.   Scratching the foundation-wall parging

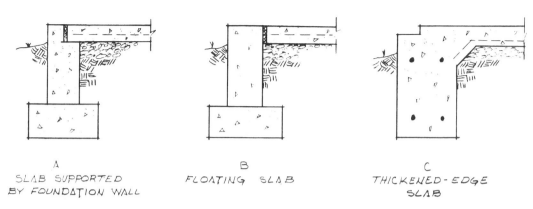

A
SLAB SUPPORTED
BY FOUNDATION WALL

B
FLOATING SLAB

C
THICKENED-EDGE
SLAB

FIGURE 3–24. Three types of slab

## SLAB-ON-GRADE

In warmer climates, where frost heave is not a serious problem, slab-on-grade construction is used successfully as a foundation. As shown in Figure 3–24, a slab can be supported by a foundation wall or supported by only the soil under the slab, or a slab and footing can be combined to form an integral unit.

Certain precautionary techniques should be followed in the construction of a slab-on-grade foundation. (See Figure 3–25.)

The slab should rest on a well-formed bed, which consists of a base course, fill, vapor barrier, and reinforcement. The base course, at least 4 inches thick, consists of sand, crushed stone, or gravel, and acts as a stop for moisture rising through the soil by means of capillary action. A polyethylene film should be placed over the base course to act as a vapor barrier, repelling the water vapor as it rises from the soil. Without the vapor barrier, the underside of the

FIGURE 3–25. Slab-on-grade

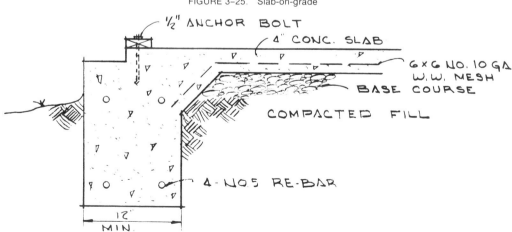

slab could absorb moisture rising from the ground.

If the slab is to receive the weight of the superstructure, is expected to receive a large live load, or is placed over cohesive soil, it should be reinforced. The reinforcing agency most often used is 6 × 6 No. 10 gauge wire mesh. This is placed over the vapor barrier and supported by metal chairs, used to keep the wire mesh elevated just below the center of the slab. If the metal chairs are not used, the wire mesh has a tendency to settle to the bottom of the concrete. If the mesh settles, the entire purpose is defeated, for there is no reinforcement embedded in the concrete. Concrete slabs have great compressive strength, but little shear resistance. Inclusion of wire mesh in the slab increases its strength. The slab should be a minimum of 4 inches thick and placed continuously. After the concrete has been placed, it should be allowed to cure properly.

## Foundation Anchorage

To prevent movement caused by the pressures of the wind, the superstructure must be firmly attached to the foundation by means of anchor bolts. The pressures created by the wind can cause translation, overturning, and rotation, as shown in Figure 3–26. *Translation* is the process by which a building is moved in a lateral direction by strong winds. *Overturning* is the uplifting of light superstructures by wind forces. To avoid overturning, both anchor bolts and framing anchors can be used. The anchors are attached to the individual framing members, usually spaced on 4-foot centers. *Rotation* is the pivoting of a light frame building on its foundation. Rotation sometimes occurs when high winds are not symmetrical. To prevent rotation, the sill plate should be firmly anchored to the foundation and the anchors should possess adequate shearing strength.

The bottom plate of the superstructure is attached to the foundation wall by means of ½-inch anchor bolts spaced on 4-foot centers. The bolts should be placed so that their threaded ends extend above the top of the sill plate. If the foundation wall is constructed of concrete blocks, the anchor bolts should be placed in a core filled with mortar. To keep the mortar

FIGURE 3-26.   Three types of structure movement caused by wind pressure

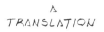

A
TRANSLATION

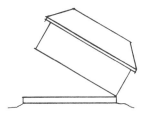

B
OVERTURNING

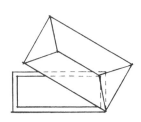

C
ROTATION

from falling from the core, a metal lath is placed below the hollow core.

## Protection Against Decay and Termites

In the design and construction of a foundation system, attention should be given to the protection of wood structural units against decay and termites. The possibility of decay and insect attack can be alleviated by proper site drainage, segregation of the structural timbers from the ground, adequate ventilation, and termite barriers.

For a building site to drain properly, a sloped grade should be placed around the perimeter of the building. If there is any danger from surface or ground water, a drainage tile should be installed and the foundation walls dampproofed. To properly segregate the structural timbers from the ground, a minimum of 18 inches should exist between the floor joist and ground; and there should be a 12-inch minimum clearance from the ground to timber girders or beams. The minimum dimensions of clearance from ground to timber allow visual inspection and provide a safe distance from possible moisture contact; in addition, the space allows room for a termite barrier. Barriers are usually termite shields, chemically treated soil, or chemically treated lumber.

There are two types of termites that feed on structural timbers—the subterranean and the drywood termite. Most termite damage is done by the subterranean termite, which lives in nests below the surface of the earth. Subterranean termites need a direct contact with some type of moisture. To maintain this contact, they build tubular structures, made from the soil, which can run from the ground to girders or joists. The tubes are usually built along the foundation wall, column, or piers. In addition to a constant source of moisture, a termite must be supplied with an adequate supply of food—wood or any cellulose material. But take food and moisture from the termite and it will cease to exist.

A termite shield is usually made of copper or other noncorrosive sheet metal, which should be no less than 26-gauge and should be placed between the foundation wall, column, or pier, and the sill plate or girder. The shields should be bent at a 45° angle, protrude a minimum of 2 inches from the edge of the wall, and placed in a full bed of mortar.

If the foundation wall is constructed of hollow concrete blocks, termites can infect the sills by crawling through the openings in the blocks. To prevent this, the blocks should be capped with solid masonry caps, or a termite shield should be placed along the top of the foundation wall.

An effective method of repelling termites is to treat the soil chemically. The best and most effective technique is to introduce a chemical—either aldrin, benzene hexachloride, gamma, or chlordane—into the soil before the slab is poured. The soil can also be treated chemically once the structure is complete.

An important aid in the protection against termites is to be sure all wood scraps are removed from the job site. Such scraps provide food for the termites and a haven for termite reproduction and growth.

Structural members decay only when there is sufficient oxygen and moisture. One of the better methods of preventing wood from decaying is to season the wood properly. If wood is going to be exposed to constant moisture, the cell structure must be impregnated with a substance to kill the fungi and ward off attacks by insects.

The wood preservatives used against decay and insects are water-soluble preservatives, creosote, and oil-soluble preservatives. Their effectiveness is dependent upon the type of chemicals used, the amounts of penetration and retention, and the uniform distribution of the preservative.

Water-soluble preservatives are relatively odorless and are used extensively in residential construction. Timber treated with such preservatives is clear and easy to paint.

Creosote, the most widely used preservative, has a strong odor and cannot be painted over. It is most often used on pilings that will come in contact with water or will be exposed to an excessive amount of moisture.

Oil-soluble preservatives are quite effective against fungi and termites unless they come in contact with salt water. A major disadvantage of an oil-soluble preservative is that it sometimes has a tendency to turn the wood a dark brown.

## CONCRETE

Concrete is a mixture of portland cement, water, and aggregates. The water should be clean and free of oil, acid, or alkali. The aggregates used in the mixture are classified as fine or coarse. Fine aggregate is sand that ranges in size from very fine to ¼ inch. If the three ingredients are properly portioned, mixed, and cured under favorable conditions, quality concrete will result.

A slump test is sometimes used to measure the consistency of concrete. It is a rough measure to determine any changes that have been made in the proportions of the aggregates, or in the water content. If there is a change, adjustments should be made to the proportion of the fine and coarse aggregates.

A slump test is made in a mold that has a base diameter of 8 inches, a top diameter of 4 inches, and a height of 12 inches. (See Figure 3–27.) Before the concrete is placed in the forms, a sample is taken and placed in the mold. It should first be filled to one-third its height and rodded 25 times with a ⅝-inch steel rod to mix and compact the concrete thoroughly. Two more layers of concrete are then poured into the mold; after the addition of each layer, the concrete is rodded 25 times, each stroke penetrating

FIGURE 3–27.   Mold for making a slump test to measure the concrete's consistency

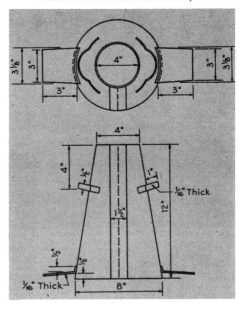

the underlying layer. When the top layer has been rodded, the concrete at the top of the mold is leveled. The mold is then gently raised and placed by the specimen. The slump is measured from the top of the specimen to the top of the mold. (See Figure 3–28.) The slump for residential construction should not exceed 5 inches.

## Handling and Placing Concrete

Prior to placing (pouring) the concrete, the forms should be treated with form oil. Diesel fuel, because of the oil base, availability, and economic advantage, is often a substitute for form oil. If a polyethylene vapor barrier is not used, the grade should be slightly moistened before the concrete is poured, to prevent the dry soil from extracting moisture from the concrete.

Concrete should be placed as near its final position as possible. It should be poured in small piles; when it is placed in large piles, the aggregates tend to separate. Too large an area should not be worked at one time, lest an excessive amount of water and mortar supersede the moving concrete, resulting in poor quality. In addition, concrete should not be poured from a height greater than 4 feet, to prevent aggregate separation. Another good practice is to discharge the concrete in a vertical position. If

FIGURE 3–28.  Measuring the slump

concrete is allowed to hit form walls and ricochet into place, segregation of the aggregates may occur.

After the concrete has been placed in the forms, voids or air pockets sometimes remain. These voids, usually referred to as honeycombs, can seriously weaken the concrete. To eliminate honeycombs, long sticks or spades are used to work the concrete around the reinforcement. This compacts the concrete and eliminates any large pockets of air. Systematic vibration is another means by which concrete can be consolidated. This method makes it possible to use a stiffer concrete mix and produces bet-

ter-quality concrete. The vibrator should be used with care because overvibrating has a tendency to cause the aggregates to separate.

## Concrete Finishing

After placing the concrete, the initial step in finishing is to strike (smooth) or screed, the surface to the proper grade. This is done by placing a long straightedge, also called a strike-off rod, on the screed and forms, moving it back and forth in a saw-like motion until the concrete is struck, or level with the screed. (See Figure 3–29.)

FIGURE 3–29.   Striking, or screeding, the surface of the concrete

FIGURE 3–30.  Hand-tamping the concrete surface

A screed is a temporary structure, sometimes made of 1 × 4s and stakes, that is used to level the concrete surface. The top of the screed should be placed level with the top of the form. Screed chairs, designed to support pipes or flat bars, can also be used to aid in striking the concrete. The strike-off rod is then moved along the top of the pipe or bar.

The distance between screeds, or between screed and form board, should never exceed 16 feet. In spanning a longer distance the strike-off rod may sag in the middle, resulting in a dished surface. To span the distance between screeds, the strike-off rod should be a minimum of 3 feet longer than the shortest distance between the screeds. In working the strike-off rod, a sawlike motion should be used, with a small amount of concrete kept ahead of the rod. The concrete preceding the rod will fill any low spots.

## Hand-Tamping

If the slump of the concrete is 1 inch or less, the concrete is stiff and the coarse aggregates must be tamped below the working surface. The tamping is done by means of a hand tamp, or jitterbug. (See Figure 3–30.) This is a metal grill about 6 inches wide and 36 inches long, with perforated openings. The openings allow the mixture of water and mortar to rise to the surface, while the coarse aggregates are pushed below the surface of the slab. In most cases hand-tamping is not necessary, and work proceeds from the striking of the concrete to the darbying process.

## Darbying

Darbying is done immediately after the concrete has been struck or hand-tamped. A darby is a wooden or aluminum finishing tool 3 to 4 inches wide and 4 to 8 feet long. (See Figure 3–31.) It is used to embed the coarse aggregate below the working surface of the slab and also to level the surface of the slab.

## Bull Float

The bull float does essentially the same job as the darby, but it enables the finisher to float a larger surface area. (A section on floating follows.) The bull float is about 10 inches wide and 4 feet in length, and is equipped with a long handle. (See Figure 3–32.) Most work with the bull float is limited to carports, drives, patios, or other areas large enough to accommodate the long handle.

## Edgers

An edger is used to produce a rounded edge on a concrete slab. (See Figure 3–33.) This increases the esthetic appeal of the slab

FIGURE 3–31.  Darbying (finishing) the concrete surface

FIGURE 3–32.   Using the bull float to finish the concrete

FIGURE 3–33.   Edging the concrete

and reduces the possibility of the edge chipping off. The radius of the edger can vary from ⅛ to ½ inch. The size of the slab and the nature of the work determine the size radius that should be used. The edging process should not be started until the concrete has started to stiffen, and care should be taken not to leave a deep impression on top of the slab. A deep impression would be difficult to remove in the floating process.

## Jointing

Joints are cut partway through a concrete slab at a specific location rather than at random intervals, to control any cracking that might occur. These cuts are usually one-eighth the thickness of the slab and are cut by a jointer (see Figure 3–34), which has a cutting edge that varies in depth from ³⁄₁₆ inch to 1 inch. In sidewalk and driveway construction, the joints are usually spaced at intervals equal to the width. For sidewalks, they should never be spaced more than 6 feet apart; and for driveways the joints should never be spaced more than 20 feet apart. The joints should be cut as soon as the concrete is firm.

## Floating

Once the water sheen has disappeared from the slab and the weight of a finisher leaves only a slight impression, the surface is

FIGURE 3–34. Jointing the concrete to control cracking

FIGURE 3-35. Using a power float to compact the concrete

ready for floating. (See Figure 3–35.) The purpose of floating is to embed the aggregate below the working surface, remove imperfections, and to compact the surface of the concrete. The floating technique may be accomplished with either wood or metal hand floats, or power floats.

## Troweling

When a smooth, dense surface is needed, the concrete is troweled after the floating operation. (See Figure 3–36.) The concrete may be subjected to several troweling operations, but the first troweling should free the surface from any defects. If more than one troweling operation is needed, the concrete should be (1) hard enough to produce a ringing sound when the trowel is passed over the surface; and (2) hard enough so that no mortar clings to the edge of the trowel. Troweling can be accomplished with either a hand or power trowel.

## Broom Finishing

Once the concrete has been troweled, the surface is very hard and smooth. To improve the traction of the surface, a broom finish is often applied. (See Figure 3–37.)

FIGURE 3–36.   Hand-troweling after floating

FIGURE 3–37.   Applying a broom finish to improve traction

This finish is achieved by drawing a broom over the surface of the slab. It leaves the surface slightly roughened and is not nearly as slippery as the troweled surface.

## Concrete Curing

The curing of concrete is one of the most important construction processes, but is also one of the most neglected. Regardless of how much care is taken in proportioning and pouring the mix, if the concrete is not properly cured it will not reach its optimum strength. Concrete should be allowed to cure slowly in a favorable temperature, and protected so that little or no moisture is lost during the first few days of hardening. Several techniques can be used to prevent the concrete from drying out. A wet burlap covering can provide an adequate source of moisture. The burlap should be kept damp throughout the curing period.

Sprinkling, as a source of moisture, is accomplished by constantly applying a fine spray of water over the exposed surface of the concrete. If this method is used, care should be taken to keep the concrete wet all the time. If the concrete is allowed to dry and then wet down, it may have a tendency to crack. Ponding can be accomplished by enclosing the perimeter of the slab with an earthen dam or a structural timber dam, and filling the enclosed space with water. This is an effective technique on flat surfaces such as slabs, sidewalks,

and drives; and it provides a uniform temperature for the concrete.

Another way to cure concrete is to prevent the loss of moisture through evaporation. This may be accomplished by sealing the surface with a plastic sheet, waterproof paper, or membrane curing compounds.

Plastic sheets are effective moisture barriers and readily adhere to simple and complex shapes. If the surface is to be exposed and discoloration of the concrete is undesirable, it may prove beneficial to use another covering, for the polyethylene film tends to discolor concrete.

Waterproof paper, also an effective moisture barrier for horizontal surfaces, does not need to be moistened and it assures suitable hydration. The paper should be placed with 4-inch lap joints secured with tap, or 2 × 4s should be placed on the joints, which must be lapped to prevent the loss of moisture. The paper should also be free of any holes or torn places.

If curing compounds are used, they should be sprayed on immediately after the finishing process, and placed uniformly. In some cases two applications of curing compound are needed.

To insure a properly cured slab, the forms should be left on as long as possible; and if wooden forms are used, they should be kept moist. The length of curing time varies, depending on size, shape, type of cement, and weather; but generally, for light construction purposes, the curing period is three to seven days.

## WOOD FOUNDATION SYSTEMS

An all-wood foundation system is a relatively new concept in the construction industry. (See Figure 3–38.) The foundation consists wholly of pressure-treated lumber

FIGURE 3-38.  A wood foundation system

and plywood placed on a bed of wash gravel. This foundation can either be constructed at the job site or fabricated in a factory. Sections of the foundation are constructed of nominal 2 × 4s or 2 × 6s covered with a plywood skin. The sizes of the studs and the thickness of the plywood skin vary with the height of the backfill, soil pressures, and vertical loads.

## Panel Construction

A typical foundation wall is constructed of 2 × 4 studs spaced 16 inches on center and sheathed with plywood. The structural requirements for vertical framing plates and footings are shown in Table 3-2. If the studs are spaced 16 inches on center, the frame can be covered with ½-inch plywood; but if they are spaced 24 inches on center, ⅝-inch plywood should be used. The structural requirements for plywood are shown in Table 3-3. The plywood skin is attached to the frame with 8d hot-dipped galvanized nails. The nails should be placed 6 inches on center along the plywood edges, and 12 inches on center at intermediate supports.

When the plywood panels are placed

## TABLE 3–2 PRESSURE-TREATED WOOD FOUNDATION: Minimum Structural Requirements for Vertical Framing, Plates, and Footings

| Height of Fill (Inch) | 25 lbs per cu ft soil pressure | | | | 30 lbs per cu ft soil pressure | | | |
|---|---|---|---|---|---|---|---|---|
| | Species & Grade of Lumber[1] | Stud & Plate Size (Nominal) | Stud Spacing (Inch) | Size of Footing (Nominal) | Species & Grade of Lumber[1] | Stud & Plate Size (Nominal) | Stud Spacing (Inch) | Size of Footing (Nominal) |
| **BASEMENT** | | | | | | | | |
| **House Width 24′ to 28′, 1 story** | | | | | | | | |
| 24 | C | 2 × 4 | 16 | 2 × 6 | C | 2 × 4 | 16 | 2 × 6 |
| 48 | B | 2 × 4 | 12 | 2 × 6 | B | 2 × 4 | 12 | 2 × 6 |
| | C | 2 × 6 | 16 | 2 × 8 | C | 2 × 6 | 16 | 2 × 8 |
| 72 | B | 2 × 6 | 16 | 2 × 8 | A | 2 × 6 | 16 | 2 × 8 |
| | C | 2 × 6 | 12 | 2 × 8 | B | 2 × 6 | 12 | 2 × 8 |
| 86 | A | 2 × 6 | 12 | 2 × 8 | A | 2 × 6 | 12 | 2 × 8 |
| **House Width 29′ to 32′, 1 story** | | | | | | | | |
| 24 | B | 2 × 4 | 16 | 2 × 8 | B | 2 × 4 | 16 | 2 × 8 |
| | C | 2 × 4 | 12 | 2 × 8 | C | 2 × 4 | 12 | 2 × 8 |
| 48 | B | 2 × 4 | 12 | 2 × 8 | C | 2 × 6 | 16 | 2 × 8 |
| | C | 2 × 6 | 16 | 2 × 8 | | | | |
| 72 | B | 2 × 6 | 16 | 2 × 8 | A | 2 × 6 | 16 | 2 × 8 |
| | C | 2 × 6 | 12 | 2 × 8 | B | 2 × 6 | 12 | 2 × 8 |
| 86 | A | 2 × 6 | 12 | 2 × 8 | A | 2 × 6 | 12 | 2 × 8 |
| **House Width 24′ to 32′, 2 stories** | | | | | | | | |
| 24 | B | 2 × 4 | 16 | 2 × 8 | B | 2 × 4 | 16 | 2 × 8 |
| | C | 2 × 4 | 12 | 2 × 8 | C | 2 × 4 | 12 | 2 × 8 |
| 48 | C | 2 × 6 | 16 | 2 × 8 | C | 2 × 6 | 16 | 2 × 8 |
| 72 | B | 2 × 6 | 16 | 2 × 8 | A | 2 × 6 | 16 | 2 × 8 |
| | C | 2 × 6 | 12 | 2 × 8 | B | 2 × 6 | 12 | 2 × 8 |
| 86 | A | 2 × 6 | 12 | 2 × 8 | A | 2 × 6 | 12 | 2 × 8 |
| **CRAWL SPACE** | | | | | | | | |
| **House Width 24′ to 28′, 1 story** | | | | | | | | |
| — | B | 2 × 4 | 16 | 2 × 6 | | | | |
| — | C | 2 × 6 | 16 | 2 × 8 | | | | |
| **House Width 29′ to 32′, 1 story** | | | | | | | | |
| — | B | 2 × 4 | 16 | 2 × 8 | | | | |
| **House Width 24′, 2 stories** | | | | | | | | |
| — | C | 2 × 6 | 16 | 2 × 8 | | | | |
| **House Width 25′ to 32′, 2 stories** | | | | | | | | |
| — | B | 2 × 4 | 12 | 2 × 8 | | | | |
| — | B | 2 × 6 | 16 | 2 × 8 | | | | |
| — | C | 2 × 6 | 12 | 2 × 8 | | | | |

[1] Species and species groups with the following minimum properties (surfaced dry or surfaced green):

| | | A | B | C |
|---|---|---|---|---|
| $F_b$ (repetitive member) psi: | 2 × 6 | 1750 | 1450 | 1150 |
| | 2 × 4 | — | 1650 | 1300 |
| $F_c$ psi: | 2 × 6 | 1250 | 1050 | 850 |
| | 2 × 4 | — | 1000 | 800 |
| $F_{c \perp}$ psi: | | 385 | 385 | 245 |
| $F_v$ psi: | | 90[2] | 90 | 75 |
| E psi: | | 1,800,000 | 1,600,000 | 1,400,000 |

[2] Length of end splits or checks not to exceed width of piece.

Reprinted with permission of National Forest Products Association.

**TABLE 3–3.    PRESSURE-TREATED WOOD FOUNDATION, PLYWOOD REQUIREMENTS:**
**Minimum Plywood Grade and Thickness for Basement Construction[1]**

| Height of Fill (Inch) | Stud Spacing (Inch) | 25 pcf Soil Pressure | | | 30 pcf Soil Pressure | | |
|---|---|---|---|---|---|---|---|
| | | Grade[2] | Minimum Thickness | Identification Index | Grade[2] | Minimum Thickness | Identification Index |
| **FACE GRAIN PARALLEL TO STUDS[3]** | | | | | | | |
| 24 | 12 | B | $\frac{1}{2}$ | $\frac{32}{16}$ | B | $\frac{1}{2}$ | $\frac{32}{16}$ |
| | 16 | B | $\frac{1}{2}$ | $\frac{32}{16}$ | B | $\frac{1}{2}$ | $\frac{32}{16}$ |
| 48 | 12 | B | $\frac{1}{2}$ | $\frac{32}{16}$ | B | $\frac{1}{2}$ | $\frac{32}{16}$ |
| | 16 | A | $\frac{1}{2}$ | $\frac{32}{16}$ | A | $\frac{5}{8}$ | $\frac{42}{20}$ |
| | | B | $\frac{5}{8}$ | $\frac{42}{20}$ | B | $\frac{3}{4}$ | $\frac{48}{24}$ |
| 72 | 12 | A | $\frac{1}{2}$ | $\frac{32}{16}$ | A | $\frac{1}{2}$ | $\frac{32}{16}$ |
| | | B | $\frac{5}{8}$ | $\frac{42}{20}$ | B | $\frac{5}{8}$ | $\frac{42}{20}$ |
| | 16 | A | $\frac{5}{8}$ | $\frac{42}{20}$ | B | $\frac{3}{4}$ | $\frac{48}{24}$ |
| | | B | $\frac{3}{4}$ | $\frac{48}{24}$ | | | |
| 86 | 12 | A | $\frac{1}{2}$ | $\frac{32}{16}$ | A | $\frac{5}{8}$ | $\frac{42}{20}$ |
| | | B | $\frac{5}{8}$ | $\frac{42}{20}$ | B | $\frac{3}{4}$ | $\frac{48}{24}$ |
| **FACE GRAIN ACROSS STUDS[3,4]** | | | | | | | |
| 24, 48, 72, 86 | 12, 16 | B | $\frac{1}{2}$ | $\frac{32}{16}$ | B | $\frac{1}{2}$ | $\frac{32}{16}$ |

[1] For crawl-space construction, use grade and thickness required for 24-inch fill depth.

[2] Minimum grade: *A* = STRUCTURAL I C-D; *B* = STANDARD C-D (exterior glue). (All panels 5-ply minimum.)

[3] Panels that are continuous over fewer than three spans (across fewer than three stud spacings) require blocking 2 feet above bottom plate. Offset adjacent blocks and fasten through studs with two 16d corrosion-resistant nails at each end.

[4] Blocking between studs required at all horizontal panel joints less than 4 feet from bottom plate.

Reprinted with permission of National Forest Products Association.

there should be a $\frac{1}{8}$-inch space between them. The space will later be filled with sealant. The panel joints should also be staggered with the joints in the pressure-treated wood footing. Once the panels are constructed and in place, a 6-mil polyethylene film should be placed below the grade portion of the treated plywood wall. Next, a treated strip should be nailed over the intersection of the plywood and polyethylene. Then the joints between the strip and plywood panel are caulked with a suitable construction adhesive.

Once the panels are constructed, they are nailed to a wood footing. The footings are $2 \times 6$s or $2 \times 8$s. If brick veneer is used, the footings are $2 \times 10$s or $2 \times 12$s. The footings are nailed to the sections with 10d corrosion-resistant nails placed 12 inches on center.

To tie adjacent wall panels together at a corner, a field-applied, untreated top plate is face-nailed to the treated top plate of the wall section. (See Figure 3–39.) The field-applied top plate of one section is allowed to extend over the intersecting section, ty-

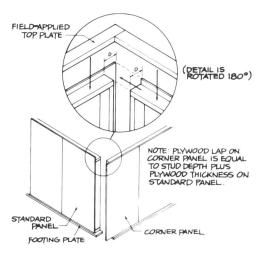

FIGURE 3–39. Corner detail, showing adjacent wall panels tied together

ing the two sections together. The two plates should be fastened together using 10d corrosion-resistant nails spaced 16 inches on center. The overlapping edges of the plywood are nailed to the field-applied top plate with 8d corrosion-resistant nails spaced 6 inches on center.

## Site Preparation

Before installation of the foundation wall and footings, the site must be excavated to the required depth. The sump and plumbing lines that are to be placed beneath the basement floor are then laid. The sump should be 24 inches in diameter or 20 inches square, and should extend a minimum of 30 inches below the top of the basement floor. (See Figure 3–40.) The sump is drained by a 4-inch drain, voiding to a storm drain or daylight.

Once the site is excavated and leveled to the proper elevation, a 4-inch-thick layer of wash gravel or crushed stone is placed over

the prepared bed. A thicker course of gravel or crushed stone is usually placed beneath the foundation wall.

For crawl-space construction, the trenches are dug to the proper elevation and filled with gravel. (See Figure 3–41.) The gravel will act as a base for the pressure-treated wood footings. The footing for a crawl space should extend below the frost line or into undisturbed soil.

## Floors

In basement construction a 6-mil polyethylene film should be placed over the bed of washed gravel. A minimum of 3 inches of concrete is then placed over the polyethylene. The concrete can be placed either before the erection of the house frame or can be delayed until the house is under roof. But the backfill around the basement walls should not be placed until the concrete floor is placed and cured. If the height of the foundation wall exceeds 4 feet, wash gravel is used to fill the lower portion.

FIGURE 3–40. Cutaway view of the basement sump

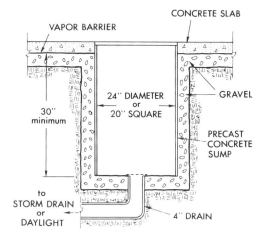

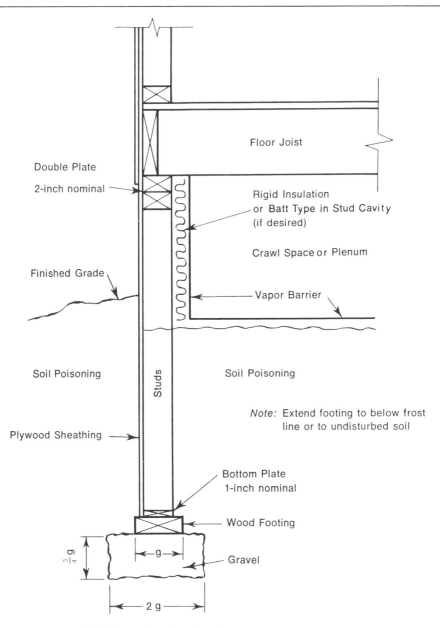

FIGURE 3–41.   A section through a pressure-treated crawl-space wall

# FRAMING

FRAMING IS USED TO SUPPORT WALLS, floors, ceilings, and roofing materials. The framing forms a skeleton, to which these materials can be attached. It must provide the proper structural strength to support the finish materials and must be properly braced to enable it to resist the racking forces of nature.

Three different framing techniques are used in light frame construction: *platform frame* construction; *balloon frame* construction; *plank-and-beam* framing.

## PLATFORM FRAME CONSTRUCTION

Platform frame construction is a popular construction technique that is used extensively for one-story construction and can be used alone or in combination with balloon construction for two-story buildings. When platform frame construction is used, the subfloor is extended to the outside edge of the building, creating a platform for the erection of exterior walls and interior partitions. (See Figure 4–1.) The wall framing is usually assembled on the floor, and the wall is then raised and placed in position.

## BALLOON FRAME CONSTRUCTION

Because of the amount of construction time and the length of framing members required, balloon frame construction is not as popular as it once was. With this technique, both the studs and first-floor joists rest on the sill. (See Figure 4–2.) The second-floor joists rest on 1 × 4-inch ribbon strips let into the edges of the studs. In some cases this type of construction is still preferred because of its resistance to movement between the wood framing and the masonry veneer. It also eliminates variations in settlement that can occur between exterior walls and interior supports.

## PLANK-AND-BEAM FRAMING

Plank-and-beam framing, in residential construction, utilizes 2-inch nominal plant subfloors and roofs supported by beams spaced up to 8 feet on center. (See Figure 4–3.) The beams are supported by posts or piers. To provide lateral bracing and a nailing base for exterior and interior wall finish materials, supplementary framing members can be placed between the posts.

One of the advantages in plank-and-beam framing is the architectural effect achieved by the higher ceilings and exposed ceiling planks. In many cases the ceiling planks are simply stained or painted. There can also be a substantial savings in labor costs, which are reduced because the posts are spaced further apart than in conventional framing, and individual pieces are larger than traditional framing members.

There are, however, certain limitations in a plank-and-beam framing system. One of the biggest disadvantages is the neces-

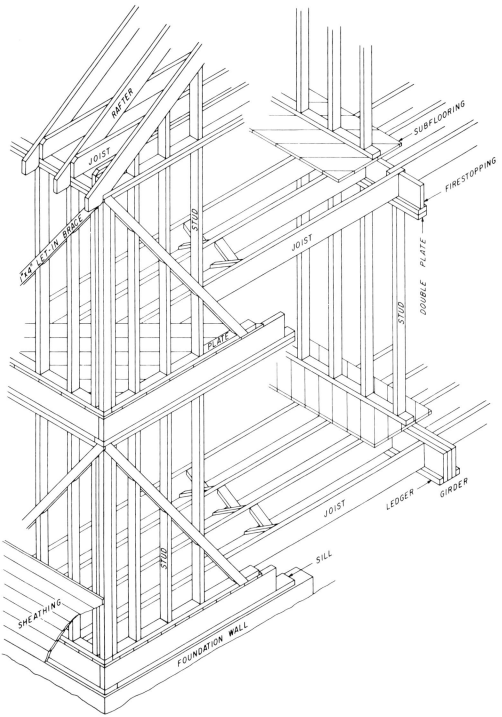

FIGURE 4-1. Platform-frame construction

59

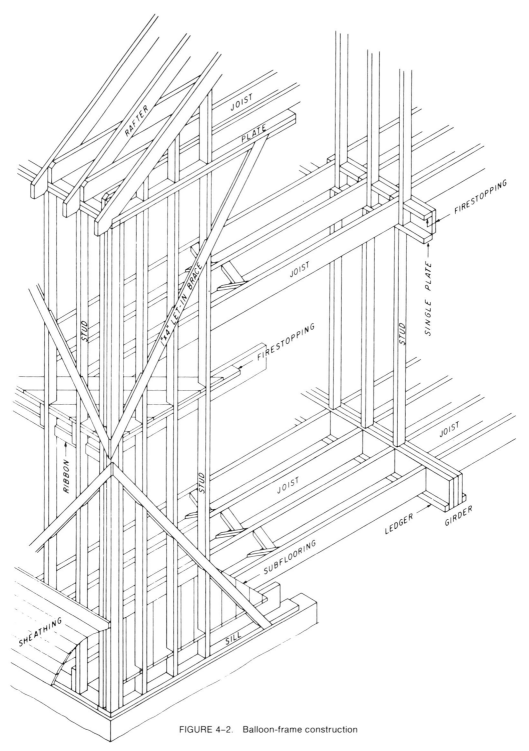

FIGURE 4–2. Balloon-frame construction

60

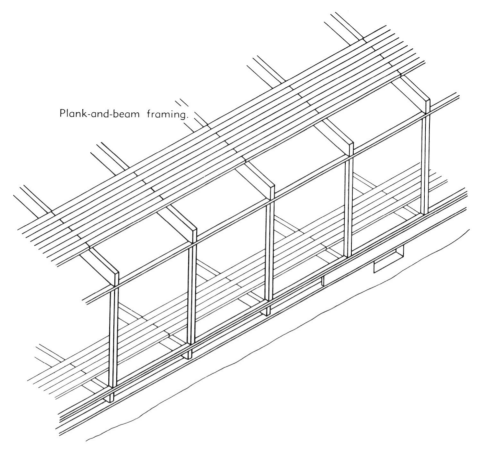

Plank-and-beam framing.

FIGURE 4–3.  Plank-and-beam framing

sity of providing additional framing under or over bearing partitions. (See Figure 4–4.) The added framing members are used to transfer the concentrated load of the bearing partition. The placement of electrical wires also poses a minor problem, but these wires can be hidden in the main support beams. The top of the beam may be routed, or the beam can be constructed of 2-inch lumber separated by short blocking. (See Figure 4–5.)

The posts are usually either $4 \times 4$ inches or are constructed of 2-inch lumber spliced together. The beams can be solid, laminated, or nominal 2-inch lumber nailed together. To create a particular architectural effect, the underside of the beams can be finished in several ways. (See Figure 4–6.) To help distribute the load, the 2-inch plank floor or roof should be tongued-and-grooved or grooved for a spline. The joints in the plank can also be constructed to achieve various architectural effects. (See Figure 4–7.)

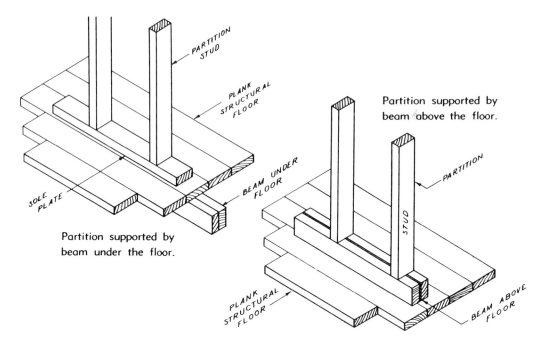

FIGURE 4–4.   Support for a nonbearing partition parallel to the planks

FIGURE 4–5.   Use of a spaced beam to accommodate electrical cable

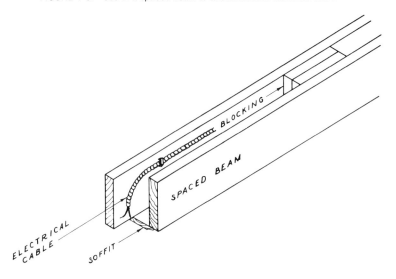

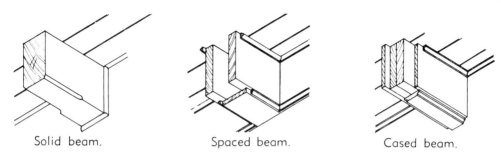

Solid beam.  Spaced beam.  Cased beam.

FIGURE 4-6.  Methods of finishing undersides of beams

FIGURE 4-7.  Methods of treating joints in exposed-plank ceilings

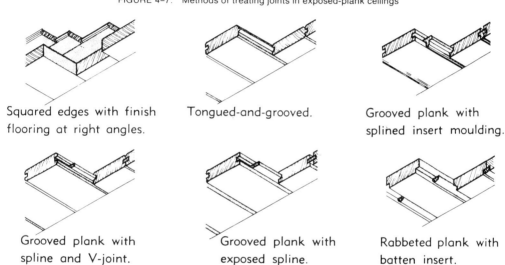

Squared edges with finish flooring at right angles.

Tongued-and-grooved.

Grooved plank with splined insert moulding.

Grooved plank with spline and V-joint.

Grooved plank with exposed spline.

Rabbeted plank with batten insert.

## FLOOR FRAMING

Once the foundation has been completed, the floor frame can be placed. It must be properly designed and constructed, for it will receive the weight of both dead and live loads. It consists of sills, girders, beams, floor joist bridging, and a subfloor.

### Sills

Sills are constructed of nominal 2-inch lumber and should be firmly anchored to the foundation wall with ½-inch anchor bolts spaced 8 feet on center. Each sill should have a minimum of two anchor bolts. If the foundation wall is constructed of concrete blocks, the anchor bolts should be embedded in 15 inches of mortar; but if the foundation wall is concrete, the anchor bolts are embedded in only 6 inches of concrete.

If sills are placed on freestanding masonry piers, they should be anchored with ½-inch bolts embedded in the piers. The sill can be either solid or built up, and should be designed to carry the load of the superstructure.

## Beams and Girders

Beams and girders are used to support the floor frame and the dead and live loads imposed on them. The beams and girders are usually constructed of solid timbers, built-up members, laminated timbers, or steel girders. To provide for visual inspections and act as a termite barrier, the top of the girder should be placed at least 18 inches above grade. When girders are framed into a foundation wall, a ½-inch air space should be provided at the end and the sides of the girder.

To determine the size that a girder should be, it is first necessary to know the span, which is the distance between girder supports. The girder load width must also be calculated. This width is the combined distance from the girder to a midpoint of a joist that rests upon it.

The last step in sizing a girder is determining the total floor load, which includes both live and dead loads. The dead load is a stationary load consisting of such things as flooring, gypsum boards, joists, and shingles. The live load is a flexible load—one

that can move—and includes things like people, furniture, wind, and snow. Some recognized loads per square foot are given in Table 4–1.

To calculate the total load of the girder, the girder span is multiplied by the girder load width by the total floor load. Table 4–2 shows the proper size girder to use for the total load.

To support the girders, posts, columns, and piers are used. The width of a column or post should be equal to the width of the girder and should be placed on 8-foot cen-

**TABLE 4–1    LOADS PER SQUARE FOOT OF FLOOR AND ROOF CONSTRUCTION**

| Type of Load | Pounds per Square Foot |
|---|---|
| Live load on roof | 30 |
| Dead load on roof | 10 |
| Live load on attic floor | 20 |
| Dead load of attic floor (not floored) | 10 |
| Dead load of attic floor (floored) | 20 |
| Dead load of partitions | 20 |
| Live load on second floor | 40 |
| Dead load on second floor | 20 |
| Dead load of partitions | 20 |
| Live load on first floor | 40 |
| Dead load of first floor | 10 |

**TABLE 4–2    ALLOWABLE TOTAL LOAD ON GIRDERS: Span**

| Size | 6 feet | 8 feet | 10 feet | 12 feet | 14 feet | 16 feet |
|---|---|---|---|---|---|---|
| 4 × 4 | 599 | 323 | | | | |
| 3 × 6 | 1659 | 936 | 582 | 387 | 267 | |
| 4 × 6 | 2323 | 1310 | 815 | 542 | 374 | |
| 6 × 6 | 3652 | 2059 | 1282 | 853 | 588 | |
| 3 × 8 | 2269 | 2154 | 1367 | 927 | 658 | 480 |
| 4 × 8 | 3177 | 3016 | 1914 | 1565 | 921 | 673 |
| 6 × 8 | 4992 | 4740 | 3008 | 2040 | 1448 | 1057 |
| 3 × 10 | 2895 | 2884 | 2795 | 1968 | 1416 | 1055 |

ters. The columns should be anchored to the girder and doweled or anchored to a base. The top of the base should not be less than 3 inches above the finished floor.

## Floor Joists

Floor joists are used to support the sub-floor, partitions, and other dead and live loads. The joists are usually nominal 2-inch lumber planks placed on edge. However, aluminum and steel joists are increasingly being used. So that conventional sheets of plywood can be used as a subfloor, the joists are usually spaced on 12-, 16-, or 24-inch centers. The spacing of the joist depends on the species and grade of the wood, the size of the joist, and whether the subfloor is glued or nailed. Table 4–3 shows the allowable clear spans for joists placed 24 inches on center.

For installation of the joists, a header joist or band is selected and laid out. The location of the joist on the header joist is indicated by two parallel lines with an X placed between them. A line is then chalked 1½ inches from the outside edge of the sill. With the chalkline used as a guide, the header joist is toenailed to the sill. (See Figure 4–8.) The joist can then be posi-

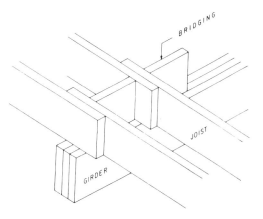

FIGURE 4–9.  Placement of the ends of joists

tioned and nailed into place, with the parallel layout lines used as guides. The joist should be positioned so that the crown is up. As natural deflection takes place, the joist will have a tendency to level out. The header joist should be nailed to the floor joist with three to five nails.

The end of the joist opposite the sill is supported by a girder. The joist can either be placed on top of the girder, as shown in Figure 4–9, or the joist can butt the girder. If the joist butts the girder, the joist should either rest on a ledger strip (See Figure 4–10) or it should be supported by a framing anchor, as shown in Figure 4–11.

FIGURE 4–8.  Placement of the header joist

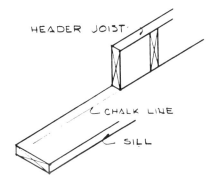

FIGURE 4–10.  Joist framed into a ledger strip

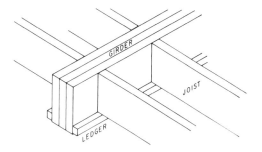

**TABLE 4–3    CLEAR SPAN FOR FLOOR JOISTS**

*Design Criteria:*
*Strength*—10 lbs per sq ft dead load plus 40 lbs per sq ft live load.
*Deflection*—Limited to span in inches divided by 360 for live load only.

| | | SPAN (feet and inches) | | | | | | | |
| | | 2 × 6 | | 2 × 8 | | 2 × 10 | | 2 × 12 | |
| *Species* | *Grade* | *Nailed* | *Glued*[1] | *Nailed* | *Glued*[1] | *Nailed* | *Glued*[1] | *Nailed* | *Glued*[1] |
|---|---|---|---|---|---|---|---|---|---|
| Douglas Fir- | 1 | 8-10 | 9-5 | 11-8 | 12-5 | 14-11 | 15-10 | 18-1 | 19-3 |
| Larch | 2 | 8-7 | 8-7 | 11-3 | 11-3 | 14-5 | 14-5 | 17-6 | 17-6 |
| | 3 | 6-7 | 6-7 | 8-8 | 8-8 | 11-0 | 11-0 | 13-5 | 13-5 |
| Douglas Fir | 1 | 8-2 | 9-1 | 10-9 | 12-0 | 13-8 | 15-4 | 16-8 | 18-8 |
| South | 2 | 7-11 | 8-3 | 10-6 | 10-10 | 13-4 | 13-10 | 16-3 | 16-10 |
| | 3 | 6-4 | 6-4 | 8-4 | 8-4 | 10-8 | 10-8 | 13-0 | 13-0 |
| Hem-Fir | 1 | 8-4 | 8-5 | 11-0 | 11-1 | 14-0 | 14-2 | 17-0 | 17-2 |
| | 2 | 7-7 | 7-7 | 10-0 | 10-0 | 12-10 | 12-10 | 15-7 | 15-7 |
| | 3 | 5-10 | 5-10 | 7-8 | 7-8 | 9-10 | 9-10 | 11-11 | 11-11 |
| Mountain Hemlock | 1 | 7-11 | 8-7 | 10-6 | 11-3 | 13-4 | 14-5 | 16-3 | 17-6 |
| | 2 | 7-9 | 7-9 | 9-11 | 10-3 | 12-8 | 13-1 | 15-4 | 15-11 |
| | 3 | 5-11 | 5-11 | 7-10 | 7-10 | 10-0 | 10-0 | 12-2 | 12-2 |
| Mountain Hemlock- | 1 | 7-11 | 8-5 | 10-6 | 11-1 | 13-4 | 14-2 | 16-3 | 17-2 |
| Hem-Fir | 2 | 7-7 | 7-7 | 9-11 | 10-0 | 12-8 | 12-10 | 15-4 | 15-7 |
| | 3 | 5-10 | 5-10 | 7-8 | 7-8 | 9-10 | 9-10 | 11-11 | 11-11 |
| Subalpine Fir | 1 | 7-0 | 7-3 | 9-3 | 9-7 | 11-10 | 12-3 | 14-4 | 14-11 |
| (Western Woods) | 2 | 6-7 | 6-7 | 8-8 | 8-8 | 11-0 | 11-0 | 13-5 | 13-5 |
| (Mixed Species) | 3 | 5-0 | 5-0 | 6-7 | 6-7 | 8-5 | 8-5 | 10-3 | 10-3 |
| Engelmann Spruce | 1 | 7-7 | 7-7 | 10-0 | 10-0 | 12-9 | 12-10 | 15-7 | 15-7 |
| (Engelmann Spruce- | 2 | 6-10 | 6-10 | 9-0 | 9-0 | 11-6 | 11-6 | 14-0 | 14-0 |
| Lodgepole Pine) | 3 | 5-3 | 5-3 | 6-11 | 6-11 | 8-10 | 8-10 | 10-9 | 10-9 |
| Lodgepole Pine | 1 | 7-11 | 8-1 | 10-6 | 10-8 | 13-4 | 13-7 | 16-3 | 16-7 |
| | 2 | 7-3 | 7-3 | 9-7 | 9-7 | 12-3 | 12-3 | 14-11 | 14-11 |
| | 3 | 5-7 | 5-7 | 7-5 | 7-5 | 9-5 | 9-5 | 11-6 | 11-6 |
| Ponderosa Pine- | 1 | 7-9 | 7-9 | 10-3 | 10-3 | 13-0 | 13-1 | 15-10 | 15-11 |
| Sugar Pine | 2 | 7-0 | 7-0 | 9-3 | 9-3 | 11-9 | 11-9 | 14-4 | 14-4 |
| (Ponderosa Pine- | 3 | 5-5 | 5-5 | 7-1 | 7-1 | 9-1 | 9-1 | 11-0 | 11-0 |
| Lodgepole Pine) | | | | | | | | | |
| Southern Pine | 1 | 8-10 | 9-5 | 11-8 | 12-5 | 14-11 | 15-10 | 18-1 | 19-3 |
| | 2 MG | 8-6 | 8-7 | 11-3 | 11-3 | 14-4 | 14-5 | 17-5 | 17-6 |
| | 3 | 6-5 | 6-5 | 8-6 | 8-6 | 10-10 | 10-10 | 13-2 | 13-2 |
| Idaho White Pine | 1 | 7-9 | 7-9 | 10-3 | 10-3 | 13-1 | 13-1 | 15-11 | 15-11 |
| | 2 | 7-1 | 7-1 | 9-4 | 9-4 | 11-11 | 11-11 | 14-6 | 14-6 |
| | 3 | 5-5 | 5-5 | 7-1 | 7-1 | 9-1 | 9-1 | 11-0 | 11-0 |
| Western Cedars | 1 | 7-6 | 7-11 | 9-11 | 10-6 | 12-8 | 13-4 | 15-4 | 16-3 |
| | 2 | 7-1 | 7-1 | 9-4 | 9-4 | 11-11 | 11-11 | 14-6 | 14-6 |
| | 3 | 5-6 | 5-6 | 7-3 | 7-3 | 9-3 | 9-3 | 11-3 | 11-3 |

[1] ¾″ UNDERLAYMENT Plywood glued to joist

Reprinted with permission of Western Wood Products Association.

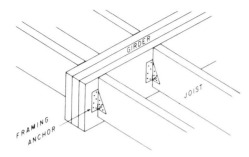

FIGURE 4-11. Joist supported by a framing anchor

## Double Joists

If a partition runs parallel to the joist, a double joist should be placed beneath the partition. (See Figure 4-12.) Double joists are also used around openings in the floor frame. The double joists are called trimmers and headers, and should be nailed together with 16d nails. The nails should be staggered and clinched to increase their holding power. The joists are doubled only if the span of the header exceeds 4 feet. If the header is longer than 6 feet, it should be supported by framing anchors or a partition, beam, or wall. If the tail joists are over 12 feet in length, they should also be supported by framing anchors. If the opening is around a fireplace, the header and

FIGURE 4-12. Placement of a double joist

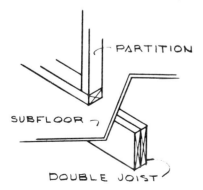

trimmers should be placed a minimum of 2 inches from the outside face of the chimney or fireplace masonry. (See Figure 4-13.)

## Notching and Boring of Joists

To maintain their optimum structural strength, the joists should not be notched or bored. But if it is necessary to notch the joists for piping, they should not be notched more than one-sixth of their depth; and the notches should never be placed in the middle third of the span. If holes are bored in the joist, their diameter should not exceed one-third the depth of the joist. Moreover, the holes should not be placed within 2 inches of the top or bottom of the joist. If the joist is notched to fit a ledger strip, the notch should not exceed one-fourth of its depth.

## Overhang of Floors

If a second-floor joist overhangs or projects beyond the first floor and the overhanging wall is at right angles to the joist, the joist may be cantilevered over the supporting wall. (See Figure 4-14.) But if the overhanging wall is parallel to the supporting joist, a double joist may be used to support the lookout joist. (See Figure 4-15.) The lookout joists are supported by framing anchors or a ledger strip. The double joist should be placed back from the supporting wall at twice the distance of the overhang.

## Bridging

Solid bridging or cross-bridging is used to increase the rigidity of the floor joist. Solid bridging is cut from joist stock and is placed between the joists. (See Figure 4-16.) Cross-bridging is usually made from

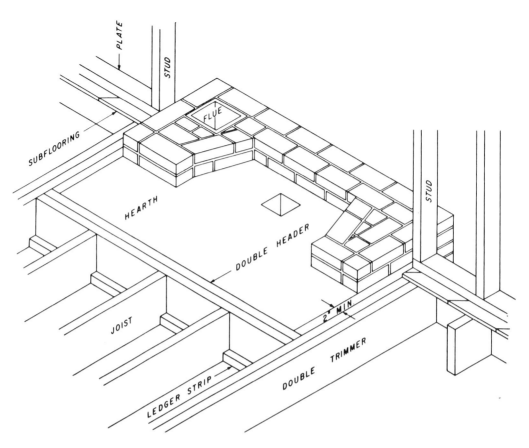

FIGURE 4–13 (above)  Floor framing around a fireplace, and 4–14 (below) Overhang of second floor

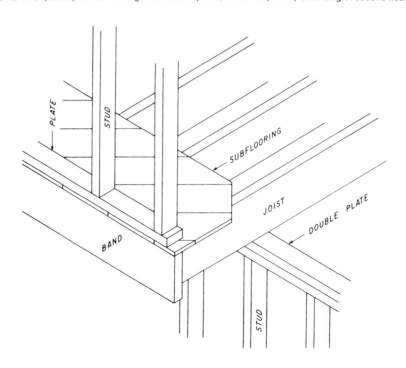

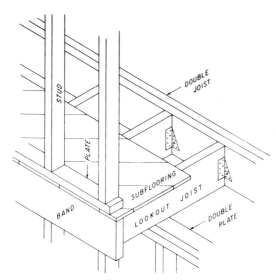

FIGURE 4-15. Lookout joist supported by a double joist

FIGURE 4-16. Solid bridging and cross bridging

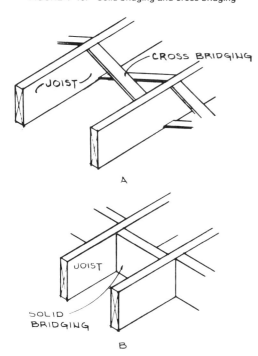

1 × 4 stock placed at an angle to the joist, as shown in Figure 4–16. When cross-bridging is used, the lower end of the bridging is not nailed until the subflooring is placed. In most cases, bridging is placed on 8-foot centers and, if the span is larger than 16 feet, two rows of bridging is required.

Although the primary purpose of bridging is to increase the rigidity of the joist and to evenly distribute the dead and live loads, some studies indicate that bridging serves no structural purpose. Nevertheless, many building codes still require bridging to be placed on 8-foot centers.

## Subflooring

To complete the floor frame, a subfloor is placed over the joist. The subfloor is usually made of tongued-and-grooved boards or plywood. The tongued-and-grooved boards are usually placed diagonal to the joists, to increase the lateral bracing and to allow the finish floor to be laid either parallel or perpendicular to the joist. The boards are nailed at each joist with two 8d nails. In most cases the boards are tightly matched, but to avoid warpage ⅛-inch cracks are sometimes left between the boards.

If plywood subflooring is used, it should be installed with the grain of the outer plies perpendicular to the floor joist. When the panels are placed, they should be staggered so that end joints in adjacent panels occur over different joists. When the plywood subflooring is being placed, a 1/16-inch space should be left between all panel end joints and a ⅛-inch space should be left between all panel edge joints. If ½-inch plywood is used as a subfloor, it should be secured with 6d common nails; 8d common nails are used for thicknesses from ⅝ to ⅞ inch; and 10d common nails are used for thicknesses from 1⅛ to 1¼ inch. The nails should be spaced at 6-inch intervals along the panel edges, and when the plywood subfloor is nailed to intermediate supports the nails should be placed at 10-inch intervals.

If resilient flooring, wood block, carpeting, or ceramic tile is to be placed over the subfloor without separate underlayment, a combination subfloor-underlayment grade of plywood should be used.

To increase the stiffness of the floor and also reduce squeaks and nail pops, the subfloor can be glued. When the plywood is bonded to the joists, the floor stiffness is 70 percent greater than for conventional floor construction. If a glued-floor system is used, the joists can span further distances, or the size of the joists can be reduced.

In laying a glued-floor system, a chalkline is snapped 4 feet from the outside edge of the wall. Glue is then spread across the joists in ¼-inch beads. The glue should be placed for only one or two panels at a time, to prevent drying. With the chalkline used as a guide, the first panel is placed with the tongue side to the wall. Once the first row of panels has been placed, glue should be inserted in the panel grooves. Additional panels are then placed adjacent to the first course. A ¹⁄₁₆-inch space should be left between all end and side joints. A

FIGURE 4–17.  Spacing the panels

gauge can be used to properly space the panels. (See Figure 4–17.) Before the glue drys, nails are placed 12 inches on center along the joist.

## WALL FRAMING

Once the subfloor has been properly placed, the exterior and interior walls can be raised. The exterior walls and load-bearing interior walls must be strong enough to support vertical loads from floors, construction materials, and roofs. The walls must also be strong enough to withstand any lateral pressures created by strong winds and earth tremors.

## Studs

The studs are the most numerous framing members in a wall. They are usually 2 × 4s placed on 16- or 24-inch centers. In the past, the studs were almost always placed on 16-inch centers, but recently there has been a trend toward use of the 24-inch centers. The apparent reason for this change is an economic one, to effect a savings on material as well as labor. But some building codes will not permit the use of 24-inch center spacing.

## Corners

At the intersection of two exterior walls, studs are nailed together to form a nailing base for the interior wall finish. The corner can be constructed before or during the erection of the wall frame. If it is constructed at the same time, it is usually built in two phases. The first phase includes the nailing of two studs to spacer blocks. The entire wall frame is then raised and the corner completed when a third stud is nailed to the previously placed stud and spacer block. (See Figure 4–18.) This technique is

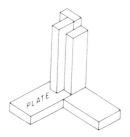

FIGURE 4–19. Corner construction

fast and is used by many contractors.

If the corner is constructed before the wall frame is built, three spacer blocks are first spiked between two full-length studs. A third stud is then nailed to the two studs and spacer blocks. It is also possible to construct a corner by first nailing two 2 × 4s together to form an L. The last stud in the adjacent wall completes the corner. (See Figure 4–19.)

## Partition Intersections

A nailing base for the interior wall finish must also be provided at the intersection of an exterior wall and an interior partition. The structural member that is placed at the intersection of an exterior wall and interior partition is called a *tee*. The tee is usually constructed of three spacer blocks and three studs. (See Figure 4–20.) The three spacer blocks are first nailed to a full-length stud. Two additional studs are then nailed to the spacer blocks.

## Framing Rough Openings

Rough openings are left in the frame wall for windows and doors. To support the

FIGURE 4–18. Completed corner

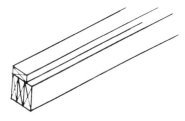

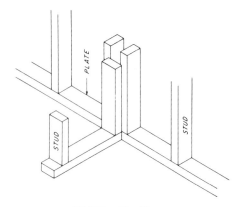

FIGURE 4-20.   The tee

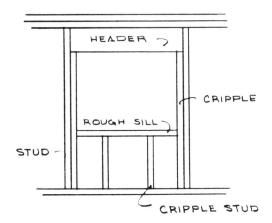

FIGURE 4-21.   Rough opening for a window

weight of the floor, ceiling, and roof, each rough opening is topped by a header, which is usually constructed from nominal 2-inch stock separated by a ½-inch shim. The shim is used to bring the headers to the thickness of the wall.

The size of the header is determined by what the span must support—one floor, ceiling, and roof, or only a ceiling and roof. Header sizes are shown in Table 4-4.

To provide a nailing base for a window, *cripples* (studs that are cut less than full length) are placed beneath the header. (See Figure 4-21.) Cripples are cut from full-length studs and should be nailed to full-length studs. A rough sill is placed parallel to the header to complete the rough open-

ing for a window. The rough sill is supported by cripple studs placed 16 to 24 inches on center. The rough opening for a door does not require a rough sill.

The size of the rough opening is determined by the height and width of the door or window. Rough window openings are usually 1 inch larger than the height and width of the window frame. This allows ½-inch clearance on each side of the window for adjusting and plumbing. The rough opening for a door is usually 3 inches wider and taller than the door, which allows for the head and side jambs, threshold, and the shims behind the hinges. When the door is placed in the rough opening, there should be a ¾-inch clearance between the cripple and side jamb.

## Plate Layout

A plate is a flat horizontal member, connected to both the top and bottom of the studs. Before the plates are laid out, the subfloor must be properly laid out. To accomplish this, chalk lines are made to indi-

**TABLE 4-4   MAXIMUM SPAN FOR HEADERS**

| Header Size | One Floor, Ceiling Roof | Ceiling and Roof |
| --- | --- | --- |
| 2-2 × 4 | 3′-0″ | 3′-6″ |
| 2-2 × 6 | 5′-0″ | 6′-0″ |
| 2-2 × 8 | 7′-0″ | 8′-0″ |
| 2-2 × 10 | 8′-0″ | 10′-0″ |
| 2-2 × 12 | 9′-0″ | 12′-0″ |

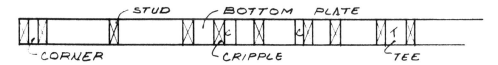

FIGURE 4–22. Corner, cripple, and tee layouts

cate the location of the bottom plate. A chalkline 3½ inches from the outside edge of the building line, or brick ledge, is used to indicate the outside frame wall. Parallel chalklines 3½ inches apart represent the location of the interior walls.

The studs will be nailed between the top plate and the bottom plate. The plates should be constructed from straight 2 × 4 stock. If the bottom plate is placed on concrete it should be pressure-treated with a wood preservative, to prevent decay and attack by insects.

If anchor bolts are used, the locations should be transferred to the bottom plate. To do this, the bottom plate must be placed around the perimeter of the building and adjacent to the anchor bolts. A square is then used to scribe two parallel lines across the bottom plate. Measuring from the chalkline, the distance is then transferred to the bottom plate and a hole is bored for the anchor bolt to slip through.

The top and bottom plates are first placed around the perimeter of the building. The plates are then laid out together, marking the stud locations and rough openings. To locate the studs, two parallel lines must be marked and an X placed between them. A corner is noted by three parallel lines and two X's. A cripple is noted by placement of a C between the two parallel lines. (See Figure 4–22.) A tee is indicated by four parallel lines and two X's.

Once the plates are laid out, the top plate is separated from the bottom plate

and studs are placed between the two plates. The top plate is then nailed to the studs with two 12d or 16d common nails.

The bottom plate can also be straight-nailed to the studs, or it can first be nailed to the subfloor, after which the wall frame is raised and the studs are toenailed to it. The straight-nailing technique is much faster than toenailing, but the toenailing technique is stronger.

As the wall frame is constructed, the rough openings are framed out and the corners and tees are placed in their proper locations. If the corners are to be braced with let-in braces, a 1 × 4 should be placed at an angle to the wall frame. (See Figure 4–23.) The 1 × 4 should form an angle to a minimum of three studs. The 1 × 4 location should then be marked and the 1 × 4 removed so that the let-in notches may be cut. When this is done, the 1 × 4 is placed in the notches and temporarily nailed.

When the wall frame is complete, a cap

FIGURE 4–23. Corner bracing

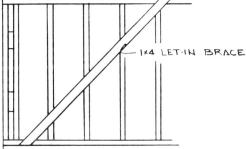

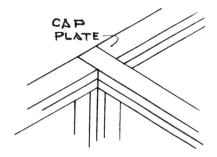

FIGURE 4-24.  Placement of the cap plate

plate is nailed to the top plate. The cap plate is used both to strengthen a load-bearing frame wall and to bring the ceiling to the proper elevation. The cap plate should not be continuous, but should have 3½-inch gaps at the intersections of interior and exterior walls. (See Figure 4–24.) The gap will allow the intersecting cap plate to lap over the top plate of the adjacent wall, tying the two walls together.

After the cap plate has been placed, the wall can be raised and temporarily braced. When all the walls, both exterior and interior, are in position the corners must be plumbed and braced.

## Corner Bracing

Before the corner can be braced, it must be plumbed. In most cases a carpenter's level is used to plumb the corner. While the level is in position and the corner is plumb, the wall is braced.

A wall can be braced with plywood, by 3 1 × 4 let-in braces, or by metal corner bracing. If 1 × 4s are used as corner bracing, they should be placed flush with the studs and should extend diagonally from the top of the cap plate to the bottom of the bottom plate. Plywood or fiberboard is becom-

ing increasingly popular as corner-bracing material. The large panels can be placed quickly, and there is no need to notch the studs. A 2-inch metal strap can also be used to prevent the wall from being racked. The strap is usually 1/16-inch thick and should extend diagonally across a minimum of three studs.

## Aligning the Walls

After the corners are plumbed, the walls must be properly aligned. A string is first stretched around the perimeter of the wall frame. The string is attached to the cap plate, but a 2 × 4 block should be placed between the string and the cap plate. (See Figure 4–25.) Once the string has been placed, a 2 × 4 can be used to determine whether or not the wall is straight. If a 2 × 4 gauge can be placed between the string and cap plate and there is no appreciable distance between the string and gauge, the wall is straight. But if the gauge

FIGURE 4-25.  Wall alignment

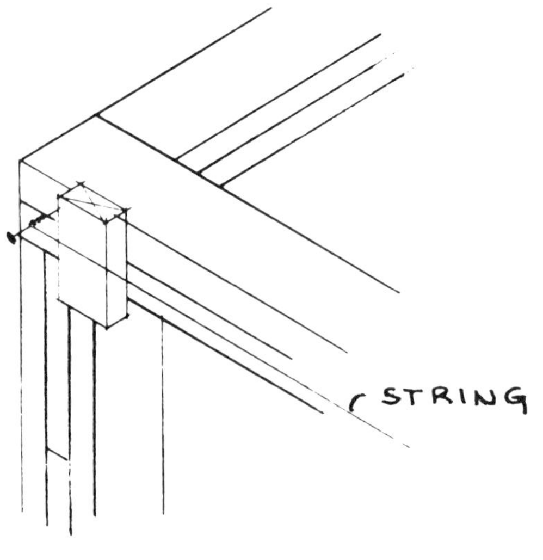

cannot be placed between the string and cap plate, or if there is a distance of more than $\frac{1}{16}$ inch between the gauge and string, the wall must be straightened. To straighten the wall a temporary brace must be nailed to the top of a stud in the exterior wall and to the bottom of a stud in an interior wall. The temporary brace should be placed on approximate 4-foot centers.

## Wall Sheathing

Wall sheathing is used to increase the strength of the wall frame, prevent air infiltration, and provide a nailing base for siding. There are four basic types of wall sheathing; wood boards, fiberboard, gypsum, and plywood sheathing.

Wood boards are usually tongued-and-grooved of shiplapped, and are covered by sheathing paper. The boards can be placed either diagonally or horizontally. Regardless of how they are placed, their end joints should fall over a stud. The boards should be nailed to the studs with two 8d nails. Wood board sheathing, not as popular as it once was, is the oldest type of wall sheathing and is still used in some areas.

Fiberboard wall sheathing is made from either wood pulp, sugarcane fibers, or cornstalks and is one of the most popular types of wall sheathing. The standard fiberboard panel is $\frac{1}{2}$ inch thick and 4 feet by 8 feet in width and length. The panels can be placed either vertically or horizontally, and are attached to the frame wall with $1\frac{1}{2}$-inch roofing nails. The nails should be placed on 3-inch centers around the edges and 6 inches on center at the intermediate supports.

Gypsum sheathing is fireproof and is highly resistant to water absorption. The sheathing is available in 8- and 9-foot

lengths and 2- and 4-foot widths. The panels are $\frac{1}{2}$ inch thick and have either "V" tongue and groove or square edges. If the panels are applied vertically, there is no need for corner bracing. The panels are attached to the wall frame with $1\frac{1}{2}$-inch roofing nails spaced 6 inches on center.

Plywood wall sheathing adds strength and rigidity to the wall frame. If plywood is used for wall sheathing, let-in bracing and building paper can be omitted. The panels can be applied either horizontally or vertically, although horizontal application is recommended when shingles are to be nailed directly to the wall sheathing. The horizontal application increases the stiffness of the wall frame under loads perpendicular to the surface. Vertical application increases the racking resistance of the wall frame. The panel thicknesses are $\frac{5}{16}$, $\frac{3}{8}$, $\frac{1}{2}$, and $\frac{5}{8}$ inch. They are nailed to the wall frame with 6d nails spaced 6 inches on the panel edges. The nails on the intermediate framing members are spaced 12 inches on center. Table 4–5 can be used to select the correct plywood wall sheathing.

**TABLE 4–5   PLYWOOD WALL SHEATHING**

| Panel Identification Index | Mimimum Thickness (inch) | Maximum Stud Spacing (inches) Exterior Covering Nailed To: | |
| --- | --- | --- | --- |
| | | Stud | Sheathing |
| 12/0, 16/0, 20/0 | $\frac{5}{16}$ | 16 | 16[1] |
| 16/0, 20/0, 24/0 | $\frac{3}{8}$ | 24 | 16 |
| | | | 24[1] |
| 24/0, 30/12, 32/16 | $\frac{1}{2}$ | 24 | 24 |

[1] When sidings such as shingles are nailed only to the plywood sheathing, or if siding joints are not over studs, apply plywood with face grain scross studs.

Reprinted with permission of American Plywood Association.

## CEILING FRAME

If roof trusses are not used, the ceiling frame is placed after the walls and partitions have been erected, plumbed, and braced. The ceiling frame is usually constructed of nominal 2-inch wide stock called ceiling joists. Ceiling joists are used to tie the exterior walls together and form a nailing surface for the ceiling finish materials. The size of the ceiling joists is determined by the span, species of wood, and their spacing. In most cases the joists are placed on 16- and 24-inch centers.

Before the joists are placed they should be crowned, and a trim cut should be made on the end of each joist. The trim cut indicates the direction of the crown and is made to prevent the joist from protruding over the top edge of the rafter. The crown should always be placed up.

Once the joists have been crowned, they can be placed on top of the cap plate. In most cases the joists span the narrowest dimension and all the joists are placed in the same direction. But the joists can also be placed in opposite directions. If a hip roof is used, the first ceiling joist is omitted and stub ceiling joists are placed perpendicular to the first ceiling joist. (See Figure 4–26.) Regular ceiling joists cannot be used in this

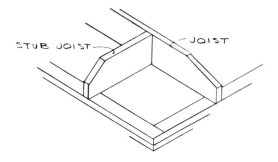

FIGURE 4–26.   Stub joist

case because the sloping rafters would hit the joists.

The joists should be toenailed to the cap plate and later nailed to the rafter. The ends of the ceiling joists rest on interior partitions and are also toenailed to the cap plate. Two intersecting ceiling joists should be nailed to each other. Ceiling joists spanning a long distance can be supported by a beam. The ends of the joists should rest on ledger strips or joist hangers.

In some cases it may be necessary to provide a nailing base for ceiling finish material, especially if the joist runs parallel to a partition. To provide the nailing base, a 2 × 6 is centered over the cap plate and nailed into place with 16d nails.

## ROOF FRAME

Once the ceiling joists have been placed, the roof frame can be cut and assembled. A typical roof frame is built of various framing members, each playing a significant role in the construction process. The more common parts of the roof frame are jack rafters, common rafters, valley rafters, ridges,

hip rafters, purlins, purlin studs, and wind beams.

There are three different types of jack rafters: hip jack; valley jack; and cripple jack. Figure 4–27 shows the first two. The hip jack extends from the cap plate to the hip rafter. The valley jack rafter extends

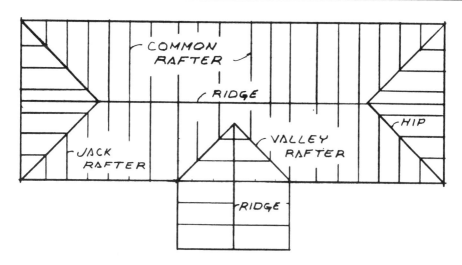

FIGURE 4–27.  Common, jack, valley and hip rafters

from a valley to a ridge, and the cripple jack extends from valley rafter to hip rafter.

A common rafter is placed at a right angle to the exterior wall frame and extends from the cap plate to the ridge, as shown in Figure 4–27. The hip is placed at a 45-degree angle to the exterior wall frame and extends from a corner of the wall frame to the end of the ridge. The valley is the lowest part of two intersecting roofs and extends from an inside corner to a ridge.

The purlin, purlin studs, and wind beams are used to brace the roof frame, as shown in Figure 4–28. A purlin is a horizontal $2 \times 4$ placed perpendicular to the rafters, and extends from a load-bearing wall to a rafter. The wind beam ties two rafters together and is placed parallel to the ceiling frame.

FIGURE 4–28.  Roof-bracing detail

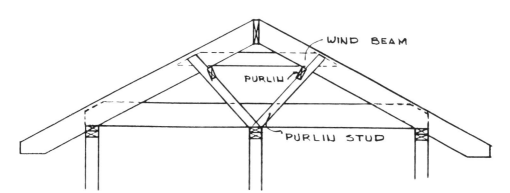

## Layout Terms

In the process of framing a roof and laying out the rafters, there are certain terms that must be understood.

*Slope* is a ratio of the vertical rise to the horizontal run. The ratio is expressed as *x* distance in 12. If the slope is 4-in-12 the roof will rise 4 inches for every horizontal foot. If the slope is not used to indicate the vertical rise of a rafter, the pitch is usually given.

The *pitch* of a roof is the ratio between the vertical rise and the span. If a roof has a rise of 5 feet and a total span of 20 feet, the pitch would be ¼.

The *span* is the horizontal distance between two walls. (See Figure 4-29.) It is considered to be the width of the building. The *run* is half the distance of the span. It is the horizontal distance covered by one common rafter.

The *rise* is a vertical distance measured from a point on the ridge that intersects a line on the common rafter. This line is projected from the outside edge of the plate.

The *unit run* is always considered to be one foot. The *unit rise* is a vertical distance measured from the end of the unit run. In light construction the unit rise is usually 3, 4, 5, or 6 inches.

## Common Rafters

Common rafters are usually laid out with a framing square. When the framing square is used, the length of the common rafter is stepped off. With the tongue held in the left hand and the blade in the right, the rafter tail is laid out by placement of the unit rise on the tongue and the unit run on the blade. (See Figure 4–30.)

If the rafter tail has an overhang of 2 feet, two steps are taken. The last point measured on the rafter tail indicates the starting point for the birdsmouth. A birdsmouth is the portion of a structural member that has been cut out so that the member may fit over a cross-timber. The birdsmouth should be laid out to a depth of 1½ inches. Measured from the birdsmouth, the rafter is stepped off with the framing

FIGURE 4–29.   The span of a building and the rise, or roof slope

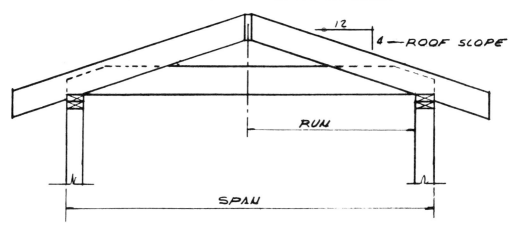

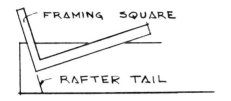

FIGURE 4–30.   Rafter-tail layout

FIGURE 4–31.   Odd-unit layout

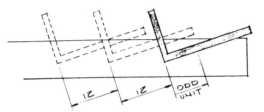

square. The number of steps should equal the run. If the run has a fraction of a foot, the unit measurement is taken on the last step, but instead of marking the unit run at 12, the mark must be placed adjacent to the inch-mark equal to the fraction of the run. (See Figure 4–31.) The last point or line laid out represents the run but, to compensate for the thickness of the ridge, a plumb-cut line should be placed to the left of the previously placed line, at a distance equal to half the width of the ridge.

## The Ridge

The ridge, the uppermost portion of a roof, is a framing member to which the rafters are nailed. On a gable roof the ridge extends the length of the building. For a hip roof the length of the ridge is equal to the length of the building minus the width of the building. Once the ridge has been cut to proper length, the location of the rafters is marked on the ridge. In most cases the raft-

ers are spaced 16 or 24 inches apart, depending on the span, size of rafter, and species of wood.

## Hip and Valley Rafters

Hip and valley rafters can be laid out in the same manner as a common rafter, but rather than using a unit run of 12 inches, one of 17 inches should be used. This is because the hip or valley rafter is the diagonal of a 12-inch square, which is 16.97 inches. The measurement is rounded off to 17 inches when a hip or valley rafter is laid out. (See Figure 4–32.) If the run has a fraction of a foot the unit measurement is taken in the last step, but instead of the unit run being marked at 17 inches, the mark is placed adjacent to the inch-mark equal to the fraction of the run.

Hip and valley rafters must also be shortened at the plumb cut, because the run was measured from a theoretical point at the intersection of hip and ridge. Since there is no way for the two framing members to meet at that exact location, the hip

FIGURE 4–32.   Diagonal of a square: the unit run of a hip rafter is the diagonal of a 12-inch square

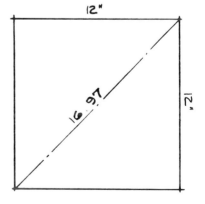

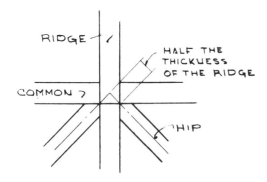

FIGURE 4–33.   Shortening a hip rafter

or valley rafter is shortened by half of the 45° thickness of the ridge. (See Figure 4–33.) The rafter tail will be stepped off the same as a common rafter, but using a unit run of 17 inches.

## Jack Rafters

Jack rafters extend from the cap plate to the hip rafter and are evenly spaced. The difference in the length of adjacent jack rafters is called the common difference in length. (See Figure 4–34.) Common difference in length can be found by first placing the framing square on a piece of rafter stock. The unit rise and unit run should be positioned correctly on the framing square.

FIGURE 4–34.   The common difference in the length of jack rafters

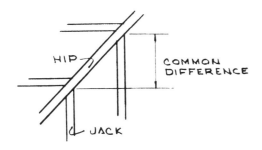

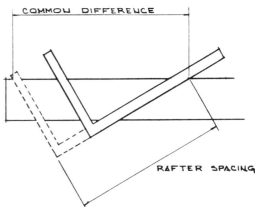

FIGURE 4–35.   Determining the common difference

Then a line should be drawn along the top of the blade, and the framing square moved along the line to a point that equals the spacing of the rafters. A horizontal measurement should then be taken between the tongue and the blade: this distance is the common difference in length for jack rafters. (See Figure 4–35.) But because of the thickness of the hip, the common difference must be shortened one-half the 45° thickness of the hip. With a common rafter used as a pattern, the common difference can be laid out the length of the rafter.

This same technique can be used for laying out valley jack rafters.

## Erecting the Roof Frame

After the roof frame has been cut, the rafters are placed around the perimeter of the building. The ridge is then placed on the ceiling frame and a rafter is nailed to each end of the ridge with 16d or 12d nails. The ridge is then lifted to its approximate height and the rafters are toenailed to the cap plate and straight-nailed to the ceiling joist. Two additional rafters are then placed and secured on the other side of the ridge, opposite the previously placed rafters. Additional rafters can then be placed, and the plumb cut nailed to the ridge and the lower portion nailed to the cap plate and joist.

When a hip or valley rafter is placed it should also be nailed to the ridge and the cap plate. Before the jack rafters are placed, a string should be placed down the center of the hip or valley rafter. The string serves as a guide to help straighten the hip or valley rafter. The jacks are then placed, with the beveled plumb cut nailed to the hip or valley rafter and the lower end nailed to the ceiling joist and cap plate.

If the roof has a gable, the ridge is sometimes allowed to extend past the building the length of the overhang, and a ladder is constructed to support the overhang. (See Figure 4–36.) The ladder should extend from one common rafter to a barge rafter placed on the ends of the lookouts. A barge rafter is a part of the roof frame that is supported by lookouts, ridge, and rough fascia. To support the ladder, the gable end is framed in with gable end studs.

## Bracing the Roof

To properly brace a roof, 2 × 4 purlins should be nailed perpendicular to the rafters, in most cases 48 inches below the ridge line. Purlin studs are cut and placed on every other rafter. The purlin stud should be nailed to a rafter and to the cap plate. In order to support the purlin and decrease

FIGURE 4–36. Roof framing for a gable end

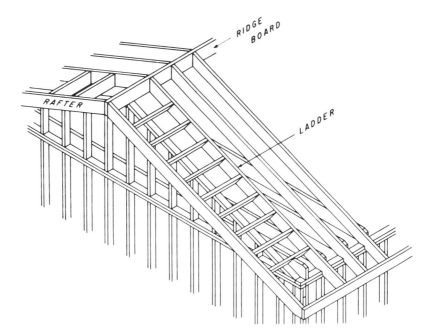

the amount of deflection in the roof, the purlin stud is notched, or a short $2 \times 4$ is placed directly below the purlin and then nailed to the purlin stud. To complete the roof bracing, $2 \times 4$ or $1 \times 8$ wind beams are nailed to the rafter. The wind beams should also be nailed to every other rafter.

To increase the rigidity and reduce the number of crowns in the ceiling joist, a stiffback is sometimes placed perpendicular to the ceiling joist. The stiffback is constructed from a $2 \times 4$ and a $2 \times 6$ nailed together in the shape of an L. The $2 \times 6$ is placed with its grain parallel to the ceiling joist, and the $2 \times 4$ is placed with its grain perpendicular to the ceiling joist.

## Roof Trusses

In light construction, roof trusses have virtually replaced the traditional joist and rafter system. The trusses are easier to erect and they are actually stronger than a joist and rafter system. Recent studies indicate that trusses spaced 24 inches on center are twice as strong as conventional framing that is placed 16 inches on center. In addition to the added strength, trusses save time and material. Greater room flexibility is also gained because trusses require no bearing partitions.

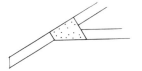

FIGURE 4–38.  Gusset plate

Trusses are available in many different styles and sizes, but the most popular truss is the standard Fink or W truss, shown in Figure 4–37. A truss is built of a top chord, bottom chord, and webs. The webs are used to tie the top and bottom chords together. The webs and chords can be joined by one of several techniques, but the most popular fasteners are split rings, metal plates, or plywood gusset plates. (See Figure 4–38.)

Trusses can be fabricated in a shop and delivered to the job site, or they can be built in the field. Regardless of where they are built, they should be accurately cut to length and angle. They have to be built in special jigs, with the members held tightly in place until the connector plates, or rings, have been properly located and placed.

After the trusses have been fabricated, they should be stored in a vertical position and should rest on temporary bearing supports. During the erection process the trusses should be carefully handled and properly braced so they will stay true and

FIGURE 4–37.  Standard fink, or "W" truss

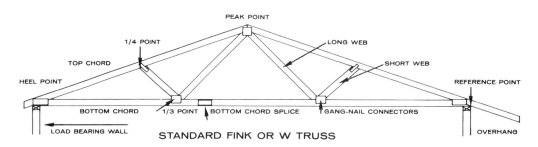

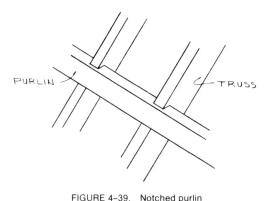

FIGURE 4-39. Notched purlin

FIGURE 4-40. Roof overhangs

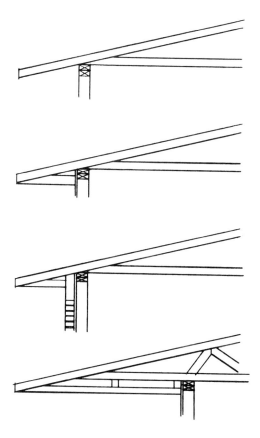

plumb. Truss hangers can be used to secure a truss to the cap plate, thus eliminating toenailing and providing a stronger attachment.

The end truss is the first truss placed, and should be plumbed and braced. Then additional trusses are placed in the truss hangers and equally spaced at the top by means of a notched purlin. (See Figure 4-39.)

In addition to the regular trusses, gable end trusses can be fully studded, or they can be studded to accommodate a triangular louver.

The plumb-cut overhang is the most popular type of roof overhang. It can be constructed with a square cut; a soffit return can be attached; an overhang return for brick veneer can be added; or a cantilevered truss with a full return can be constructed, as shown in Figure 4-40. In addition, half truss systems, hip louver, and jack truss systems can be fabricated.

## Roof Sheathing

Roof sheathing is used to provide a covering or skin over the roof frame. If also increases the structural rigidity of the roof frame, provides a nailing base for roofing materials, distributes live loads to the framing members, and can act as a vapor barrier and provide insulation. There are three basic types of roof sheathing used in light construction: spaced sheathing, plywood sheathing, and nominal 1-inch-thick boards.

Spaced sheathing is used primarily when wood shakes are selected as a roof covering. The sheathing consists of 1 × 4s spaced on centers equal to the weather exposure of the wood shakes. (See Figure 4-41.) The

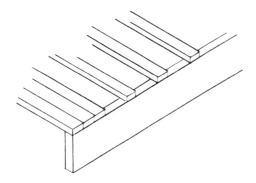

FIGURE 4–41.   Spaced sheathing

sheathing should not be placed on centers greater than 10 inches. The boards are secured to the roof frame with two 8d nails placed at each framing member.

Plywood sheathing is used extensively in light construction, for the large panels go down fast and cover a large area. The panels also form a smooth base with few joints.

The plywood panels should be placed with the face grain perpendicular to the framing members. The thickness of the plywood and the spacing of the framing members should conform to the specifications given in Table 4–6. As the panels are placed, the joints should be staggered. A $\frac{1}{16}$-inch space should be left at all end joints, and a $\frac{1}{8}$-inch space at all edge joints. The panels should be nailed 6 inches on center along the edges and 12 inches on center at intermediate supports. The nails should be 6d common for panels $\frac{1}{2}$-inch or less, and 8d for panels thicker than $\frac{1}{2}$ inch. The edges of the panels can be supported by clips, tongue-and-groove panels, or lumber blocking between joist.

Nominal 1-inch-thick boards are usually

### TABLE 4–6   PLYWOOD ROOF DECKING

| Identification Index | Plywood thickness (inches) | Maximum span (inches) | Unsupported edge-max. length (inches) | Allowable live loads (psf) | | | | | | | | | |
|---|---|---|---|---|---|---|---|---|---|---|---|---|---|
| | | | | Spacing of supports center to center (inches) | | | | | | | | | |
| | | | | 12 | 16 | 20 | 24 | 30 | 32 | 36 | 42 | 48 | 60 |
| 12/0 | $\frac{5}{16}$ | 12 | 12 | 150 | | | | | | | | | |
| 16/0 | $\frac{5}{16}, \frac{3}{8}$ | 16 | 16 | 160 | 75 | | | | | | | | |
| 20/0 | $\frac{5}{16}, \frac{3}{8}$ | 20 | 20 | 190 | 105 | 65 | | | | | | | |
| | $\frac{3}{8}$ | | 20 | | | | | | | | | | |
| 24/0 | $\frac{1}{2}$ | 24 | 24 | 250 | 140 | 95 | 50 | | | | | | |
| 32/16 | $\frac{1}{2}, \frac{5}{8}$ | 32 | 28 | 385 | 215 | 150 | 95 | 50 | 40 | | | | |
| 42/20 | $\frac{5}{8}, \frac{3}{4}, \frac{7}{8}$ | 42 | 32 | | 330 | 230 | 145 | 90 | 75 | 50 | 35 | | |
| 48/24 | $\frac{3}{4}, \frac{7}{8}$ | 48 | 36 | | | 300 | 190 | 120 | 105 | 65 | 45 | 35 | |
| 48/24[1] | $\frac{3}{4}, \frac{7}{8}$ | 48 | 36 | | | | 225 | 125 | 105 | 75 | 55 | 40 | |
| 2-4-1 $1\frac{1}{8}''$ | $1\frac{1}{8}$ | 72 | 48 | | | | 390 | 245 | 215 | 135 | 100 | 75 | 45 |
| Grp. 1 & 2 $1\frac{1}{4}''$ | $1\frac{1}{8}$ | 72 | 48 | | | | 305 | 195 | 170 | 105 | 75 | 55 | 35 |
| Grp. 3 & 4 | $1\frac{1}{4}$ | 72 | 48 | | | | 355 | 225 | 195 | 125 | 90 | 65 | 40 |

[1] Loads apply only to C-C EXT-APA, STRUCTURAL I C-D, INT-APA, and STRUCTURAL I C-C EXT-APA. Check availability before specifying.

Reprinted with permission of American Plywood Association.

6 to 8 inches wide and should be placed perpendicular to the framing members. To provide adequate support for the roofing, they should be laid close together. The boards are usually shiplapped or tongued-and-grooved. To secure the boards two 8d nails should be placed at each framing member.

## METAL FRAMING

In recent years many residential and light commercial structures have used metal framing members. (See Figure 4–42.) The framing members offer long-term stability, fire safety, adaptability, and a high strength-to-weight ratio.

In most cases the framing members are placed much the same as in wood construction practice. The metal studs and roof framing members are usually spaced on 2-foot centers to accommodate wall sheathing materials. To provide for a proper bearing

FIGURE 4–42. Metal framing

surface for transfer and distribution of loads from the studs, the studs are placed in metal runner tracks.

If an extra load capacity is needed, the studs can be nested together to form tubular sections. This particular technique is often used at door and window openings. The header over the opening can be a typical X-bracing header, plywood-composite beam, or an engineered metal channel 8 to 12 inches in depth.

The metal-to-metal joining is done by spot-welding, crimping, or with self-tapping screws. The exterior and interior finish materials are attached with self-drilling screws.

# CHAPTER 5

# CORNICE
# CONSTRUCTION

THE CORNICE (SEE FIGURE 5–1) IS A PART OF the exterior trim that is used to complement a particular architectural style, insulate the eave line, and in some cases to protect the sidewall from the elements.

The cornice may be either open or closed, but a closed cornice is usually preferred. The closed cornice is subdivided into two basic classifications—the horizontal soffit and the sloping soffit. If a closed cornice has a horizontal soffit, the soffit is nailed to framing members that are placed perpendicular to the frame wall. But if a sloping soffit is used, it is nailed directly to the sloping rafter tails. The soffit is sometimes referred to as the *plancier* and is used to cover the exposed rafter tails. It is usually constructed of hardboard, plywood, or aluminum, but gypsum board or plaster can also be used. Regardless of the material used, most soffits are constructed with some type of opening—either small perforations or larger openings cut for ventilation.

In addition to the soffit, the cornice includes the fascia, drip edge or shingle strip, rough fascia, lookouts, ledger, deadwood, frieze, and molding. The fascia is usually a $1 \times 6$ and is used to close the ends of the rafter tails. The fascia can be plain, or it can be slotted to receive the soffit. The slotted type is more desirable, for it often eliminates the necessity of scribing and cutting the soffit. Because the fascia will be exposed to an excessive amount of moisture, a material that is highly resistant to decay should be used—cedar, cypress, or redwood is usually recommended.

A drip edge is placed under the underlayment and over the fascia, to help shed water at the roof's edge. There are many different styles of drip edges, but most are formed from 26-gauge galvanized steel. They extend on the roof sheathing approximately 3 inches and are bent over the fascia, causing the water to drip free of the fascia. A shingle strip is used for the same purpose as a drip edge, but it is usually constructed from a $1 \times 2$. If a $1 \times 2$ is used, it is nailed to the finished fascia with 6d hot-dipped galvanized nails, aluminum nails, or stainless steel nails.

A rough, or false, fascia is sometimes nailed directly to the ends of the rafter tails. A rough fascia is not always neces-

FIGURE 5–1.   The cornice

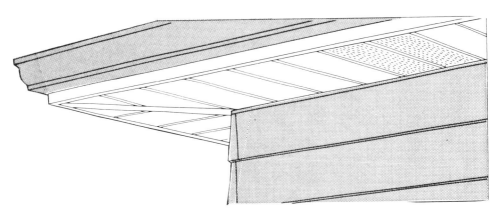

sary, but it provides a straight and level nailing base for the finish fascia. If a rough fascia is used it should be secured to each rafter tail with two 12d nails.

A 2 × 4 ledger is nailed directly over the wall frame, serving as a nailing base for the lookouts. On a hip roof the 2 × 4 ledger runs the perimeter of the building, but on a gable roof the ledger extends past the building at a distance equal to the roof overhang.

Lookouts are usually constructed of 2 × 4s and are used to support the soffit. Lookouts are placed perpendicular to the wall frame, and extend to the rough or finish fascia. To provide adequate support for the soffit, the lookouts are usually placed on 16- to 48-inch centers.

Deadwood serves no structural purpose other than serving as a nailing base for the frieze and molding. Deadwood is usually constructed of 2 × 4s and is only needed for brick veneer buildings. If plywood is used as siding, the frieze can be nailed directly to the siding.

Moldings are thin strips of trim that are used to cover the intersection of two surfaces. They are also used to enhance and complement a particular architectural style. There are three basic types of moldings that are used in cornice construction: cove, crowns/bed, and quarter-round. (See Figure 5–2.) Cove molding has a concave profile and is probably the most widely used of the three. Crowns/beds are used to cover large angles and are considered to be more decorative than cove or quarter-round. The basic difference between crown and bed molding is that crown molding is always sprung, while bed molding may be sprung or plain. The backside of a sprung piece of molding is beveled to allow for a close fit. Quarter-round is less decorative than most molding, but is often used to cover the intersection of the frieze and soffit.

FIGURE 5–2. Close cornice with a horizontal soffit

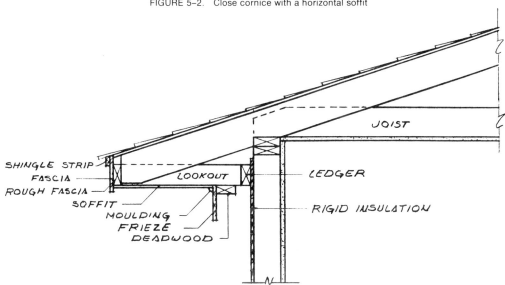

## CORNICE CONSTRUCTION (HORIZONTAL SOFFIT)

The first step in cornice construction is to place a chalkline against the rigid insulation, or building paper. To allow for a slight variance in the overhang of the rafters, the line is placed ½ inch below the rafter tail. With the chalkline used as a guide, the ledger is nailed to the wall frame with 12d nails. Then a rough fascia, if it is used, is nailed to the rafter tails. Once the ledger and rough fascia have been placed, the lookouts are nailed between the ledger and rough fascia. The addition of the lookouts creates a ladder and a supporting agency for the soffit. Many contractors prefer to assemble the ladder on the ground, and then raise the entire assembly into position. This technique is fast and eliminates the need to handle many small pieces of wood on a scaffold. If brick veneer is to be placed over the wall frame, a piece of deadwood should be nailed to the ladder. The front edge of the deadwood should be placed 5½ inches, or a distance equal to the width of the brick ledge, from the frame wall.

Before the soffit is placed, the rough fascia should be straightened. To properly straighten the rough fascia, a string must be nailed on the outside edge and projected away from the rough fascia with a ¾-inch block. (See Figure 5–3.) With the string as a guide and the block as a gauge, the rough

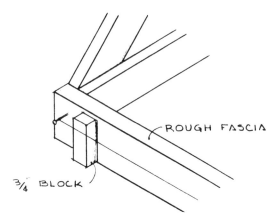

FIGURE 5–3.  Straightening procedure for the rough fascia

fascia can be tapped in or out to produce a straight nailing base. Once the rough fascia is straight, the soffit should be nailed or screwed to the ladder with rust-resistant fasteners. A fascia is then nailed to the rough fascia to cover the rough fascia and finish closing the cornice. In most cases the fascia is allowed to project ½ to ¾ inch below the soffit. To facilitate the placing of the fascia, a gauge block is often used to measure the correct distance from the soffit to the bottom of the fascia. After the fascia has been placed, the drip edge, or shingle molding, can be put in place. At that point, the cornice, the frieze and molding can be cut and fitted to the cornice.

## GABLE END TRIM

There are many variations of gable trim, but in most cases the construction process is similar to that for cornices. (See Figure 5–4.) To terminate the gable trim, a box is framed along the eave line, then covered

FIGURE 5–4.  Gable trim

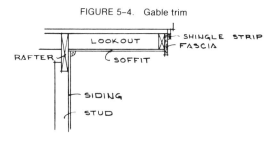

with nominal 1-inch siding. The side of the box can be finished in many ways, but a vertical or scribed cut is usually used to dress the side board.

## PREFABRICATED CORNICE MATERIAL

To eliminate the time-consuming process of cornice construction, many contractors have turned to prefabricated cornice materials such as aluminum, vinyl, or PVC (Poly Vinyl Chloride) soffit systems. All of these allow the soffit to be placed without wood framing members. (See Figure 5–5.) First the J-channel frieze must be secured to the building. The soffit is then slipped into the J-channel frieze and fastened at the fascia. Once the fascia has been placed, the fascia cap is slipped into the trim at the roof line—its lower leg covers the outer edge of the soffit.

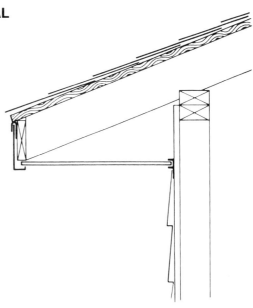

FIGURE 5–5 (right)   Prefabricated cornice material

## OPEN CORNICE

The open cornice has exposed rafter tails and exposed roof sheathing. The roof sheathing should be thick enough to prevent roofing nails or staples from protruding out of the exposed underside. If the sheathing over the open cornice is thicker than the roof sheathing, the intersection should be shimmed at each rafter to provide a flush joint at the change of the plywood thickness. (See Figure 5–6.)

FIGURE 5–6.   Open soffit

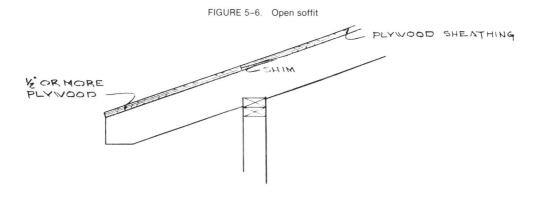

PLYWOOD SHEATHING

SHIM

½" OR MORE PLYWOOD

The open cornice may or may not have a fascia, but a trim board must be placed between the rafters at the wall line. (See Figure 5–7.) The trim board reduces the possibility of air infiltration and it is used to close or finish the siding.

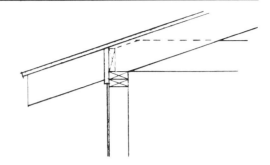

FIGURE 5–7 (right)   Open cornice

## GUTTER SYSTEMS

Gutters are used to collect the water falling from the roof and divert it away from the

FIGURE 5–8.   Gutter parts

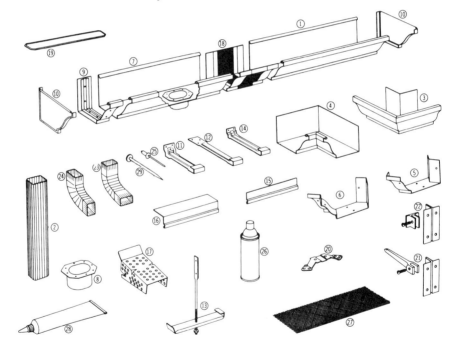

| | | | |
|---|---|---|---|
| **1** Free Floating Gutter | **6** Miter Strip Inside | **11** Regular Hanger | **18** Expansion Joint | **23** 2″x3″ Square Elbow Style A 60° - 75° |
| **2** 2″x3″ Square Downspout | **7** Section With Drop Outlet Tube | **12** Roof Hanger | **19** Downspout Pipe Band | **24** 2″x3″ Square Elbow Style B 75° |
| **3** Miter Section Outside | **8** Wide Flange Outlet Tube | **13** Adjustable Roof Hanger | **20** Downspout Pipe Cleat | **25** Rivets |
| **4** Miter Section Inside | **9** Slip Joint Connector | **14** Fascia Hanger | **21** Downspout Pipe Bracket, Masonry | **26** Touch Up Paint |
| **5** Miter Strip Outside | **10** Rivet Type End Cap | **15** Fascia Apron | **22** Downspout Pipe Bracket | **27** Gutter Cover |
| | | **16** Roof Apron | | **28** Joint Sealer |
| | | **17** Perforated Strainer | | **29** Aluminum Nail |

foundation system. If an excessive amount of moisture is allowed to accumulate around foundation walls, seepage into the basement might occur. Gutters are made in a wide variety of designs and styles, but most are constructed from aluminum, galvanized sheet metal, or vinyl. The entire system consists of many different parts, but some of the more common parts are the gutter, downspout, outside miter, inside miter, joint connectors, and end caps. (See Figure 5–8.)

In gutter installation, first the downspouts, then the midpoint between the downspouts, are located. From these two points a line is chalked along the fascia. The line should be sloped toward the downspout at a rate of $\frac{1}{32}$ inch per foot. The hangers are then screwed to the fascia, with the chalkline as a guide, and finally the gutters are placed in the hangers. The different lengths of gutters can be connected by joint connectors or expansion joints.

Rivets are used to assemble most gutter systems, although other fasteners can be used. Corners in the system are turned by inside and outside miter sections, and the ends are terminated by end caps. When the system has been hung, the joints are finished with gutter seal.

Some of the gutter systems manufactured today are virtually maintenance-free, while others require periodic maintenance. Gutters and downspouts constructed from aluminum and vinyl require very little maintenance; those constructed from galvanized sheet metal usually require periodic repainting and repairs.

# CHAPTER 6

# ASPHALT SHINGLES

## MATERIALS

Asphalt shingles are one of the most popular types of roof covering used, accounting for more than 80 percent of the roof covering on American homes. They are economical, durable, and attractive. They are also fire-resistant, wind-resistant, and are available in a wide variety of colors and styles. The most popular style is the square butt strip shingle. It is rectangular in shape and is available with one, two, or three tabs. The less popular asphalt shingles are the hexagon and the individual shingles.

Asphalt shingles are made of a base mat of either organic (cellulose fibers) or inorganic (glass or asbestos fibers) material. The base mat is saturated with asphalt and surfaced with opaque mineral granules. The asphalt waterproofs the roof and the granules protect the shingles from the sun.

Shingles are sold by the square and the approximate weight per square. A square of shingles will cover 100 square feet of roof surface. The approximate weight ranges from 205 to 380 pounds per square. The most popular weight of asphalt shingles is 240 pounds per square.

## ROOF SLOPE LIMITATIONS

To provide adequate drainage, a roof must be properly sloped. (See Figure 6–1.) If "free tab" square butt strip shingles are used, the degree of slope, or pitch, should be not less than 4 inches per foot. However, a self-sealing type square butt strip shingle

FIGURE 6–1.   Minimum pitch requirements for asphalt roofing products

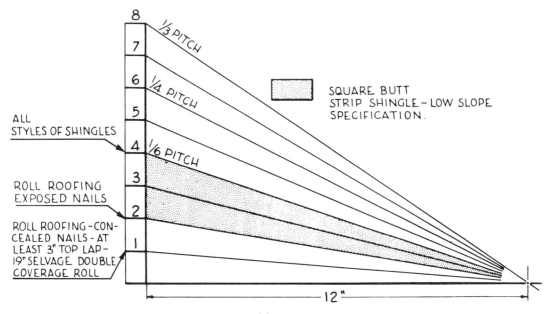

can be placed on a roof with a slope down to 2 inches per foot. "Free tab" shingles can also be placed on a slope of 2 inches per foot if the tabs are cemented down. If roll roofing is lapped 2 inches, it can be used on pitches down to 3 inches. If the pitch is 1 inch per foot—but not less—double-coverage roll roofing may be used.

## DECK PREPARATION

Asphalt shingles can be placed on a new deck or an old deck. If they are to be placed over an old deck, the old shingles should first be removed. (Some roofers place the new shingles directly over the old shingles, claiming that there is double insulation value. This is true, but with this technique the old decking cannot be properly inspected. If the shingles need replacing, chances are that some areas of the deck may also need replacing.)

The old shingles and roofing nails can be removed by means of a square-tip shovel, and starting from the ridge line. If any nails that are holding the sheathing in place have worked loose, they should also be pulled and new nails hammered in at new locations. After the shingles are removed, the deck must be swept clean of all debris, and the deteriorating places of the deck removed and replaced. Any large knotholes should be covered with small pieces of metal. The metal prevents an excessive amount of resin from coming into contact with the underlayment and shingles. Once the deck is properly prepared, the underlayment can be placed.

In new construction, solid lumber sheathing should be well-seasoned, a minimum of 6 inches wide, and 1-inch-thick tongued-and-grooved or shiplapped. The boards should be laid close together and secured with 7d annularly threaded nails or 8d common wire nails at each framing member.

Plywood roof sheathing should be of the proper interior or exterior construction grade. If the plywood is $\frac{3}{8}$ or $\frac{1}{2}$ inch thick it can be placed on rafters spaced up to 24 inches on center. The plywood should be secured with 5d annularly threaded nails or 6d common wire nails.

## UNDERLAYMENT

In new construction, if the roof is pitched higher than 3 inches per foot, 15-pound asphalt-saturated felt should be placed directly over the roof sheathing. The felt is available in 36-inch-wide rolls, weighs approximately 15 pounds per 108 square feet, and will cover approximately 400 square feet per roll. A felt heavier than No. 15 or a covering of some other material should not be used. A heavy felt acts as a vapor barrier, allows accumulation of moisture between roof sheathing and underlayment, and will wrinkle over an extended length of time. The accumulation of moisture between roof sheathing and underlayment could lead to the decay of the roof sheathing and cause structural damage to the framing members.

Underlayment serves a threefold purpose:

- It acts as a primary barrier against moisture penetration until the shingles can be placed
- It serves as a secondary barrier against moisture penetration once the shingles have been placed
- The felt acts as a buffer between the resinous areas on the roof sheathing and the asphalt shingles

Underlayment is laid flush with the rake edge and eave line. If a drip edge is used, the felt should be placed over the drip edge at the eave line, and at the rake edge the felt should be placed under the drip edge. (See Figure 6–2.) To properly place the underlayment, three nails should be placed in a triangular pattern in the center of the felt and near the rake edge. Then the felt should be rolled out approximately 15 feet and aligned with the eave line. As the felt is positioned, care should be taken to avoid "fishmouths," or wrinkles, which will later telegraph through the shingle, causing a distortion in the roof. The felt is nailed to the roof sheathing with $\frac{7}{8}$- or $\frac{3}{4}$-inch roofing nails spaced 6 inches on center at the eave line and 12 inches on center at the top

of the felt. If shingles are to be applied immediately after the application of the felt, it is not necessary to place any nails in the center of the felt. But if there is a time lapse between the application of the underlayment and the shingles, two rows of nails should be placed in the field of the underlayment. The first row of nails should be 12 inches above the bottom of the felt and placed 12 inches on center. The second row of nails must also be placed 12 inches on center and staggered between the previously placed nails.

The second and succeeding courses of underlayment are laid in the same manner as the first course. Each succeeding course is lapped 2 inches horizontally and has 4-inch end laps. The felt should be lapped 6 inches from both sides of all hips and ridges.

Valleys (intersections of two sloping roofs) should be lined with a 36-inch-wide strip of underlayment. The horizontal strips of underlayment are projected 6 inches on the valley felt and are trimmed parallel to the intersection of the two sloping roofs. The placement of the nails is very important in valley felt, because if they are incorrectly placed, the roof could leak. To obtain a waterproof valley, no nails should be placed within 12 inches of the intersection of the two sloping roofs.

Stack vents and vent stacks protrude through most roof decks and require special attention during the application of the underlayment. When the underlayment is to be placed around a stack, it should be raised and gently forced over the top of the stack. If too much force is used, a ragged joint is created which must then be properly sealed. To do this, a piece of 15-pound felt should be cut and laid on the top of the

FIGURE 6–2.  Placement of metal drip edge

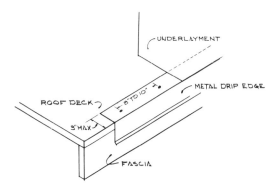

stack. Then the felt is gently tapped with a hammer, creating an impression of the stack on its underside. When the impression is cut out of the piece of felt, the felt is slipped over the stack and nailed into position, with care that the top is placed under the edge of the felt that is above the stack. The nails should be placed 3 inches on center around the perimeter of the piece of felt.

The intersection of a vertical wall and a sloping roof is a vulnerable location for water penetration and should be treated properly. The underlayment should be allowed to project 6 inches up a vertical wall, and placed so that descending water will be carried over the joint and not into it.

Roofs that have a pitch of 2 to 4 inches per foot should be provided with a double coursing of 15-pound asphalt-saturated felt. The first course is an 18-inch-wide strip placed along the eave line. A full 36-inch sheet is then placed over the 18-inch strip and nailed in place. Succeeding courses are placed, with 18 inches overlapping the preceding course.

## NAILS

There is no single step in roofing more important than the nails and nailing procedure, for improper nailing will decrease the roof's weather resistance. The three most important features to be considered in nailing are placing the nail in the correct location, using the correct-size nail, and using the correct number of nails.

A minimum of four nails must be used to secure a strip shingle to the roof deck. The nails for a three-tab shingle should be located $5/8$ inch above the tops of the cutouts and 1 inch from each end. For a two-tab shingle the nails should be placed 1 inch from each end and $5/8$ inches from the bottom of the shingle. Two additional nails are placed $5/8$ inches from the bottom of the shingle and 5 inches to each side of the center cutout. The nailing pattern for a no-cutout shingle is $5/8$ inches from the bottom of the shingle; the ends are nailed 1 inch from each side; and the other two nails are placed 12 inches from each end.

The nails used for strip shingles should be large-head galvanized or aluminum roofing nails. They should be no smaller than 12-gauge and have $3/8$-inch-diameter heads. The nails should be long enough to penetrate the roofing material and at least $3/4$ inch into wood deck lumber. They should be driven straight and flush with the surface of the shingle. If a roofing nail is driven at an angle, the head has a tendency to cut the shingle. If it is driven into a crack or does not go into solid sheathing, it should be pulled and another nail placed adjacent to the hole.

Staples are sometimes used in place of roofing nails. Care should be taken that the legs are long enough to penetrate the roof decking and go in at least $3/4$ inch into solid lumber sheathing. The staples should be driven flush and not through the shingle. They are placed parallel to the bottom of the shingle, in the same location as nails. Most staples are inserted with pneumatic staplers, a method valued for its speed. However, speed should not replace craftsmanship; properly combined, they should produce a high-quality roof at a lower cost.

## APPLICATION OF ASPHALT STRIP SHINGLES

Various techniques are used in the application of asphalt strip shingles, each affording certain advantages and disadvantages. The pyramid, step, and racking techniques are the most widely used.

### Pyramid Technique

The pyramid application of asphalt strip shingles is often used when the roof exceeds 75 feet in length. (See Figure 6–3.) The four advantages of the pyramid technique are:

■ Maintenance of a good color pattern because the shingles are distributed more evenly
■ Ease in keeping the cutouts of the shingles in proper alignment
■ Easier and faster nailing
■ Prevention of improper nailing

The first step before the actual application of the shingles is to lay out a series of parallel and perpendicular lines on the underlayment, as shown in Figure 6–4. The first line is chalked parallel to and 6 inches from the eave line, and will be used to align the starter shingle. A second line is chalked

11 inches from the eave line, and is used to keep the first course of shingles straight. From the second chalkline a series of lines 10 inches or 20 inches apart are marked. These parallel lines are also used to keep the shingles in proper alignment. Some contractors prefer to check alignment every two courses, while some check the shingles on every fourth course. This is the reason for the 10- or 20-inch spacing of chalklines.

After the parallel lines are chalked, a series of seven chalked lines must be marked perpendicular to the parallel lines. The initial step is to locate the center of the roof on the chalkline that is 6 inches from the eave line. From the center a mark is placed 16 inches on either side. A 6-foot rule is extended its full length and a nail is placed through a small hole at the end of the rule. The holes are ⅛ inch in diameter and are drilled on each end of the rule. The nail that has been placed through the hole is put on the mark that is 16 inches from the center point. Another nail is placed in the hole at the other end of the rule and a downward sweeping motion is made, marking the felt in the process. (See Figure 6–5.) The rule is then transferred to the other marked point and the procedure is repeated. A 1 × 2 with a nail in each end can

FIGURE 6–3.  Pyramid application

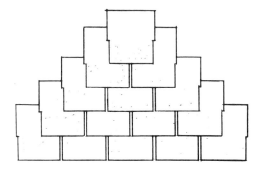

FIGURE 6–4.  Layout for pyramid application

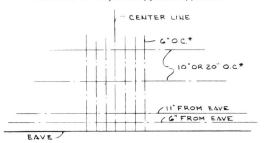

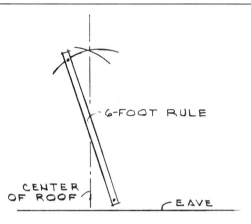

FIGURE 6-5. Locating the center line

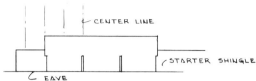

FIGURE 6-7. Placement of shingles

also be used to scribe the arcs, but it is sometimes difficult to find suitable stock on the job. A roofer always carries a 6-foot rule, and the only preparation needed before its use is to drill the small holes in the ends of the rule.

A line is chalked from the intersection of the two arcs to the center of the roof on the first parallel line. The result is a chalkline perpendicular to the starter line. (See Figure 6-6.) To create the pyramid effect six other perpendicular lines must be chalked, three on each side of the center line. The perpendicular lines are used to align the cutouts and should be spaced 6 inches apart. (See Figure 6-7.)

Once the perpendicular and horizontal lines are chalked, the roof should be stocked, with the shingles strategically located on the roof. Because of their weight, they should not be placed over hips and

ridges, nor should they be placed in one large pile.

Before the placement of the shingles, starter shingles must be cut and placed. Starter shingles are 7 inches wide and are cut from a regular strip shingle. (See Figure 6-8.) Any color shingle may be used for a starter, because the color is cut off. But starters must be of the self-sealing type. They are placed with the factory edge along the starter line and both edges butting the outside perpendicular lines. A second and third starter shingle are placed on each side of the first starter shingle, each butted against the first.

Many roofing mechanics use an inverted full shingle for a starter course. But with this technique the thermoplastic dots are not placed at the bottom of the tabs, so the first course of shingles will not be wind-resistant.

After the starter shingles have been

FIGURE 6-8. Step technique

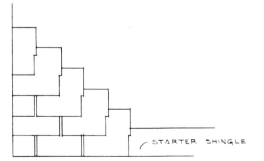

FIGURE 6-6. Starter shingle

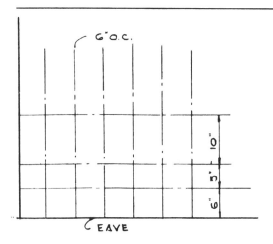

FIGURE 6–9. Layout for step technique

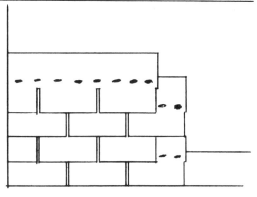

FIGURE 6–10. Racking technique

placed, a whole strip shingle is put in, with the chalkline 11 inches from the eave line as a guide. (See Figure 6–9.) The end of the shingle should be placed on the second perpendicular line nearest the rake edge. A two-tab shingle should then be cut and placed next to the previously placed whole shingle. The second course of shingles is composed of a whole shingle and a one-tab shingle. Using the third perpendicular line from the rake edge, the whole shingle is placed in position. To complete the second course a one-tab shingle must be placed next to the whole shingle. The third course is comprised of a whole shingle. This is placed on the fourth perpendicular line. The fourth and fifth courses are made of a two-tab and a one-tab shingle. The two-tab shingle is centered above the whole shingle and the one-tab is centered above the two-tab. When the one-tab shingle is put in place; the pyramid is complete and the courses can be finished. (See Figure 6–10.) The shingles are placed from the eave line up, with as many as six roofers working from the pyramid. (This is another advantage of the pyramid technique.)

## Step Technique

The step technique of laying asphalt strip shingles is very similar to the pyramid technique. (See Figure 6–11.) The difference is that the pyramid technique is started in the center of the building and the step technique is started at the rake edge. The step technique is most often used on roofs that do not exceed 75 feet in length. The biggest disadvantage of this technique is that it only allows a limited number of workers at one time. However, good color distribution is achieved and the shingles along one rake edge do not need trimming.

FIGURE 6–11. Layout for racking technique

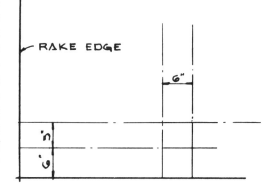

In the application of asphalt strip shingles horizontal and vertical chalklines should always be used to insure good shingle alignment. The first horizontal chalkline should be 6 inches from the eave line and will serve as a guide for the starter course. A second chalkline is then made 11 inches from the eave line; it will align with the first course of shingles. With the second horizontal line as a starting point, a series of lines should be chalked 10 or 20 inches apart.

Vertical lines are then chalked along the rake edge. The first should be 5 inches from the rake edge; five additional lines are then chalked 6 inches apart and parallel to the rake.

Once the horizontal and vertical lines are chalked, a 7-inch-wide starter shingle is placed along the eave line and is butted to the sixth chalkline. The starter shingle should be laid with the thermoplastic dots at the eave line. A full shingle, minus a half of a tab, is placed over the starter shingle and aligned with the second horizontal chalkline and the fifth vertical chalkline. The second course is started with a full shingle, minus a tab, and is aligned with the fourth vertical chalkline. The third course is started with a full shingle minus one and one-half tabs, and is aligned with a 10-inch horizontal line and the third vertical line. The fourth shingle in the step is a full shingle, minus two tabs, and is aligned with the second vertical chalkline. The last shingle needed to complete the step is a full shingle minus 30 inches. The 6-inch shingle is allowed to overhang the rake edge by 1 inch and is aligned with the chalkline 5 inches from the rake edge. When the step is completed, full shingles are butted to the steps and aligned with the horizontal chalklines.

## Racking Techniques

The racking technique of applying asphalt strip shingles is often used by roofers, but is discouraged by some building codes. (See Figure 6–12.) This technique leads to color variations in the roofing material and also promotes improper nailing. The popularity of the technique, however, lies in its simplicity and speed.

Before the placement of the shingles, horizontal and vertical lines are chalked on the underlayment. The first horizontal line is 6 inches from the eave line and the second is 11 inches from the eave line. The remaining horizontal lines are chalked 10 or 20 inches apart, measuring from the 11-inch line. Two vertical lines are chalked parallel to the rake edge, the first line 29 inches and the second 35 inches from the rake edge. Then starter shingles are cut and placed. The first starter shingle is placed next to the line that was chalked 35 inches from the rake edge. The top of the starter shingle is aligned, with the chalkline 6 inches from the eave line used as a guide. Four nails should be used to secure the starter shingle to the roof sheathing. A whole shingle is then placed adjacent to the

FIGURE 6–12. Cutting the shingles at a vertical wall

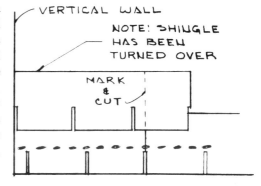

line that was chalked 29 inches from the rake edge. The whole shingle is secured with only three nails; the fourth nail will be added on the second run of shingles. (Many roofers, however, do not add the fourth nail, and most manufacturers of asphalt shingles will not guarantee a roof if only three nails per shingle are used.) Another whole shingle is placed, with the 35-inch chalkline as a guide. The shingle is nailed with three nails, leaving the end next to the vertical line unnailed; a shingle will be slipped under this free end. The remaining courses are carried to the ridge, alternating from the 29-inch line to the 35-inch line. Once the run is complete, the rake edge must be trimmed with a pair of snips. In most cases the shingles extend 1 inch past the rake edge.

After the first run of shingles is laid, a second run is placed next to the first run. The starter is placed, aligned, and nailed. A whole shingle is then slipped under the second shingle of the first run and butted to the first shingle of the first run. The shingle is aligned and fastened, with four nails. An additional nail is placed on the end of the second shingle of the first run (a procedure that is omitted by many roofing mechanics). A second shingle is then placed and secured with only three nails. The remainder of the run is completed by alternating the placement of the shingles.

## Checkover for Strip Shingles with Cutouts

To keep the shingles in proper alignment, it is sometimes necessary to make a checkover. This procedure is needed only when a ridge intersects a sloping roof. The first step in a checkover is to locate the nearest chalkline above the intersecting ridge, and continue this line to the rake edge. Additional chalklines are then placed down the roof at 10- to 20-inch intervals. The last chalkline should be 11 inches from the eave line. With the uppermost chalkline as a guide, a horizontal course of shingles is laid. This course of shingles should be carried past a point that corresponds with the intersection of the valley and eave lines. To maintain a straight line with the cutouts, a second course of shingles is placed over the first course. Then a vertical line is chalked perpendicular to the last strip shingle; at the same time it may be necessary to align the cutouts of the shingles. One of the easiest methods is to align them with a framing square. An additional vertical chalkline is then placed parallel and 6 inches from the previously placed line.

After the placement of the two vertical lines, the starter shingle is placed at the eave line. The placement of the starter shingle is an important step, for if it is not placed on the correct vertical line the checkover will not work. The starter shingle should be placed adjacent to the vertical line that is opposite to the end of the last shingle placed. The first course of shingles is then started by placement of the end of the shingles on the next vertical line. Care should be taken to keep the shingles in proper alignment, or they will not have any head lap when the top course is reached. The checkover should be placed in the racking fashion, thus allowing for inspection. After the first rack is placed, the checkover is filled with several other racks.

The checkover is a relatively simple procedure if the roof is square. But if it is not, the chalklines must be adjusted. They should never be placed any closer than 4 inches or any further apart than 5⅛ inches. If adjustments are made, care should be taken in the shingle placement, because if a line is added or subtracted the courses will be different.

## Cutting the Rake Edge

When certain application techniques are used the rake edge must be cut after the shingles have been placed. One of the best techniques is to mark a 1-inch overhang near the eave line and a 1-inch overhang near the ridge. A line is then chalked between the two points and the shingle is cut with a pair of snips. This technique gives a smooth and neat cut.

The shingles can be cut before placement, but this leaves the rake edge uneven and rough. Some contractors feel that this technique is faster, but actually it is slower. It is easier, but the easiest method is not always the best.

Another method is to nail a row of shingles along the rake edge with the desired overhang. The shingles are placed like starter shingles and serve as a guide in the cutting process. Once the rake shingles are placed, a knife can be used to trim them with, using the starter shingles as a guide.

## Cutting the Shingles at the Intersection of a Vertical Wall

When shingles are cut along a vertical wall, there is a simple procedure that can be used to facilitate the cutting. The shingle is first turned over and placed against the vertical wall. The tabs should now project down the slope of the roof. The shingle is then marked and cut where it overlaps the previously placed shingle. The point is cut and then the shingle is flipped over, aligned, and nailed in place.

## Hips and Ridges

Hips and ridges can be covered by an individual hip or ridge unit, or a 12- by 12-inch unit can be cut from a strip shingle. There are two techniques used to cut the units from a strip shingle. They can be cut straight across from the cutout, as in Figure 6–13A, or two cuts can be made, each at an angle to the cutout, as in Figure 6–13B. The latter technique is usually more satisfactory because the edges of the shingles do not show once the units are nailed. The first shingle placed on the hip is bent lengthwise and mitered to fit the roof line. A nail is then placed on each side of the hip unit. The shingles are placed 5½ inches up from the exposed edge and 1 inch from the edge. The remaining hip units are placed up the hip, leaving a weather exposure of 5 inches. In some cases chalklines are placed 6 inches on either side of the hip and used to align the hip units.

Ridge units are placed in the same manner as hip units, but there are two different techniques used in the termination of the units. To terminate the ridge unit at the in-

FIGURE 6–13. Hip units

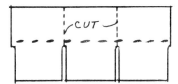
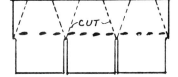

tersection of a hip and ridge, a ridge unit is cut down the center. One edge of the cut unit is then tucked under and a nail is placed over the folded end.

To terminate a ridge on a gable roof, a cap unit must be used. The ridge units are started on one end of the ridge and are placed down the ridge to a point 2 feet from the end of the ridge. A second run of ridge units is then started and is laid until it intersects with the first run. A bed of roofing cement is placed over the intersection and a cap shingle is placed over the cement.

For added protection the cap shingle is secured with four exposed nails.

## FLASHING

During the application of strip shingles, the roof must be properly flashed in areas around chimneys, soil stacks, intersections of other roofs, and projections through the deck. Flashing is an essential element in any roof and proper application techniques should be followed.

### Valley Flashing

A valley is the intersection of two sloping roofs. It must be flashed to provide runoff along the joint. In the *open-valley method,* the valley is lined with sheet metal or roll roofing. There are two types of metal valleys—"stomp" or roll valleys, and formed valleys. The latter is the preferred metal valley.

The roll valley is formed by rolling 28- or 30-gauge sheet metal in the center of the valley. The galvanized sheet metal is 12 to 20 inches wide and is nailed along one side. In order to form the flashing to the configuration of the valley, the metal is stomped into place. (This stomping can cause disfiguration of the valley and breakage in the galvanized coating. Eventually the break will rust and the valley will leak.) The flashing is allowed to project over the ridge 1½ inches. If there is a valley on each side of the ridge, the other valley flashing is projected over the ridge and lapped 1½ inches over the other piece of valley flashing. After the flashing is stomped into place, the other side of the flashing is nailed to the roof sheathing.

The preferred type of metal valley is the formed valley. The valley flashing is 18 inches in width, 8 to 10 feet in length, and is formed in a metal brake. (See Figure 6–19, p. 109.) Before applying this flashing, a line is chalked down the center of the valley. Starting at the bottom, the first piece of flashing is placed with the formed center falling over the chalkline. The valley flashing is projected over the eave line until the corners are nailed into place. The nails are spaced 8 to 10 inches apart, and are placed 1 inch from the edge of the valley flashing. Pressure is then applied to the center of the flashing and the other side is nailed into place. A second piece of flashing is then placed with a 3-inch head lap. It is placed in the same manner as the first piece of valley flashing.

#### Roll Roofing Valley Flashing

If roll roofing is used to flash the valley, a contrasting or matching color of roofing is selected. (See Figure 6–14.) An 18-inch

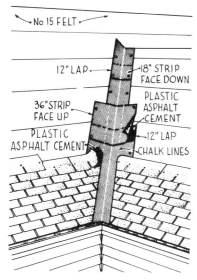

FIGURE 6–14.   Use of roll roofing for open-valley flashing

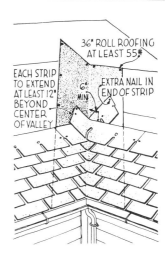

FIGURE 6–15.   Closed, or woven valley

strip of roll roofing is then cut and placed, surfaced side down, in the valley. This strip is first nailed along one side. Pressure is then applied to the center of the strip to make the flashing conform to the shape of the valley. The other side of the flashing is then nailed 1 inch from the edge. Then the lower edge is trimmed flush with the eave line. To complete the flashing, a 36-inch strip of roll roofing is placed over the 18-inch strip, surfaced side up, and secured in the same fashion as the 18-inch strip of flashing.

### Woven Valleys

A woven valley is created by weaving adjacent courses of shingles over a valley. (See Figure 6–15.) This type of valley construction is sometimes preferred because of its double coverage. The first step in making this type of flashing is to center a strip of 50-pound roll roofing in the valley. The first course of shingles is laid along the eave line and the last shingle should project 12 inches beyond the center of the valley. The first course of shingles on the adjoining roof is then laid, and the last shingle is allowed to overlap the valley and the previously laid shingle. Next, additional courses of shingles are laid, alternating from one roof to the other. The shingles should be pressed firmly into the valley and should not be nailed within 6 inches of the valley.

### Closed-Cut Valleys

In the construction of a closed-cut valley, the shingles on one roof are allowed to project over the valley and up the adjoining roof. (See Figure 6–16.) The shingles should project up the adjoining roof at least 12 inches and should not be nailed within 6 inches of the valley. The shingles are then applied on the intersecting roof and allowed to "run wild" in the valley. A line is chalked down the center of the valley and the overlaid course of shingles is cut along the chalkline. As they are being cut, the upper corner of each shingle that

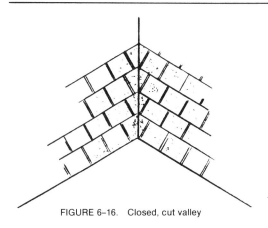

FIGURE 6–16.   Closed, cut valley

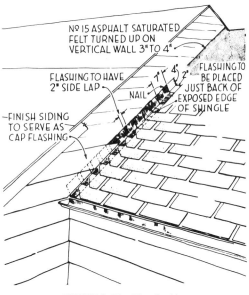

FIGURE 6–17.   Step flashing

projects into the valley should be cut at a 45-degree angle. This procedure will prevent water from penetrating the different courses.

## Wall Flashing

Wall flashing is placed at the intersection of a sloping roof and a vertical wall. Its primary objective is to keep water from penetrating the roof. There are three basic types of wall flashing: composition flashing; step flashing; the solid-water guide.

### Composition Flashing

Composition flashing consists of a special flashing cement and fiberglass membrane that is applied to the wall and shingles. After the shingles are cut up the sloping roof, an application of flashing cement is applied with a trowel. The membrane is then embedded in the base coat of flashing. A second coat is then parged over the membrane. This is the least expensive and also the least effective type of wall flashing.

### Step Flashing

Step flashing is made from sheet metal and is installed after each shingle is cut and placed. (See Figure 6–17.) The flashing shingle is rectangular in shape, usually 6 inches long and 7 inches wide. The flashing is bent so that it extends up the vertical wall at least 4 inches and has a minimum of 2 inches of flashing extending out over the roof deck. The first piece of flashing is placed over the starter shingle. Next, the first course of shingles is laid over the flashing. The remaining courses of shingles are then laid, with the flashing and the shingles alternated. The flashing is secured with one nail in the top corner and is placed above the exposed edge of the shingle. Step flashing is expensive and is not always the most effective type of flashing.

### The Solid-Water Guide

The solid-water guide is a solid piece of metal flashing 8 to 10 feet long and 4 inches wide. It is bent down the center and a $\frac{1}{2}$-inch lip is turned up on one edge. This flashing is nailed directly to the studs

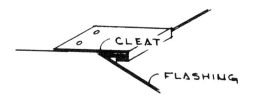

FIGURE 6–18. Securing flashing with cleats

before the finish siding is placed. The base of the flashing is secured with cleats, which are made of ½-inch horizontal strips of flashing. (See Figure 6–18.) The lip of the cleat is placed over the lip of the flashing, and the tab of the cleat is nailed to the roof sheathing. The cleats should be installed 12 inches on center. As the shingles are placed, they are cut to fit ½ inch from the wall. To keep water from running down the shingles and under the flashing, the top corner of each shingle that is laid over the flashing should be cut at a 45-degree angle 1 inch from the top. This is called cutting the "tips" or "points."

## Chimney Flashing without a Cricket

Chimney flashing is complicated and expensive. It consists of wall flashing, apron flashing, counter flashing, step flashing, and back flashing. The material that is most often used for chimney flashing is galvanized sheet metal or copper.

*Apron flashing* is the first piece of flashing placed. (See Figure 6–19.) It lays 4 inches over the shingles and turns up the chimney 4 inches. The ends should be allowed to extend 4 inches past the corner of the chimney; they are then bent around the corner.

At this time *vertical wall flashing* is placed on both sides of the chimney. (See Figure 6–20.) It extends to the bottom of

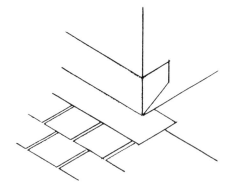

FIGURE 6–19. Apron flashing

the apron flashing and 4 to 6 inches behind the back of the chimney. Next, shingles are placed over the flashing up the roof to a point at which the color of the shingle is even with the back of the chimney.

*Back flashing* is then placed behind the chimney, extending 18 inches up the roof and 4 inches up the chimney.

*Counter flashing* is placed around the chimney when the back flashing has been completed. Straight counter flashing is placed over the apron flashing and is allowed to turn the corners of the chimney for 2 inches. The top of all counter flashing is placed and nailed into a raked mortar joint. Mortar is then placed over the junction of the flashing and chimney.

FIGURE 6–20. Straight-counter flashing

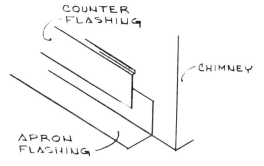

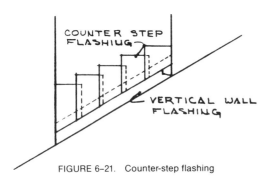

FIGURE 6-21. Counter-step flashing

*Counter step flashing* is then placed on the sides. (See Figure 6–21.) The width of the chimney determines the number of pieces of step flashing, but in most cases three or four pieces are required. The last piece of flashing is a straight piece of counter flashing placed over the back flashing. It should project around the corners 2 inches. Once the flashing is complete, shingles may be placed over the back flashing and the remainder of the roof.

## Eave Flashing

In areas where there is a chance of an ice buildup along the eave line, the eave should be properly flashed. (See Figure 6–22.) If the roof has a normal slope of 4 inches per foot, or if the slope is greater, the eave is flashed by placement of a course of 90-pound mineral-surfaced roll roofing along the eave line. The roll roofing should overhang the metal drip edge or shingle strip by ¼ to ⅜ inch and should extend up the roof to a point 12 inches inside the interior wall line. The double course is cemented together and should overhang the metal drip edge ¼ to ⅜ of an inch.

## Stack Flashing

Flashing must be placed over and around protruding stacks (pipes that protrude through the roof). This flashing is usually factory-made and of lead, galvanized sheet metal, copper, or rubber. Stack flashing has two basic parts—the flange and the barrel. The flange rests on the roof deck and the barrel is the portion of the flashing that fits around the stack. In some cases a third piece called the cap is used—this fits on top of the barrel.

In flashing a stack, the shingles must be placed under the front of the flange and brought up to a point at least three-

FIGURE 6-22. Eaves flashing

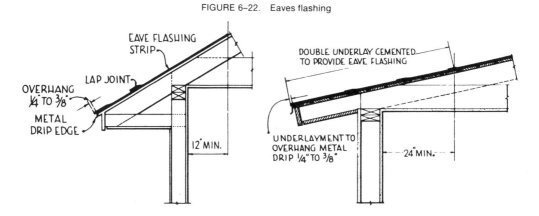

quarters of the way to the top of the exposed stack. In some cases it will be more than this, because some vents are smaller. In this case the shingles are carried to a point where the nails will be covered and only the color will be showing.

After the shingles are nailed in place in front of the stack, flashing is placed over the stack. It is aligned and nailed in place, with the nails on the outer edge of the flange. (The nails should be placed in such a manner that the next shingle placed will cover the nails.) Then a shingle is placed next to the stack, marked, and cut to fit the stack. The shingle is then placed in the correct position and properly secured.

## REROOFING EXISTING ROOFS WITH STRIP SHINGLES

If an overlay is needed, the existing deck should be sound and free from any structural defects. Any loose or curled shingles must be nailed down or cut away. The stack flashing must be removed, and the hip and ridge units removed. Loose and protruding nails must be removed; old overhangs on both the eave and rake edges must be broken or cut, and the roof should be cleaned of any debris.

The first step in the overlay procedure is to cut 2 inches off the bottom of a 7-inch starter shingle and place it along the eave. The top of the starter butts the bottom of the old shingle. The nails used in an overlay should be long enough to penetrate ¾ inch into a solid wood deck or through a plywood deck. The first course of shingles has 2 inches cut off the width and is placed with a 1-inch rake overhang. The first course of shingles is butted to the bottom of the third course of old shingles. This technique keeps the roof from being uneven and irregular. The third course and all succeeding courses are laid with full-width shingles and are butted to the bottom edges of the old shingles. The exposure will be automatic and will coincide with the old roof.

If the shingles have cutouts, they must never be in line with the cutouts on the old roof. To eliminate any possibility of alignment, chalked lines can be placed between the cutouts.

## GIANT INDIVIDUAL SHINGLES

Giant individual shingles are sometimes used in place of strip shingles. They are made of the same material as strip shingles, but cut to a different size. Each one is usually 16 inches long and 12 inches wide. There are two application techniques that can be used for giant individual shingles—the American method, used for new construction, and the Dutch lap method, used over old roofing.

Before the shingles are placed, a drip edge and eave flashing should be placed. If the shingles are applied over a new deck, a layer of 15-pound asphalt-saturated felt should be laid. (If the shingles are to be placed over old roofing the underlayment may be omitted.)

## Application by the American Method

To keep the individual shingles in proper alignment, a horizontal chalkline is placed 11 inches from the eave line. A second line is marked 4 inches from the first line. Additional horizontal lines are then chalked 11 inches apart. (See Figure 6–23.)

Once the horizontal lines are chalked, the starter course is placed, with the first chalkline as a guide. The starter-course shingles are laid horizontally and butted to each other. The first course is started with a full shingle laid flush with the second chalkline. As the individual shingles are placed, a ¾-inch gap is left between the shingles. The shingles are nailed 6 inches up from the bottom of the shingle and 1½ inches from the side. The nails should be long enough to penetrate at least ¾ inches into solid sheathing and completely through plywood sheathing.

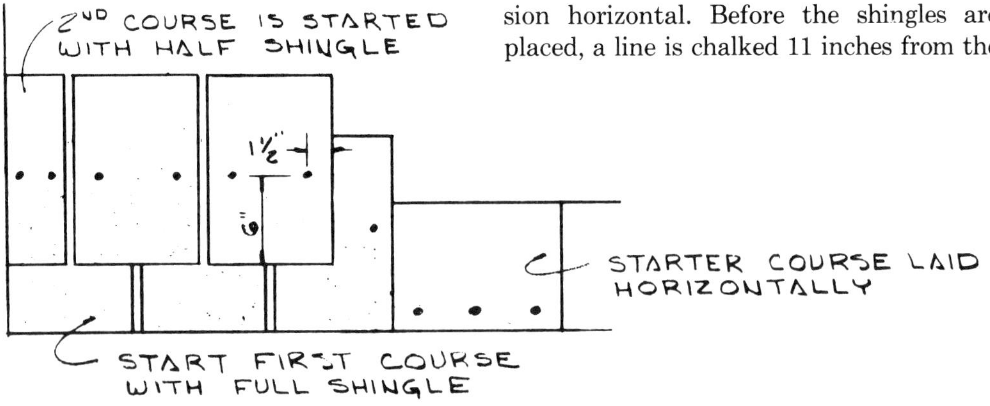

The second course is started with a shingle that is 8 inches wide and continues with full shingles, leaving 5 inches to the weather. The third course is started with a shingle 4 inches wide, and the fourth course starts with a full shingle. The sequence is repeated until the roof is covered.

If the shingles are placed in areas that are subjected to high winds, shingle tab cement can be placed under each tab 1 inch up from the butt; the glue spot should be 1 inch in diameter.

The hips and ridges can be finished with the giant unit shingle, or individual hip and ridge units can be used. The shingles are bent down the center to fit over the hip or ridge. If the weather is cold, the shingles should first be warmed. The first shingle placed on a hip should be trimmed to fit the roof line. The shingle is then placed at the bottom of the hip and secured with one nail on each side. The nails should be placed 5½ inches up from the exposed end, and 1 inch up from the edge.

## Application by the Dutch Lap Method

In using the Dutch lap method the individual shingles are laid with the long dimension horizontal. Before the shingles are placed, a line is chalked 11 inches from the

FIGURE 6–23. Application of giant individual shingles by the American method

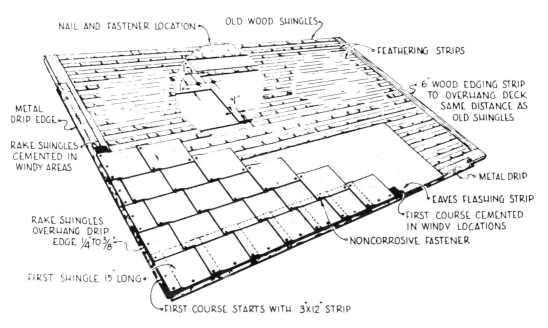

NAIL AND FASTENER LOCATION

OLD WOOD SHINGLES

FEATHERING STRIPS

6" WOOD EDGING STRIP TO OVERHANG DECK SAME DISTANCE AS OLD SHINGLES

METAL DRIP EDGE

RAKE SHINGLES CEMENTED IN WINDY AREAS

METAL DRIP

EAVES FLASHING STRIP

FIRST COURSE CEMENTED IN WINDY LOCATIONS

RAKE SHINGLES OVERHANG DRIP EDGE 1/4 TO 3/8"

NONCORROSIVE FASTENER

FIRST SHINGLE 15" LONG

FIRST COURSE STARTS WITH 3"X12" STRIP

FIGURE 6–24.    Application of Dutch-lap shingles

eave line. Additional horizontal lines are then chalked 9 inches apart. (See Figure 6–24.)

The first course is started with a 3- by 12-inch strip laid flush with the chalkline. The small piece of shingle should be secured with a nail in each corner, 1 inch in from the edges. A full-size shingle is then placed over the 3-inch strip and secured with four nails, one in each corner. The remaining shingles of the first course are then placed, lapping each previously laid shingle by 3 inches. The shingles should be fastened with two nails and a noncorrodible wire staple fastener. Along the eave line and rake edge, the shingles should be face-nailed. When the shingles are laid left to right, nails are placed in the upper left-hand and the lower right-hand corner of the shingle. The wire staple is placed in the lower left-hand corner. The second course of shingles is started with a full unit, lapped 2 inches over the top of the first course, and butted against the exposed vertical edge of the shingle below. Additional shinglés are then placed, maintaining a 2-inch top lap and 3-inch side laps. The remaining courses are self-aligning and are laid in the same manner as the second course. Once the roof is covered, the rake edge must be trimmed.

## Interlocking Shingles

Interlocking shingles are sometimes used in place of strip shingles and giant individual shingles. They vary greatly in design, but they all have the same purpose—to give adequate protection and resistance in strong winds without the use of adhesives. These shingles are most effective over old asphalt-shingled roofs or old wood shakes and shingles.

Before the interlocking shingles are placed over an existing roof, any loose or curled shingles should be nailed or cut away. If the existing roof is laid with asphalt shingles, they should be broken back at the eave line and rake edge and replaced with a 1 × 4 eave strip.

The first step is to mark lines for the starter strip. The chalkline should be 8 inches from the eave line. A chalkline is then placed for the starter shingle. The distance from the eave line is determined by cutting a shingle horizontally at the point where the locking device couches the shingle. The vertical distance is measured and a 1-inch overhang is allowed; then the chalkline for the starter shingle is placed. The coverage of the shingle is determined by locking the two shingles together, aligning them, and measuring the distance from the top of the bottom shingle to the top of the top shingle. Chalked lines are then placed for every fourth course of shingles.

Once the horizontal lines are chalked, a vertical line next to the rake edge will be

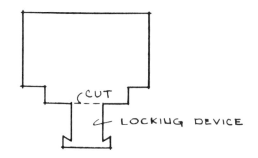

FIGURE 6–25. Cutting the starter shingle

equal to half the width of a shingle plus 1 inch. The 1 inch represents the overhang of the shingle.

Next, starter strips are cut at a point just below where the shingle locks. (See Figure 6–25.) With the first chalkline as a guide, the starter strips are placed from rake to rake. Starter shingles are then placed over the starter strips. The first shingle is placed, top flush with the second chalkline, and the middle of the shingle is aligned with the vertical line along the rake edge. (See Figure 6–26.) To facilitate alignment with the vertical line, most shingles have a small mark in the middle.

The next starter shingle is then placed beside the first, leaving a small gap of ¼

FIGURE 6–26. Application of interlocking shingles

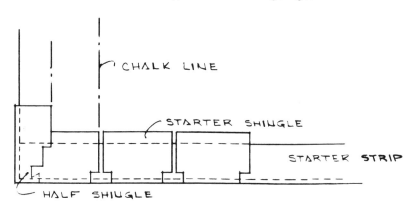

inch between the two shingles. The remaining starter shingles are then placed in the same manner.

A rake shingle (the vertical half of a whole shingle) is then hooked at the rake edge and aligned with the vertical chalkline. It is aligned by pulling up the previously placed shingle and putting the side of the rake shingle next to the vertical chalkline. The second course of shingles is then completed, leaving a ¼-inch gap between the shingles. Additional courses are placed in the same manner as the second course. The half shingles on the rake edge and the starter shingle must be cemented down with a small amount of cement placed under each shingle.

The step technique is one of the best methods of application and a crew of three workers is considered to be the most efficient means of applying the interlocking shingles. If a crew of three is used, one can hook, one can nail, and the other can hand the shingles to the roofers.

## ROLL ROOFING

Roll roofing is manufactured in 36-inch widths and can be placed on roofs that have a rise as little as 1 inch per foot. The roofing has a 19-inch lap and a 17-inch mineral-surfaced exposure. It can be applied with hot asphalt, cold asphalt adhesives, or it can be nailed down. (See Figure 6–27.)

Before the roofing is laid, a metal drip edge should be placed along the eave and rake edge. This edge should extend 3 inches onto the roof deck and should be secured with roofing nails spaced 8 to 10 inches on center.

The first step in the application of roll roofing is to cut a 19-inch starter strip from a roll of roofing. (The starter strip is the 19-

FIGURE 6–27. Application of double-coverage roll roofing parallel to the eaves

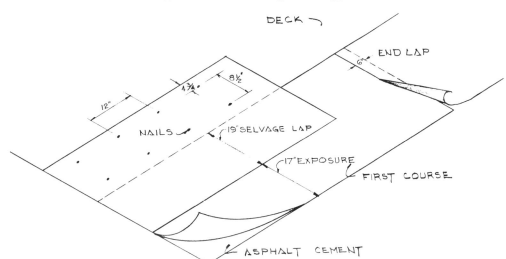

inch salvage portion of the roofing.) The starter strip is then placed along the eave line and secured with 11- or 12-gauge galvanized roofing nails with heads at least ⅜ inch in diameter and shanks ⅞ to 1 inch in length. The starter strip is secured with three rows of nails, one near the top of the strip, one near the eave line, and one in the middle.

The first course of roll roofing is then placed and aligned with the starter strip. After it is properly aligned, two rows of nails should be placed 4¾ inches from the upper edge of the roofing. A second row of nails is then 8½ inches down from the first row. The nails should be placed 12 inches on center and staggered in the salvage portion.

Additional courses are placed, each maintaining a 19-inch salvage. To maintain proper alignment of the courses, chalklines can be marked at 19-inch intervals. Asphalt roofing cement is then applied to the salvage portion of each course. To properly apply it, the mineral-surfaced portion must be lifted and the cement applied according to manufacture's specifications. To bond the plies together, slight pressure should be applied with a light roller or broom. End laps should be at least 6 inches wide and secured with a row of nails 4 inches on center. (See Figure 6–28.) In addition to the nails, cement should also be placed under the lapped area.

Hip and ridge units are 12 inches by 36 inches and are cut from a piece of roll roofing. A starter piece is cut from the salvage. The starter piece is bent lengthwise and placed over the hip. Rows of nails are then used to secure it. The first hip unit is cemented to the starter and a row of nails placed on each side of the 19-inch salvage. This process is continued until the hip is completely covered.

Once the roofing has been laid, it should be inspected to make sure none of the laps have pulled away. If it is necessary, recement any laps that have pulled away.

FIGURE 6–28.   Construction of the end lap

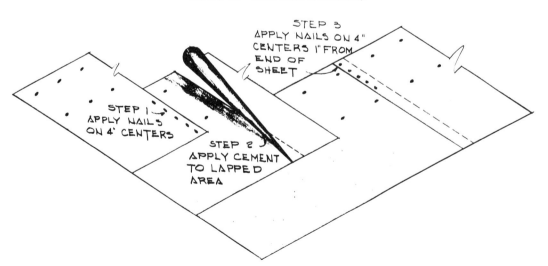

# CHAPTER 7

# WOOD SHAKES AND WOOD SHINGLES

WOOD SHAKES AND WOOD SHINGLES HAVE successfully been used as a roof-covering material for many years. Wood shakes are split and wood shingles are sawed. Shakes have at least one textured side, while shingles have relatively smooth surfaces. Both are strong, provide good insulation, and are resistant to winds; the appearance of each is pleasing.

In recent years, however, the popularity of wood shakes and shingles has dropped. One of the reasons is the increased cost and the fire hazard created by their use. But in recent years, a wood shake and shingle has been developed that carries a U.L. Class "B" rating, or a U.S. Class "C" rating. To obtain a U.L. Class "C" rating, the cell structure is impregnated by fire-retardant chemicals. A Class "B" rating is obtained when the wood shakes and shingles are placed over an underlayment of plastic-coated steel foil.

The underlayment must be placed over a ½-inch minimum plywood deck or 2-inch minimum tongued-and-grooved lumber.

The three most popular species of wood used for wood shakes and shingles are Western red cedar, redwood, and cypress. Cypress shingles carry the grades of No. 1, best, prime, and economy. Redwood shingles are graded as No. 1s and No. 2s. Red cedar shingles are graded as No. 1 (blue label), No. 2 (red label), and No. 3 (black label). (See Table 7–1.) Red cedar shakes are graded as No. 1 handsplit and resawn; No. 1 tapersplit, and No. 1 straightsplit. (See Table 7–2.) Wood shakes and shingles are manufactured in lengths of 16, 18, and 24 inches. They vary in width from 3 to 14 inches, but if a shingle is wider than 9 inches it should be split before it is placed. The wood shakes and shingles are packaged in bundles. It usually takes four or five bundles to cover a square. The actual area covered is determined by both the weather exposure and the length of the shingle, as shown in Tables 7–3 and 7–4.

## SLOPE LIMITATIONS

To provide adequate drainage, the roof must be properly sloped and the wood shakes and shingles must have the proper exposure. To increase the efficiency of the roof, the exposure is reduced as the slope of the roof decreases. If the roof pitch is 5 inches in 12 inches or steeper, a 16-inch shingle should have a 5-inch exposure; an 18-inch shingle, a 5½-inch exposure; and a 24-inch shingle should have a 7½-inch exposure. (See Table 7–5.) If the roof pitch is less than 5 inches-in-12 inches but not below 3 inches-in-12 inches, a 16-inch shingle would have a 3¾-inch exposure; an 18-inch shingle a 4¼-inch exposure; and a 24-inch shingle a 5¾-inch exposure.

## SHEATHING

Wood shakes and shingles may be placed over spaced or solid sheathing. Spaced sheathing usually consists of 1 × 4s or 1 × 6s spaced on centers equal to the exposure of the shake or shingle. Regardless of the exposure, the sheathing materials

**TABLE 7–1.   RED CEDAR SHINGLE**

| Grade | Length | Thick-ness (at Butt) | No. of Courses Per Bundle | Bdls/Cartons Per Square | Description |
|---|---|---|---|---|---|
| No. 1: Blue Label | 16″ (Fivex) 18″ (Perfections) 24″ (Royals) | .40″ .45″ .50″ | 20/20 18/18 13/14 | 4 bdls. 4 bdls. 4 bdls. | The premium grade of shingles for roofs and sidewalls. These top-grade shingles are 100% heartwood. 100% clear and 100% edge-grain. |
| No. 2: Red Label | 16″ (Fivex) 18″ (Perfections) 24″ (Royals) | .40″ .45″ .50″ | 20/20 18/18 13/14 | 4 bdls. 4 bdls. 4 bdls. | A good grade for many applications. Not less than 10″ clear on 16″ shingles, 11″ clear on 18″ shingles and 16″ clear on 24″ shingles. Flat grain and limited sapwood are permitted in this grade. |
| No. 3: Black Label | 16″ (Fivex) 18″ (Perfections) 24″ (Royals) | .40″ .45″ .50″ | 20/20 18/18 13/14 | 4 bdls. 4 bdls. 4 bdls. | A utility grade for economy applications and secondary buildings. Not less than 6″ clear on 16″ and 18″ shingles, 10″ clear on 24″ shingles. |
| No. 4: Under-coursing | 16″ (Fivex) 18″ (Perfections) | .40″ .45″ | 14/14 or 20/20 14/14 or 18/18 | 2 bdls. 2 bdls. 2 bdls. 2 bdls. | A utility grade for undercoursing on double-coursed sidewall applications or for interior accent walls. |
| No. 1 or No. 2: Rebutted-Rejointed | 16″ (Fivex) 18″ (Perfections) 24″ (Royals) | .40″ .45″ .50″ | 33/33 28/28 13/14 | 1 carton 1 carton 4 bdls. | Same specifications as above for No. 1 and No. 2 grades but machine-trimmed for exactly parallel edges with butts sawn at precise right angles. For sidewall application where tightly fitting joints are desired. Also available with smooth sanded face. |

Reprinted with permission of Red Cedar Shingle & Handsplit Shake Bureau.

should never be spaced more than 10 inches apart. In areas that experience wind-blown snow and where the outside design temperature is 0° F or lower, solid sheathing is recommended.

The biggest advantage of spaced sheathing is that it costs less, and allows the shingles to dry out quickly.

Solid sheathing is either notched or unnotched 1-inch boards or plywood. The

## TABLE 7–2 RED CEDAR HANDSPLIT SHAKES

| Grade | Length and Thickness | 18″ Pack | | Description |
|---|---|---|---|---|
| | | # Courses Per Bdl. | # Bdls. Per Sq. | |
| No. 1: Handsplit & Resawn | 15″ Starter-Finish | 9/9 | 5 | These shakes have split faces and sawn backs. Cedar logs are first cut into desired lengths. Blanks or boards of proper thickness are split and then run diagonally through a bandsaw to produce two tapered shakes from each blank. |
| | 18″ × ½″ to ¾″ | 9/9 | 5 | |
| | 18″ × ¾″ to 1¼″ | 9/9 | 5 | |
| | 24″ × ⅜″ | 9/9 | 5 | |
| | 24″ × ½″ to ¾″ | 9/9 | 5 | |
| | 24″ × ¾″ to 1¼″ | 9/9 | 5 | |
| No. 1: Tapersplit | 24″ × ½″ to ⅝″ | 9/9 | 5 | Produced largely by hand, using a sharp-bladed steel froe and a wooden mallet. The natural shingle-like taper is achieved by reversing the block, end-for-end, with each split. |
| | | 20″ Pack | | |
| No. 1: Straight-Split | 18″ × ⅜″ True-Edge | 14 Straight | 4 | Produced in the same manner as tapersplit shakes except that by splitting from the same end of the block, the shakes acquire the same thickness throughout. |
| | 18″ × ⅜″ | 19 Straight | 5 | |
| | 24″ × ⅜″ | 16 Straight | 5 | |

Reprinted with permission of Red Cedar Shingle & Handsplit Shake Bureau.

## TABLE 7–3 APPROXIMATE COVERAGE OF ONE SQUARE OF SHINGLES

| Length and Thickness | Approximate coverage of one square (4 bundles) of shingles based on following weather exposures | | | | | | | | | | | | |
|---|---|---|---|---|---|---|---|---|---|---|---|---|
| | 3½″ | 4″ | 4½″ | 5″ | 5½″ | 6″ | 6½″ | 7″ | 7½″ | 8″ | 8½″ | 9″ | 9½″ |
| 16″ × 5/2″ | 70 | 80 | 90 | 100[1] | 110 | 120 | 130 | 140 | 150[2] | 160 | 170 | 180 | 190 |
| 18″ × 5/2¼″ | — | 72½ | 81½ | 90½ | 100[1] | 109 | 118 | 127 | 136 | 145½ | 154½[2] | 163½ | 172½ |
| 24″ × 4/2″ | — | — | — | — | — | 80 | 86½ | 93 | 100[1] | 106½ | 113 | 120 | 126½ |

| Length and Thickness | 10″ | 10½″ | 11″ | 11½″ | 12″ | 12½″ | 13″ | 13½″ | 14″ | 14½″ | 15″ | 15½″ | 16″ |
|---|---|---|---|---|---|---|---|---|---|---|---|---|---|
| 16″ × 5/2″ | 200 | 210 | 220 | 230 | 240[3] | — | — | — | — | — | — | — | — |
| 18″ × 5/2¼″ | 181½ | 191 | 200 | 209 | 218 | 227 | 236 | 245½ | 254½[3] | — | — | — | — |
| 24″ × 4/2″ | 133 | 140 | 146½ | 153[2] | 160 | 166½ | 173 | 180 | 186½ | 193 | 200 | 206½ | 213[3] |

[1] Maximum exposure recommended for roofs.

[2] Maximum exposure recommended for single-coursing No. 1 and No. 2 grades on side walls.

[3] Maximum exposure recommended for double-coursing No. 1 grades on side walls.

Reprinted with permission of Red Cedar Shake & Handsplit Shake Bureau.

**TABLE 7–4  APPROXIMATE COVERAGE OF ONE SQUARE OF SHAKES**

| Shake Type, Length, and Thickness | Approximate sq. ft. coverage of one square based on these weather exposures: | | | | | | | | |
|---|---|---|---|---|---|---|---|---|---|
| | 5½ | 6½ | 7 | 7½ | 8½ | 10 | 11½ | 14 | 16 |
| 18″ × ½″ to ¾″ Handsplit-and-Resawn | 55[1] | 65 | 70 | 75[2] | 85[3] | 100[4] | — | — | — |
| 18″ × ¾″ to 1¼″ Handsplit-and-Resawn | 55[1] | 65 | 70 | 75[2] | 85[3] | 100[4] | — | — | — |
| 24″ × ⅜″ Handsplit | — | 65 | 70 | 75[5] | 85 | 100[6] | 115[7] | — | — |
| 24″ × ½″ to ¾″ Handsplit-and-Resawn | — | 65 | 70 | 75[1] | 85 | 100[8] | 115[7] | — | — |
| 24″ × ¾″ to 1¼″ Handsplit-and-Resawn | — | 65 | 70 | 75[1] | 85 | 100[8] | 115[7] | — | — |
| 24″ × ½″ to ⅝″ Tapersplit | — | 65 | 70 | 75[1] | 85 | 100[8] | 115[7] | — | — |
| 18″ × ⅜″ True-Edge Straight-Split | — | — | — | — | — | — | — | 100 | 112[9] |
| 18″ × ⅜″ Straight-Split | 65[1] | 75 | 80 | 90 | 100[7] | — | — | — | — |
| 24″ × ⅜″ Straight-Split | — | 65 | 70 | 75[1] | 85 | 100 | 115[7] | — | — |
| 15″ Starter-Finish Course | Use supplementary with shakes applied not over 10″ weather exposure. | | | | | | | | |

[1] Maximum recommended weather exposure for three-ply roof construction.

[2] Maximum recommended weather exposure for two-ply roof construction; 7 bundles will cover 100 sq. ft. roof area when applied at 7½′ weather exposure.

[3] Maximum recommended weather exposure for side wall construction; 6 bundles will cover 100 sq. ft. roof area when applied at 8½′ weather exposure.

[4] Maximum recommended weather exposure for starter-finish course application; 5 bundles will cover 100 sq. ft. when applied at 10″ weather exposure.

[5] Maximum recommended weather exposure for application on roof pitches between 4-in-12 and 8-in-12.

[6] Maximum recommended weather exposure for application on roof pitches steeper than 8-in-12.

[7] Maximum recommended weather exposure for single-coursed side wall construction.

[8] Maximum recommended weather exposure for two-ply roof construction.

[9] Maximum recommended weather exposure for double-coursed side wall construction.

Reprinted with permission of Red Cedar Shingle & Handsplit Shake Bureau.

thickness of the plywood varies with the spacing of the rafters, but in most cases ½-inch plywood with an exterior glue is used.

The use of solid sheathing increases the insulation properties of the roof and reduces air infiltration.

**TABLE 7–5  MAXIMUM EXPOSURE OF WOOD SHAKES**

| Pitch | Maximum exposure recommended for roofs: | | | | | | | | |
|---|---|---|---|---|---|---|---|---|---|
| | No. 1 Blue Label | | | No. 2 Red Label | | | No. 3 Black Label | | |
| | 16″ | 18″ | 24″ | 16″ | 18″ | 24″ | 16″ | 18″ | 24″ |
| 3-IN-12 to 4-IN-12 | 3¼″ | 4¼″ | 5¾″ | 3½″ | 4″ | 5½″ | 3″ | 3½″ | 5″ |
| 4-IN-12 and Steeper | 5″ | 5½″ | 7½″ | 4″ | 4½″ | 6½″ | 3½″ | 4″ | 5½″ |

Reprinted with permission of Red Cedar Shingle & Handsplit Shake Bureau.

## UNDERLAYMENT AND INTERLAYMENT

An underlayment is usually not recommended with wood shingles, because many roofing contractors feel that asphalt-saturated felt causes condensation trouble. Condensation subsequently leads to sheathing damage and roof leakage. If underlayment is used to reduce air infiltration, a rosin-sized building paper is usually used. If wood shakes are placed in areas that experience wind-driven snow, layers of 30-pound felt may be placed between the courses of shakes. (See Figure 7–1.) Before placement of the shakes, a 36-inch-wide strip of 30-pound asphalt-saturated felt is placed along the eave line, and then a double or triple starter course is placed over the strip. After each course of shakes is placed, an 18-inch-wide strip of felt is placed over them. The distance from the butt of the shake to the interlayment should equal twice the weather exposure. For example, if an 18-inch shake is laid with an 8-inch weather exposure, the distance from the butt of the shake to the bottom of the interlayment is 16 inches.

To secure the interlayment to the roof

FIGURE 7–1.   Placement of interlayment

sheathing, ⅞-inch or ¾-inch roofing nails are used. These should be spaced 12 inches on center at the top of the felt. It is not necessary to place any nails at the bottom of the felt.

Valleys should be lined with a 36-inch-wide strip of 30-lb asphalt-saturated felt. The horizontal strips of interlayment are projected 6 inches on the valley felt, and are trimmed parallel to the intersection of the two sloping roofs.

## FLASHING

### Valley Flashing

Valley flashing can be made of galvanized iron, lead, copper, or aluminum sheets. It should be 24-gauge and extend 10 inches on both sides of the valley. The valley should be center-crimped, and the edges should have formed guides to direct the flow of water and prevent an accumulation of water under the wood shake or shingle.

Before the application of the valley flashing, a line should be chalked down the center of the valley. Starting at the bottom of the valley, the first piece of flashing is placed with the crimped center on the chalkline. One side of the valley is then nailed into place. The nails are spaced 8 to 10 inches apart and are placed 1 inch from the edge of the valley tin. Pressure is then applied to the center of the valley flashing, and the other side is nailed into place. A second piece of flashing is placed over the first piece, creating a 3-inch head lap. The second piece is secured in the same manner.

## Eave Flashing

In areas that experience extreme freezing temperatures, eave flashing is recommended. (See Figure 7–2.) If the roof has a slope of 4 inches per foot or greater, two courses of 30-pound felt are placed along the eave line. The felt should overhang the eave line by ¼ to ⅜ inch, and extend up the roof to a point 24 inches inside the interior wall line. If the flashing is wider than 36 inches, there should be a 2-inch cemented horizontal lap located above the cornice.

If the roof has a slope less than 4 inches per foot, a double course of asphalt-saturated felt that is cemented together should extend 24 inches inside the interior wall line. The cement should be applied at a rate of 2 gallons of cement per 100 square feet, and should be uniformly spread with a comb trowel so that the two pieces of underlayment are not allowed to touch. When the cement is spread, the overlying portion

of the eave flashing should be pressed firmly into place.

## Sidewall Flashing

Sidewall flashing is placed at the intersection of a sloping roof and a vertical wall, mainly to keep water from penetrating the roof. The type of sidewall flashing often used by roofing contractors is step flashing. (See Figure 7–3.) Step flashing, made from galvanized sheet metal, is installed after each adjacent shingle to the vertical sidewall is placed. The flashing shingle is rectangular in shape, usually 10 inches long and 7 inches wide. The flashing, bent to extend up the vertical wall at least 4 inches, is placed between the courses of wood shakes or shingles, with flashing and the shingles alternated. The flashing is secured with one nail in the top corner and is placed just above the butt edge of the wood shake or shingle.

## Chimney Flashing

A chimney is flashed by using sidewall flashing, apron flashing, counter flashing, and step flashing. The material that is

FIGURE 7–2.  Eaves flashing

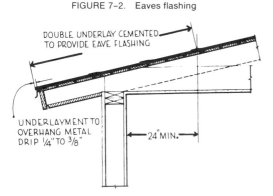

FIGURE 7–3.  Step flashing

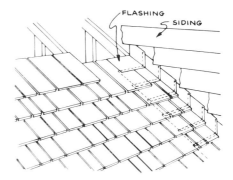

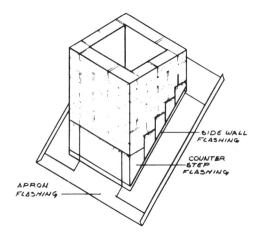

FIGURE 7-4.   Apron, counter, and counter-step
flashing

most often used is galvanized sheet metal
or copper.

Apron flashing is first placed next to the
chimney and over the previously placed
shingles. (See Figure 7-4.) The apron flashing should extend over the wood shakes or
shingles a minimum of 4 inches and should
also turn up the chimney 4 inches. The
ends of the apron flashing should be allowed to extend 4 inches past the corner of
the chimney. It is then bent around the
corner of the chimney. Next, vertical wall
flashing is placed on the other three sides of
the chimney. The right and left side vertical wall flashing should be placed flush
with the front of the chimney and allowed
to extend 6 inches behind the back of the
chimney. The ends of the flashing are then
bent around the chimney, and flashing is
placed for the back side of the chimney.

Straight counter flashing is then placed
over the apron flashing. (See Figure 7-4.)
The ends of the counter flashing extend 2
inches past the corner of the chimney and
are bent around the corner. The top of all

counter flashing is placed and nailed into a
raked mortar joint. Mortar is then applied
over the junction of the flashing and chimney. Counter step flashing is then placed on
the remaining sides. (See Figure 7-4.) The
width of the chimney determines the
number of pieces of step flashing, but in
most cases three or four pieces are required.

Once the flashing is complete, shingles
may be laid.

## Stack Flashing

A factory-built piece of flashing must be
placed over a vent stack. The flashing is
usually made of lead, galvanized sheet
metal, copper, or rubber. The manufactured unit has two basic parts—the flange
and the barrel. The flange rests on the roof
deck and the barrel is the portion of the
flashing that fits around the stack. In some
cases a third piece, called the cap, is used.
It fits on top of the barrel.

In flashing a stack, the shingles should
be placed under the front of the flange and
brought up to a point at least three-
quarters of the distance to the top of the
exposed stack. In some cases it will be more
than this because some vents are smaller
than others. In such cases the shingles are
carried to a point where the nails will be
covered.

After the shingles are nailed in place in
front of the stack, the stack flashing is
placed over the stack. The flashing is
aligned and nailed in place, on the outer
edge of the flange, and positioned so that
the next shingles placed will cover the nails.
Once the flashing is on, a shingle is placed
next to the stack, marked, and cut to fit the
stack. The shingle is then placed in the correct position and properly secured.

## NAILS

Wood shakes and shingles should be secured with rust-resistant nails, usually hot-dipped zinc-coated nails or aluminum nails. If the nails are driven into plywood, threaded nail shanks should be used—these have better holding power. For new construction, 1¼-inch nails should be used, 16- and 18-inch shingles require a 1¾-inch nail; and a 24-inch shingle requires a 2-inch nail. For nailing hip and ridge units, a nail two pennies larger than the nail used on the roof should be used. Double coursing also requires a larger nail, usually 1¾ inches long.

## APPLICATION OF WOOD SHAKES (NEW CONSTRUCTION)

To keep the shakes in proper alignment along the eave line, a shake is placed next to both rake edges and a string is stretched between the two shakes. The shakes should be allowed to project 1½ inches over the eave and rake edge. If the roof is long, it may be necessary to place a "trig" shingle at its midpoint. The "trig" shingle will support the string and keep it from getting out of proper alignment.

The first course of shingles is then placed, with the stretched string as a guide. Each shingle should be secured with only two nails, placed ¾ inch from the edge of the shingle and up from the butt edge enough so that the next course of shingles will cover them by ¾ inch. (See Figure 7–5.) The nails should be driven flush with the shake, but not below the surface. As the individual units are placed, a ¼-inch gap should separate the units. This allows for expansion and prevents buckling. A further precautionary technique is to soak the shingles before their installation. This causes them to swell, reducing the chance that they will later swell and buckle. This technique is not often used, however.

Once the first course is placed, a second or double course is added over the first course; for extra texture, it can be tripled. The joints of the double course should be broken a minimum of 1½ inches. The butt ends of the course should align with the butt ends of the first course of shakes.

An 18-inch-wide strip of 30-pound roofing felt is placed over the top portion of the shakes in such a way that the distance from the butt of the shake to the interlayment is equal to twice the weather exposure.

The second course of shakes is started by marking a chalkline along the shakes. The distance of the chalkline from the butt edge depends on the desired weather exposure. Using the chalk-line as a guide, the shakes are then placed. If any of the individual units are wider than 9 inches, they should be split. Another way to align the

FIGURE 7–5. Nail placement and correct spacing of wood shakes

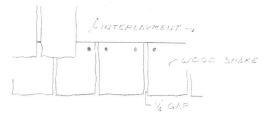

shakes is to tack a straightedge, usually a 1 × 4, temporarily in place and butt the shakes to it.

The succeeding courses of shakes are placed in a similar fashion: spacing the shakes ¼ inch apart; leaving side laps of 1½ inches; never having two vertical joints in direct alignment if they are separated by only one course of shakes; and placing an interlayment of 30-pound roofing felt between the courses.

The shakes can be placed with staggered butt lines, with the butt ends slightly above or below a horizontal chalkline. Longer shakes at random intervals can create an extremely staggered butt line.

## APPLICATION OF WOOD SHINGLES (NEW CONSTRUCTION)

The application of wood shingles is similar to the method used for wood shakes. (See Figure 7–6.) The primary difference is that wood shingles do not require interlayment or underlayment, as wood shakes do.

To keep the shingles in proper alignment along the eave line, a string is stretched parallel to the fascia. The string should be 1½ inches from the fascia, and will serve as a guide for the first course of shingles, which is then placed, each individual unit secured with two nails. The nails should be placed ¾ inch from the edge and 1 to 2 inches above the butt line of the next course. A second layer of shingles is then nailed over the first course, providing a minimum side lap of 1½ inches. Adjacent shingles in every course should be separated by a ¼-inch gap. To assure a watertight roof, the joints in adjacent courses should not be in direct alignment.

The second course of shingles is then placed over the double first course. They are kept in proper alignment by use of a chalkline or straightedge. If the roof has a pitch of 4-in-12 or steeper, a 5-inch, 5½-inch, or a 7½-inch weather exposure is used for 16-inch, 18-inch, and 24-inch shingles, respectively. If the roof has a flatter pitch, the exposure must be reduced.

The succeeding courses of shingles are placed in a similar fashion: spacing the shingles ¼ inch apart; leaving side laps of 1½ inches; and alternating course joints.

### Alternate Techniques

Wood shingles can also be placed with alternate application techniques, such as the thatch, serrated, Dutch weave, and pyramid methods. In the thatch technique the

FIGURE 7–6.  Application of wood shingles

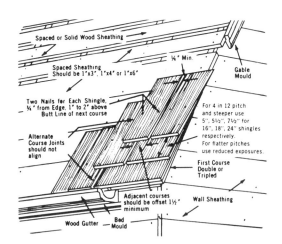

Spaced or Solid Wood Sheathing

¼" Min.

Spaced Sheathing
Should be 1"x3", 1"x4" or 1"x6"

Gable Mould

Two Nails for Each Shingle,
¾" from Edge, 1" to 2" above
Butt Line of next course

For 4 in 12 pitch
and steeper use
5", 5½", 7½" for
16", 18", 24" shingles
respectively.
For flatter pitches
use reduced exposures.

Alternate
Course Joints
should not
align

First Course
Double or
Tripled

Adjacent courses
should be offset 1½"
minimum

Wall Sheathing

Wood Gutter

Bed
Mould

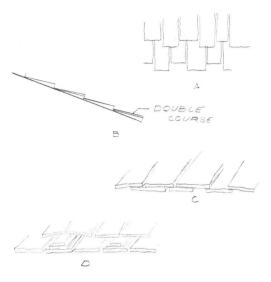

FIGURE 7-7. Alternate techniques of wood-shingle application

shingles are placed with a staggered butt line. (See Figure 7-7A.) The butt ends are staggered above or below an imaginary line, not to exceed 1 inch. The serrated technique incorporates the doubling of every third, fourth, fifth, or sixth course. (See Figure 7-7B.) In the Dutch weave method, shingles are sporadically doubled throughout the roof. (See Figure 7-7C). In the pyramid method, two narrow shingles are superimposed over a larger shingle. (See Figure 7-7D.) The pyramids are placed at random throughout the roof area.

## Hips and Ridges

Before the application of the hip and ridge units, a strip of 30-pound asphalt-saturated felt should be placed down the ridge or hip line. The strip should be 12 inches wide and placed with a 6-inch exposure on each side. A straightedge is then tacked onto both sides of the ridge or hip, 6 inches from the crown. It will serve as a guide in laying the shingles. The first wood shake or shingle is placed adjacent to the straightedge and nailed in place with two 8d nails, long enough to penetrate the underlying sheathing. The placed shake is then trimmed to fit the hip or ridge line. Next, the shake or shingle opposite the one previously placed is then positioned and cut back on a bevel. The starting course is then doubled and the remainder of the shakes are placed, with the direction of each lap alternated. This procedure is called the "Boston lap." (See Figure 7-8.) Preformed factory-made hip and ridge units, which reduce the amount of work and speed up the job, may be used in place of site-applied units.

A ridge line that runs from gable to gable should be started at each end and terminated in the middle. Each end should have a double starter course. The shakes or shingles are then applied, using the same weather exposure as the roof. When the ridge units meet in the middle, a cap unit should be face-nailed over the intersection of the two lines.

FIGURE 7-8. Boston lap

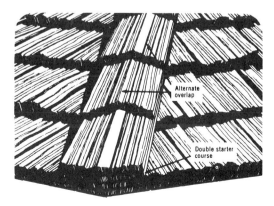

## REROOFING EXISTING ROOFS

Placing wood shakes or shingles over an existing roof provides several advantages; it affords a double roof with added insulation and storm protection, and the interior is protected from sudden rains during the construction period.

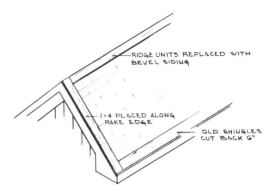

FIGURE 7–9.   Reroofing preparations for existing roofs

Before any wood shakes or shingles are placed, the roof sheathing should be inspected for any structural defects. In order to have a smooth application, any loose or curled shingles should be nailed down or cut away. In most cases if a roof requires an overlay, the stack flashing must also be replaced; therefore, it is necessary to remove all stack flashing. The stacks will require new flashing as the overlay is placed. To assure a sound deck and a pleasing roofline, the old shingles should be cut back 6 inches from the rake edge, and a 1 × 4 nailed on top of the sheathing adjacent to the rake edge. (See Figure 7–9.) The old shingles should also be cut back one course from the eave line. The ridge units should be removed and replaced with a thin strip of bevel siding, bevel side down. To keep the old valley flashing and new valley flashing segregated, a 1 × 4 must be nailed in the valley.

The new wood shakes or shingles are then applied, with the same techniques as those described for new construction. The only difference is that 5d rust-resistant nails should be used.

## PANELIZED SHINGLES AND SHAKES

Wood shakes and shingles bonded to 46¾-inch by 8-foot plywood panels are gaining wide acceptance. (See Figure 7–10.) The large panels reduce the handling of many small pieces, are self-aligning, and can be applied with color-matched, annular-threaded nails. The wood shakes and shingles are bonded to a ½-inch strip of 8-foot plywoood with phenolic resin adhesives and modified urea resin. The phenolic resin adhesive meets test requirements for exterior plywood; the modified urea resin meets those for Type II hardwood plywood.

The panels are available in two-, three-, and four-ply panels. The two-ply panels can be placed over spaced or solid sheathing, using two nails per shingle or shake. To assure a watertight roof, a layer of 15-pound asphalt-saturated felt is required between panels. If a four-ply roof is placed on a roof with a 4-in-12 pitch, or if the pitch is steeper, a 6-inch exposure exists and no interlayment is required. But if the roof has a 10-inch exposure, an 18-inch strip of interlayment must be placed between the panels.

FIGURE 7–10. Panelized wood shingles

## Application Techniques for Panelized Shingles and Shakes

Before the panels are placed, a chalkline or starter strip should be placed parallel to the eave line. A starter panel is then nailed to the rafter with two 6d rust-resistant or zinc-coated nails. (See Figure 7–11.) The ends of each panel should fall on a rafter. Once the starter-panel course has been placed, the roof-shingle panels can be applied directly to the rafters. The panels should be nailed to each rafter with two 6d nails. The first nail should be placed 1 inch below the top of the plywood strip, and the second nail approximately 2 inches up from the lower edge of the strip. The succeeding courses of panels should be cut so the shingles will have a minimum of 1½ inches offset. The vertical joints in alternate courses

should vary slightly. (See Figure 7–12.) To apply flashing around a vent, the shingles must first be cut with a sabre saw. The flashing is then placed over the vent, and the top panel is nailed in place. At the intersection of a vertical wall, flashing is applied after each course of panels. (See Figure 7–13.) The vertical wall flashing should be nailed near its outside edge and above the exposure line.

To finish the rake edge, the panel is placed flush with the outside edge of the verge rafter. The last shingle on the panel is removed and rake molding is positioned to cover the exposed plywood edge. (See Figure 7–14.). Additional shingles are then placed on the panel to the desired overhang.

Once all the panels are placed, the hips and ridges are finished in a "Boston lap."

FIGURE 7-11. Application of panelized wood shingles

The hip and ridge units can be made from the remnants of the panels, or factory-assembled units can be used. The weather exposure should be the same as for the roof shingles. Two nails (usually 8d) should be used on each side of the hip or ridge unit.

To flash valleys, 1 × 6s are placed parallel to the valley rafter. (See Figure 7-15.)

FIGURE 7-12. Alignment of vertical joints

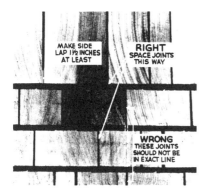

FIGURE 7-13. Stack flashing

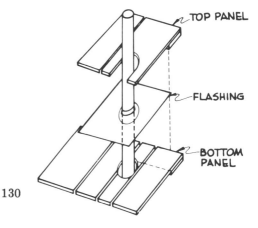

130

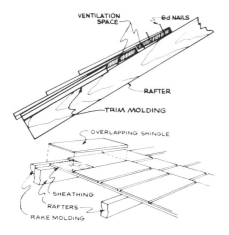

FIGURE 7-14.   Placement of rake molding

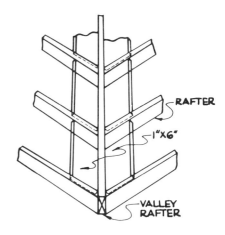

FIGURE 7-15.   Placement of 1x6s parallel to the valley

The 1 × 6s are cut to fit between the valley jack rafters, and are placed flush with the tops of these rafters. This technique provides a solid base for the flashing and permits an open valley of the desired width.

To cover the strip of plywood, the flashing must be applied after each course of panels. (See Figure 7–16.) The flashing should be nailed above the exposure line and near the outside edge of the flashing.

FIGURE 7-16.   Application of flashing

# CHAPTER 8

# METAL AND
# BUILT-UP ROOFING

METAL HAS BEEN USED FOR DECADES TO provide durable roofing. A metal roof can reflect as much as 85 percent of the sun's heat, and its low radiation rate cuts the heat transfer from the exterior to the interior. Metal roofs made of aluminum are corrosion-resistant, noncombustible, and nonsparking.

Metal roofing is also available in a wide range of matching and complimentary colors. The roofing is lighter in weight than most other types of roof covering and will not deteriorate, warp, or split.

The most popular types of metal roofing are simulated wood shakes, corrugated sheets, and sheet metal.

## SHEATHING

Corrugated sheets are usually nailed to $2 \times 4$ purlins, although they can be placed over solid sheathing or insulation board. If the sheets are to be placed over solid sheathing, the sheathing should first be covered with underlayment; but if they are placed over insulation board, $2 \times 4$ purlins should first be placed over them. An air space should always be provided between any metal roofing and porous or absorbent

insulations. Insulation board acts as a vapor barrier, preventing the migration of moisture from the inside of a building or the absorption of outside condensation.

Aluminum shingles or shakes may be installed over solid sheathing or metal purlins. Purlins are properly spaced on the deck and secured with a No. $8 \times \frac{1}{2}$ screw at each standing seam. The sheathing can be tightly matched 1-inch stock or plywood.

## UNDERLAYMENT

If simulated wood shakes are used, a layer of 30-pound asphalt-saturated felt should be placed over the roof sheathing. The underlayment should be held in place with $\frac{7}{8}$-inch aluminum felting nails. If a roof has a 2- to 3-inch rise per foot, the overlapping plies of felt and the nail heads should be sealed in hot asphalt or asphalt cement. Fifteen-pound felt can be used as underlayment for aluminum shakes, but it should be

placed with head laps of 19 inches and end laps of 6 inches.

If corrugated metal roofing or rolled sheet metal is placed over solid sheathing, the sheathing must first be covered with a layer of 15-pound asphalt-saturated felt. The felt should be placed with a 2-inch top lap and 4-inch side laps. The felt should be lapped 6 inches from both sides of all hips and ridges.

## FASTENERS

Copper sheets are held in place by copper cleats. (See Figure 8–1.) The cleats should be fastened to the roof sheathing with cop-

per or approved bronze nails. All accessories used on copper roofing should be made of copper, bronze, or brass. Wood screws

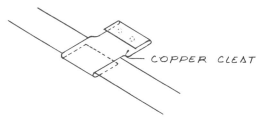

FIGURE 8–1.  Copper cleat

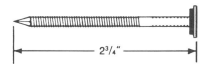

FIGURE 8–2.  Screw-thread roofing nails

should be No. 12 gauge and the machine screws or bolts should be ¼ inch in diameter. Cleats are used to secure copper sheets to roof sheathing. They should be 1½ inches wide, and long enough to make at least a ½-inch interlock with the sheet copper.

For corrugated aluminum roofing, screw-thread roofing nails with a flat washer seal should be used. (See Figure 8–2.) If the roofing is nailed to purlins or solid sheathing, the nails should be 2 inches long and have a head diameter of ½ inch. If the corrugated roofing is placed over insulation or an old roof, the nails should be 2¾ inches long.

Aluminum shakes should be installed with aluminum screw-shank nails, long enough to penetrate ½ inch into the sheathing. Eave and gable trim are attached with 1¼-inch screw-shank aluminum nails and No. 8 × ⅜-inch sheet metal pan-head slotting screws.

## APPLICATION OF ALUMINUM SHAKES AND SHINGLES ON WOOD SHEATHING

The first step in the application of aluminum wood shakes is to install the eave trim. (See Figure 8–3.) The trim is first nailed to the roof sheathing with 1¼-inch screw-shank aluminum nails placed 12 inches on center. Sheet metal pan screws are then screwed through the exposed face of the trim. The screws should be placed 18 inches on center. When the trim is first placed, it should be allowed to extend 1½ inches past the rake edge.

Gable trim is then placed along the rake edge. (See Figure 8–4.) It should be butted to the trim at the eave line, coping the trim

FIGURE 8–3.  Installation of eaves trim

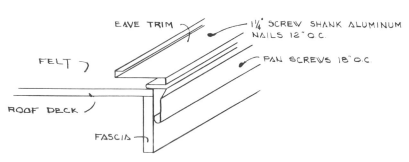

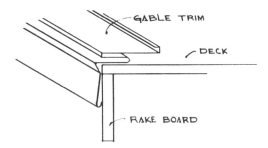

FIGURE 8–4.   Gable trim

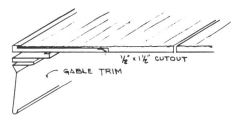

FIGURE 8–6.   Shingle placement

to form a butt joint. A second piece of gable trim is then placed over the first line, and is secured to roof sheathing with nailing tabs. The tabs should be placed 12 inches on center and secured with 1¼-inch screw-shank aluminum nails. The gable trim is allowed to extend 2 inches over the ridge line. The trim is then bent over and secured to the sheathing with a nailing tab. (See Figure 8–5.) The first shingle is placed at the lower left corner of the roof. (See Figure 8–6.) For the first shingle to fit the gable trim is then placed over the first piece, flush with the eave line, and is secured to roof sheathing with nailing tabs. The tabs should be placed 12 inches on center and secured with 1¼-inch screw-shank alumi- num nails. The gable trim is allowed to extend 2 inches over the ridge line. The trim is then bent over and secured to the

shingles are started with a three-quarter, one-half, and one-quarter shingle. When the course of shingles reaches the opposite rake edge, a 1½-inch × ½-inch section is cut from the butt of the last shingle. The shingle is then locked into position and secured with nailing tabs. (See Figure 8–7.)

If aluminum shingles are placed over a metal deck, a metal purlin is first laid along the eave line. The purlin is attached to the roof deck with one No. 8 × ½ self-drilling screw at each standing seam. The first pur- lin is placed to lock the first course of shin- gles. Additional courses of shingles require properly spaced purlins to also lock them into position.

When the shingles reach the ridge line, they should be bent over the ridge 2 inches and secured with 1¼-inch aluminum nails. The nail heads are then caulked with a sealant. The first ridge unit is butted into the gable trim and secured to the sheathing

FIGURE 8–5.   Placement of nailing tab

FIGURE 8–7.   Securing shingles with 4 nailing clips

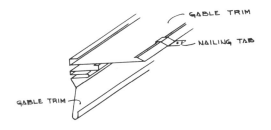

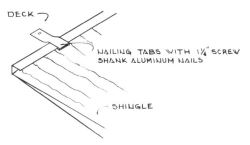

with two nailing tabs. As the caps are placed, they should be locked together and then secured into position. (See Figure 8–8.) This procedure is followed until the units reach half the distance of the ridge. Additional ridge units are then started from the other rake edge. The two rows of ridge units should be allowed to overlap a minimum of 2 inches. The joint is then caulked and a matching aluminum piece is placed over the joint.

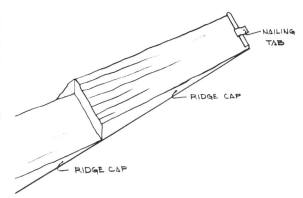

FIGURE 8–8. Cutting and securing the last shingle

## FLASHING ON A METAL ROOF

### Vent-Pipe Flashing

To properly flash a vent pipe, a hole is first cut in the shingle and mastic placed over the bottom of an aluminum vent cone. The vent cone is then placed over the stack. It should be placed under the shingles that fall above the stack and over the shingles that are below the stack. (See Figure 8–9.) If the butt of a shingle runs under the flange of the flashing, the butt should be hammered flat. The bottom of the flange is

secured with aluminum screws and the edges are chalked.

### Valley Flashing

The first step in valley flashing is to place aluminum flashing down the center of the valley. The flashing should have 10-inch lap joints or 1-inch lock seams in the direction of flow. Two pieces of eave and gable trim are then placed back to back in the valley. The trim should be placed within 1 inch of the valley and secured with nailing tabs placed 12 inches on center. (See Figure 8–10.) Once the two pieces of trim are in position, the shingles can be mitered to fit

FIGURE 8–9. Placement of ridge units

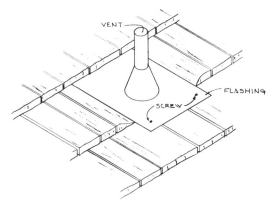

FIGURE 8–10. Vent-pipe flashing

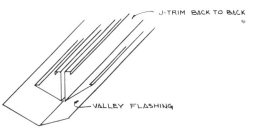

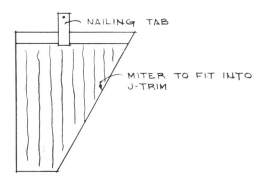

FIGURE 8–11.   Valley flashing

## Hips

Hips are covered by first mitering a full shake short of the hip line by 1 inch. (See Figure 8–12.) After both sides of the hip are covered, the back end of the unit is slipped up under the top flange of the adjacent shingle. After the front edge of the hip unit locks over the butt end of the adjacent shakes, it is then secured with 1¼-inch aluminum nails.

## Vertical-Wall Flashing

the trim. (See Figure 8–11.) The nails should be 1¼ inches long and should have their heads caulked.

When a sloping roof intersects a vertical wall, a gable end strip is used to flash the

FIGURE 8–12.   Mitered shingles

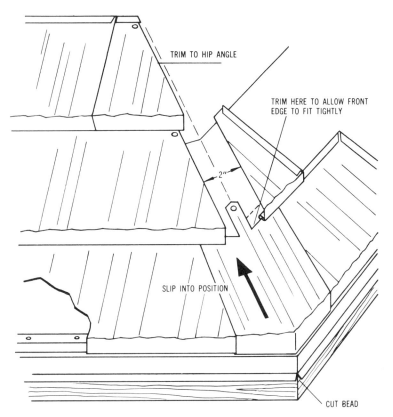

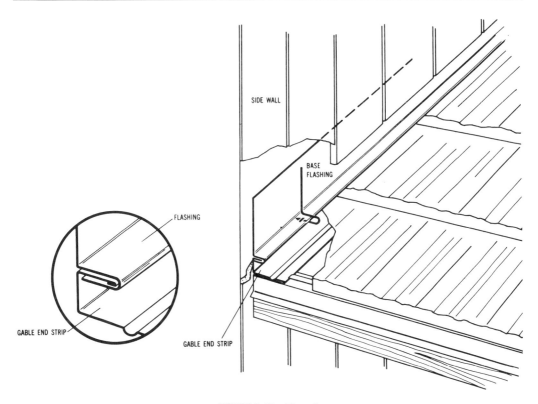

FIGURE 8–13. Hip units

intersection. (See Figure 8–13.) The joints should be lapped 4 inches and the strip nailed on 6-inch centers. The flashing should be placed under the sidewall cover. If the sidewall is brick, however, counter flashing is placed in a mortar joint. The counter flashing should extend over the gable end strip, and should be caulked in the mortar joint.

## Chimney Flashing

First, base flashing should be placed around the chimney. (See Figure 8–14.) The base on the sides of the chimney should fold around the front of the chimney 4 inches. The base flashing in the back of the chimney should fold around the sides of the chimney 4 inches, and be placed so that the water drains from the top of the flashing to the layer below. As the base flashing is installed, it should be placed in a bed of caulking.

Gable end strips should then be placed over the base flashing. They should be located 4 inches from the side of the chimney, nailed on 6-inch centers, and caulked to prevent leakage. The shingles can then be cut and fitted around the chimney. To start the shingle in the back, an eave strip must be placed next to the chimney. After the base flashing is complete and the shingles have been placed, counter flashing is placed over the base flashing.

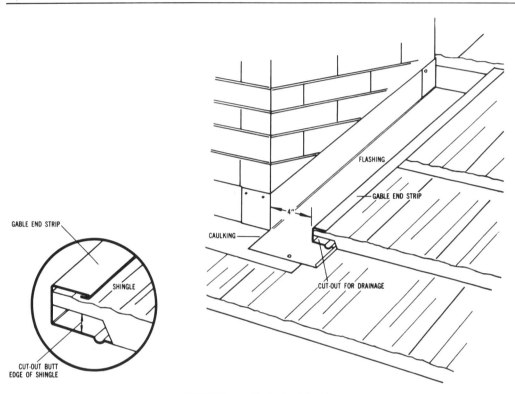

GABLE END STRIP

SHINGLE

CUT-OUT BUTT
EDGE OF SHINGLE

FLASHING

GABLE END STRIP

4"

CAULKING

CUT-OUT FOR DRAINAGE

FIGURE 8–14.   Vertical-wall flashing

## APPLICATION OF CORRUGATED SHEET ROOFING

Before the corrugated roofing is placed, a vertical line should be chalked parallel to the rake edge. Additional lines are also chalked parallel to the eave line. The spacing of the lines depends on the width and length of the roofing. The first sheet is placed flush with the chalklines and is nailed into position. To completely seal the roof, 48-inch rubber filler strips can be placed under each panel at the eave and ridge lines. (See Figure 8–15.)

The nails are placed on the top of a ridge and should penetrate at least 1 to 1½ inches into solid wood. They should make snug contact with the roofing. (See Figure 8–16.) If the nail does not seat the washer,

FIGURE 8–15.   Base flashing

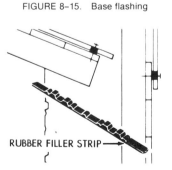

RUBBER FILLER STRIP

**CORRECTLY** – seated nail makes good snug contact between nail head, washer and roofing, sealing out moisture.

FIGURE 8-16.   Correctly placed nail

the roof might leak, or the roofing may be moved by the wind. (See Figure 8–17.) If the nail is overdriven, the surrounding metal may be pulled down, forming a depression that will hold water. (See Figure 8–18.) At the eave line, the roof panel should be placed with a 2-inch overhang. The panel should be flush with the rake edge and covered with gable trim. As additional sheets are placed, a nail should be

FIGURE 8-17.   Underdriven nail

**UNDERDRIVEN**—nail does not seat the washer, creates a potential leak around the nail shank, may allow roofing or siding to move with wind, further enlarging hole.

**OVERDRIVEN**—nail can pull down surrounding metal, forming a dimple which holds water, may break seal. Cover this potential leak with sealer.

FIGURE 8-18.   Overdriven nail

placed through the side laps and adjacent crown.

To facilitate the application and to insure maximum weather-tightness, the nail holes should be prepunched where more than two thicknesses occur. The end laps are usually 6 inches, but if the slope of the roof is less than 4 inches in 12 inches, 12-inch laps should be used.

The ridge is finished by the installation of a ridge cap down the center of the ridge. (See Figure 8–19.) The ridge cap should be nailed on 12-inch centers.

## Sidewall Flashing

There are three basic techniques that can be used to flash the intersection of a sloping

FIGURE 8-19.   Ridge cap

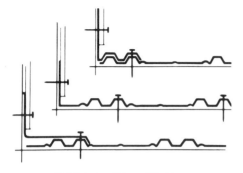

FIGURE 8-20.  Side-wall flashing

roof and a vertical wall. (See Figure 8–20.) A piece of corrugated roofing can be bent at the job site to extend 5 inches up under the bottom of the siding. The bottom portion of the flashing should lap two ribs and bend to extend 5 inches under the bottom edge of the siding. The flashing can also be constructed to fit two ribs and fit 5 inches under the siding. The flashing should be nailed at the rib farthest from the wall, at 12-inch intervals. Roofing can also be bent to fit up under the bottom of the siding.

## Valley Flashing

If valley flashing is required, it should be constructed of galvanized iron, lead, or aluminum, depending on the material used for the roofing. The valley should be center-crimped and the edges should have formed water guides to direct the flow of water and prevent its accumulation under the roofing.

Before the valley flashing is placed, a line should be chalked down the center of the valley. Then, starting at the bottom of the valley, the first piece of flashing is placed with the crimped center at the chalkline. One side of the valley is then nailed into place, with the nails 8 to 10 inches apart. They should also be placed 1 inch from the edge of the valley flashing. Pressure should be applied to the center of the valley flashing, and the other side then nailed into place. A second piece of flashing is placed over the first piece, creating a 3-inch head lap. The second piece is then secured in the same manner as the first.

## BUILT-UP ROOF

A built-up roof consists of alternate coatings of bitumen and layers of felt over decking (called the membrane), and is used on roofs that have relatively low slopes. The decking of a built-up roof, which is the exterior roof covering, falls into two basic categories—nailable and non-nailable. These two categories can be further subdivided into steel, concrete, wood, gypsum, and structural wood fiber decks. The primary purposes of the roof deck are to support the roofing membrane, to partially prevent heat gain or heat loss, to control sound, and in some cases to provide a desirable inside appearance.

One of the most important qualities of a roof deck is its ability to shed water. Decks, however, are all too often constructed dead level; when they are, a depression develops. Water can collect in the depression and the freeze-thaw action of the ponding water will eventually deteriorate the roofing membrane. Moreover, if left standing, water will invariably find its way through small openings in the membrane and penetrate the exposed organic felts by capillary action or absorption. Therefore, the roof deck should be designed to provide a minimum $\frac{1}{4}$-inch slope per foot so that it may drain freely.

Before the roofing membrane is placed, the roof deck should be properly inspected, because the success of the roof is largely dependent upon the deck. Incorrect nailing in a wood deck could cause the boards to warp and the roof to crack; nails improperly placed might work loose and come through the membrane.

## Wood Decks

Board decks and plywood decks are the two basic types of wood decks. They are used extensively in light construction, and because of their popularity will be the only type discussed in this chapter.

Board decks must be constructed of well-seasoned lumber that is at least ¾ inch thick. Shiplap and tongued-and-grooved boards are the most popular types used for decking; they should be secured with two nails per 6-inch width. The nails should be long enough to penetrate the framing members and hold the decking in position without working loose. When the decking is placed, the ends should be centered over the rafter and the board nailed to each rafter or purlin. If the boards are warped or split, they should be discarded for more suitable stock. But if the boards have small cracks or knotholes, these imperfections should be covered with sheet metal nailed in place.

If plywood is used as a roof decking it should be a minimum ½ inch in thickness with 24-inch rafter spacing, and should be an exterior grade. The plywood panels should be applied at right angles to the framing members and should be continuous over two or more rafters with either solid backing or with a ply clip placed between each rafter. To allow for the possibility of dimensional change, there should be a ¹⁄₁₆-

inch joint at the ends of the panels and a ⅛-inch joint at the sides of the panels. The panels should be nailed to the framing members, both at the edges and throughout the main body of the panel. As the panels are being nailed in place, the mechanic should avoid placing any direct weight on a panel, for this might nail in a deflection, which could cause a depression that will accumulate water.

## Coal-Tar Pitch and Asphalt

Coal-tar pitch and asphalt are two agencies that bind the plys of roofing felt together and are integral parts of the roofing membrane. Both coal-tar pitch and asphalt are bituminous material and similar in appearance. They are often used for the same purpose; but the two materials are basically different. Asphalt is manufactured from a variety of residuals left over from distillation of petroleum oil, and consists primarily of equal parts of paraffinic and naphtenic hydrocarbons. In contrast, coal-tar pitch is a chemical composition of aromatic hydrocarbons of the benzenering type.

If coal-tar pitch is used, there are several things that should be considered:

- *Slope:* Coal-tar pitch can normally be placed on a slope up to 1 inch per foot, and under certain conditions it can be placed on slopes up to 2 inches per foot.
- *Aggregate Surfacing:* Coal-tar must be covered by a surfacing of slag or gravel. The slag or gravel screens out ultra-violet light, helps prevent the possibility of a fire starting by burning brands, and improves the stability of the roofing membrane.
- *Roof Deck:* The roof deck on which the

coal-tar pitch is placed should be clean, dry, smooth, and free of depressions. If coal-tar pitch is placed over a damp surface, blisters might form under the roofing membrane. If the substrata are not clean, there will be an inadequate bond.

■ *Heating of Coal-Tar:* Coal-tar pitch should be heated to a maximum temperature of 400° F until after the felt is applied.

### Asphalt

Roofing asphalts are available in the grades of Low Melt, Medium Melt, High Melt, and Extra High Melt. The softening point range of the various asphalts are: Low Melt 135° to 150° F; Medium Melt 160° to 175° F; High Melt 180° to 200° F; and Extra High Melt 205° to 225° F. Asphalt should be heated to a maximum temperature of 450° F. The temperature should be raised slowly and kept relatively stable by the addition of fresh asphalt. To maintain a constant temperature, most asphalt kettles are equipped with thermostatic controls.

## Felts

Felts used in a built-up roof can be classified as asbestos or organic. Asbestos felts are composed primarily of asbestos fiber, a nonrotting inorganic mineral fiber which provides longer life, greater resistance to weathering, and lower maintenance costs.

Organic felts are made of fibrous organic materials and are subject to deterioration unless heavily saturated with a bituminous saturant. The organic felt for a roof should weigh about 500 to 600 pounds per square.

## Surfacing Materials

Surfacing materials for a built-up roof may be water-worn or river-washed gravel, crushed rock, or crushed blast-furnace slag. The surfacing material should be hard, dry, opaque, and free of dirt, dust, and foreign matter. The size of the aggregate should range from $\frac{1}{4}$ to $\frac{5}{8}$ inch. When the surfacing aggregate is placed, it should be thoroughly embedded in a poured flood coat of hot asphalt or coal-tar pitch.

## APPLICATION OF ROOFING

One of the first steps in the application of a built-up roof is to cover the board deck with a ply of rosin-sized sheathing paper. This paper is used to keep the asphalt or coal-tar pitch from bleeding through the felt and onto the decking. This bleeding can cause the membrane to split, due to the differential movement of the decking.

Once the sheathing paper has been placed, a layer of 30-pound asphalt-saturated felt is placed, starting at the low edge and working up the slope. The felt should be placed perpendicular to the slope and

lapped 2 inches for each course. The laps should be nailed on 6-inch centers and down the longitudinal center of each run, with the two rows of nails spaced approximately 11 inches apart. The nails down the center of the felt should be spaced on 18-inch centers. The nails should be $\frac{1}{2}$-inch-diameter head, galvanized roofing nails at least $\frac{7}{8}$ inch long.

After the asphalt-saturated felt has been placed, an 18-inch-wide strip of asbestos finishing felt should be placed along the lower edge. Next, a full 36-inch-wide strip

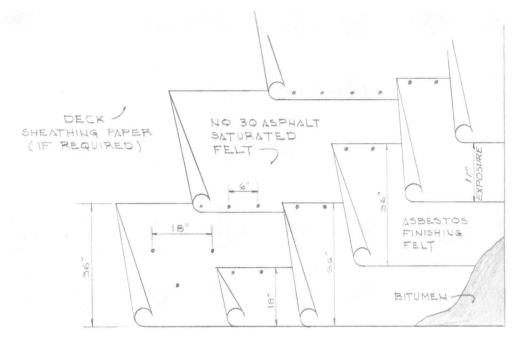

FIGURE 8–21.  Felt placement

is placed over the previously placed half-strip. Additional rows of felt are then placed full width, overlapping the preceding felt by 19 inches. (See Figure 8–21.) When the felt is placed in this manner, at least two plies of felt cover the base at any point. If the slope is greater than 2 inches per foot, it should be nailed on 9-inch centers along the back edge. As each strip of felt is placed, it should be set in a uniform layer of hot asphalt that has been spread at a minimum weight of 23 pounds per square. To assure a good bond, the felt should be rolled and broomed into the hot asphalt.

Once the felt has been placed, the entire roofing surface must be flooded with a uniform flood coat of roofing asphalt at a weight of 60 pounds per square. While the asphalt is hot, surfacing aggregate should be evenly spread over the roof surface.

## FLASHING ON A BUILT-UP ROOF

To prevent the migration of moisture, vulnerable locations should be properly flashed. Some of the areas that need flashing are the intersection of the roof and a vertical wall; collars for large stacks and flagpoles; vent-pipe connections; and at the eave line of the roof.

### Base Flashing

To properly flash the intersection of a vertical wall and a built-up roof, concrete primer should be applied to the wall and allowed to dry before the roofing felt is placed. (See Figure 8–22.) A cant strip with

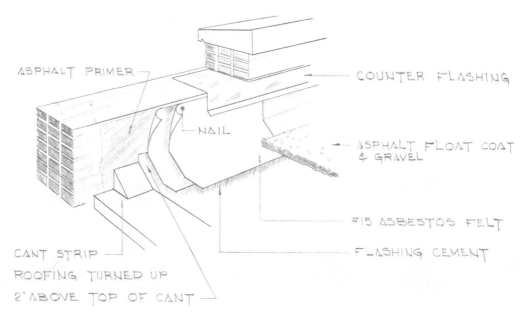

ASPHALT PRIMER

COUNTER FLASHING

NAIL

ASPHALT FLOAT COAT & GRAVEL

#15 ASBESTOS FELT

FLASHING CEMENT

CANT STRIP
ROOFING TURNED UP
2' ABOVE TOP OF CANT

FIGURE 8–22.   Base flashing

4-inch vertical and horizontal sides should then be placed at the intersection of the horizontal and vertical surfaces. After the wall has been primed and the cant strip positioned, the roofing felt can be placed on the roof. It should be allowed to extend up the vertical wall for a distance of 6 inches. Next, a flashing assembly consisting of one layer of No. 15 asbestos felt and one layer of asbestos-reinforced flashing sheet is placed over the cant strip. The flashing should extend well above the turned-up roofing and out onto the deck proper at least 4 inches. The different layers of flashing should be mopped both to the vertical wall and to each other with special flashing cement or hot High Melt asphalt, and also nailed to the vertical wall with case-hardened concrete nails, to be placed a minimum of 1½ inches from the top of the flashing, on 12-inch centers. The lengths of

the base flashing should be a minimum of 12 feet, and have 3-inch end laps. After the flashing has been nailed, a uniform coat of flashing cement should be troweled over the entire surface of the base flashing.

## Vent-Pipe Connections

When a vent pipe protrudes through the roof deck, the area around the vent should first be flooded with hot asphalt. (See Figure 8–23.) The flange of the vent is then placed in the hot asphalt, and nailed to the roof deck. Next, two squares of No. 15 asbestos felt are cut and placed over the flange. The felt should extend 4 inches and 6 inches beyond the flange, and should be set in hot asphalt. To complete the flashing of the vent pipe, a flood coat of hot asphalt and gravel is placed over the two layers of felt.

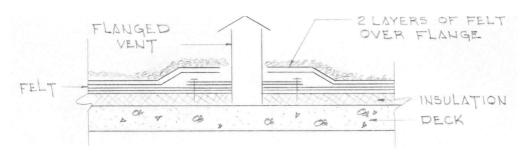

FIGURE 8–23.   Vent-pipe flashing

## Gravel Stops

To properly flash the roof's edge, the first two layers of felt should extend a minimum of 6 inches beyond the edge. (See Figure 8–24.) After the remaining felts have been placed, the edges of the previously placed felts should be folded over and embedded in hot asphalt. When this technique is used, the asphalt cannot bleed down the fascia. Next, the gravel stop is set in a thick bed of flashing cement and nailed to the deck. The nails should be ½-inch-diameter head galvanized roofing nails at least ⅞ inch long and spaced on 6-inch centers. The ends of the flashing should be lapped a minimum of 3 inches and packed in flashing cement. The flange of the gravel guard should then be sealed with two layers of No. 15 asbestos felt embedded in hot asphalt. The bottom piece of felt should be 6 inches wide and the top layer should be 12 inches wide.

FIGURE 8–24.   Gravel stop

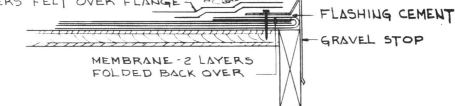

# CHAPTER 9

# CLAY AND CONCRETE MASONRY CONSTRUCTION

CLAY AND CONCRETE MASONRY CONSTRUC-
tion has been used successfully in light construction for decades. The masonry units
are permanent, provide good insulation,
and offer design flexibility; and in most
cases very little maintenance is required.

To make clay masonry units, mined
shale or clay is pulverized and tempered
with water. This process is called "grinding
and pugging." The clay is then formed into
bricks and dried. Next, the bricks are
placed in a kiln to be burned. When they
have cooled, they are ready for shipment.

For many years bricks were manufactured in three basic sizes, but in recent
years manufacturers have developed new
styles, characteristics, and sizes of brick
units, as shown in Table 9–1. But the three
brick unit sizes considered to be standard
are called *standard,* Roman, and Norman.

Concrete masonry units are manufactured from heavyweight or lightweight aggregates mixed with water and portland
cement. Depending on the type of aggregate used in the manufacture of the individual units, they are classified as either
*heavyweight* or *lightweight.* The aggregates
in the heavyweight units can be sand,
gravel, crushed stone, or air-cooled slag;
those in the lightweight units are coal cinders, expanded shale, slag, pumice, or volcanic cinders. A concrete masonry unit
8″ × 8″ × 16″ that is constructed from
heavyweight aggregate will weigh from 40

**TABLE 9–1   NOMINAL MODULAR SIZES OF BRICK**

| Unit Designation | Thickness Inches | Face Dimension | | |
|---|---|---|---|---|
| | | Height Inches | Length Inches |
| Conventional brick | 4 | 2⅔ | 8 |
| Roman brick | 4 | 2 | 12 |
| Norman brick | 4 | 2⅔ | 12 |
| Engineer's brick | 4 | 3⅕ | 8 |
| Economy brick | 4 | 4 | 8 |
| Jumbo brick | 4 | 4 | 12 |
| Double brick | 4 | 5⅓ | 8 |
| Triple brick | 4 | 5⅓ | 12 |
| "SCR brick" | 6 | 2⅔ | 12 |

Reprinted with permission of Portland Cement Association.

to 50 pounds, while a unit of the same size
made with lightweight aggregates will
weigh from 25 to 35 pounds.

Individual concrete masonry units are
available in a variety of shapes and sizes,
each designed for a specific job. (See Figure
9–1.) Some of the more common units are
stretchers, corners, jambs, full-cut headers,
and solid tops. The individual units are
usually referred to by their nominal dimensions. Therefore, a typical stretcher has
nominal dimensions of 8″ × 8″ × 16″, but
the actual units of measurement are
7⅝″ × 7⅝″ × 15⅝″. When the unit is laid
in a ⅜-inch mortar bed, and has a ⅜-inch
head joint, the jointed assembly will measure 8″ × 8″ × 16″.

## WALLS

The three basic types of wall used in light
construction are veneered, solid masonry,
and cavity walls.

## Veneered Walls

Brick veneer is typically placed over and
anchored to wood-frame construction with
corrosion-resistant metal ties.

150

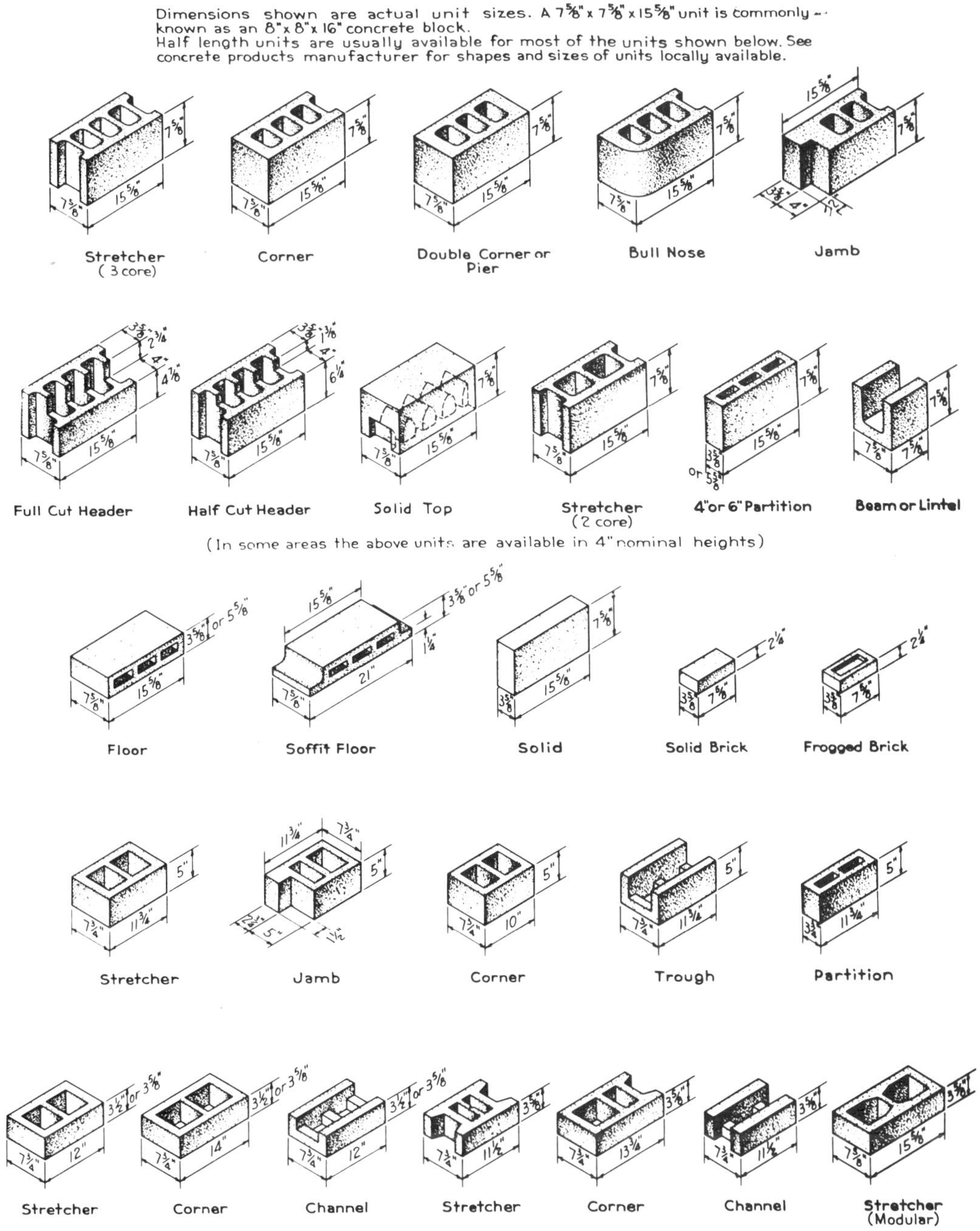

Dimensions shown are actual unit sizes. A $7\frac{5}{8}$" x $7\frac{5}{8}$" x $15\frac{5}{8}$" unit is commonly known as an 8"x 8"x 16" concrete block.
Half length units are usually available for most of the units shown below. See concrete products manufacturer for shapes and sizes of units locally available.

Stretcher
( 3 core)

Corner

Double Corner or
Pier

Bull Nose

Jamb

Full Cut Header

Half Cut Header

Solid Top

Stretcher
( 2 core)

4"or 6" Partition

Beam or Lintel

(In some areas the above units are available in 4" nominal heights)

Floor

Soffit Floor

Solid

Solid Brick

Frogged Brick

Stretcher

Jamb

Corner

Trough

Partition

Stretcher

Corner

Channel

Stretcher

Corner

Channel

Stretcher
(Modular)

FIGURE 9–1. Typical shapes and sizes of concrete masonry units

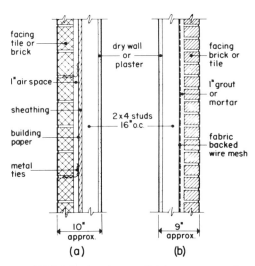

FIGURE 9–2.   Attachment of brick veneer to a wall
frame

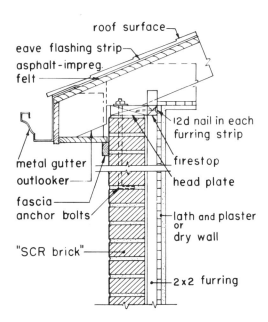

Two techniques are used to secure the
veneer to the wall frame. (See Figure 9–2.)
In the one most widely used, one corrosion-
resistant wall tie is installed every 2 square
feet of wall area. The wall ties should be
corrugated, at least ¾ inches wide, and 6⅝
inches long. It is also possible to attach veneer to the frame wall by grouting it to
paperbacked, welded wire mesh. The mesh
is attached directly to the frame wall, thus
eliminating the need for wall sheathing.
This technique is sometimes called reinforced masonry veneer. Regardless of the
technique used, there should be at least a 1-
inch space between the brick and the wall
frame. In the case of the reinforced masonry veneer the space is fitted with mortar
or grout. Satisfactory performance from a
brick veneer wall requires an adequate
foundation, strong and well-braced frame
wall, sound anchorage system, and the use
of good materials and craftsmanship.

FIGURE 9–3.   (right) Single-wythe clay masonry
wall

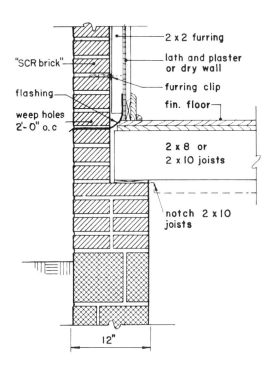

## Cavity Walls

A cavity wall consists of two wythes (single vertical sections of masonry) of masonry separated by a continuous air space not less than 2 inches wide. The individual wythes can be constructed of clay tile, brick, concrete masonry, or a combination of two of the materials. To connect the two wythes, metal ties are firmly anchored in the mortar. The ties should be $\frac{3}{16}$-inch-diameter steel rods or have equivalent stiffness. One wall tie should be placed for each $4\frac{1}{2}$ square feet of wall area and staggered on alternate courses. The exterior wythe is always a nominal 4 inches in thickness, while the interior may be 4, 6, or 8 inches wide. The cavity should be free of mortar droppings and bridgings and should also be built with weep holes, so that the cavity can drain.

## Solid Masonry Walls

The two most common types of solid masonry walls used in light construction are concrete masonry walls and single-wythe clay masonry walls. Concrete masonry walls are usually constructed from nominal $8 \times 8 \times 16$ units laid in a full bed of portland cement mortar. To increase the strength of the wall, the masonry units are laid in a lap bond and reinforcement is added to the wall. The reinforcement can include continuous reinforced concrete bond beam, reinforced concrete studs tied to the footing, reinforced concrete footing, and reinforcement placed in the horizontal mortar joints.

Single-wythe clay masonry walls are usually constructed from 6-inch clay masonry units. (See Figure 9–3.) The 6-inch units allow for a maximum wall height at the eave line of 9 feet and 15 feet to the peak of the gable. The units are laid in common (half) bond with full head and bed joints. Single-wythe clay masonry walls are designed for use with furring strips, which provide a barrier to moisture penetration, permit the installation of plumbing lines and electrical fixtures, and allow the use of blanket insulation.

## BONDS

The term *bond,* when used in reference to masonry, can have three different meanings:

- *Mortar bond:* Refers to the ability of the mortar to adhere to the masonry units or to reinforcing steel
- *Structural bond:* Refers to the technique in which individual masonry units are interlocked or tied together to form an integral unit
- *Pattern bond:* Refers to the development of a particular pattern through placement of individual masonry units

### Structural Bonds

Structural bonding of masonry walls can be accomplished by overlapping the individual masonry units; by the use of metal ties embedded in the mortar joints of the two adjacent wythes; or by the grouting of adjacent wythes of masonry. The individual masonry units can be placed in different positions to effect a particular pattern or bond. Thus an ordinary masonry unit may take on a different name.

The overlapping of the individual masonry units is based on a variation of two traditional techniques of bonding—the English bond and the Flemish bond. The English bond consists of alternating courses of headers and stretchers, while the Flemish bond consists of alternating headers and stretchers in every course. To bond the wall transversely, the headers are placed with their long axis across the wall. The headers should not be spaced further apart than 24 inches, and should comprise a minimum of 4 percent of the total wall area.

If metal ties are used to bond the wall, one metal tie should be used for each 4½ square feet of wall area. The ties should be placed on equal centers not to exceed 24 inches, and the ties should be staggered in alternate courses. Because of the flexibility of metal ties, stresses may be relieved, thus preventing the possibility of cracking.

## Pattern Bonds

A pattern bond is the arrangement of the individual masonry units to produce a particular pattern. Five basic pattern bonds are used (See Figure 9–4):

- *Running bond:* The simplest of the five. This type utilizes all stretchers and is used primarily in cavity-wall construction and veneered walls.
- *Common or American bond:* Constructed with a course of full-length headers placed at regular intervals. The header course provides structural bonding, and is usually placed every fifth or sixth course. The courses other than the header course are usually placed in a running bond.
- *Flemish bond:* Consists of alternating headers and stretchers in every course. Either full-length headers can be used to provide a structural bond, or the header units can be broken into to provide a pattern bond.
- *English bond:* Consists of alternating courses of headers and stretchers. The

FIGURE 9–4.   Basic pattern bonds

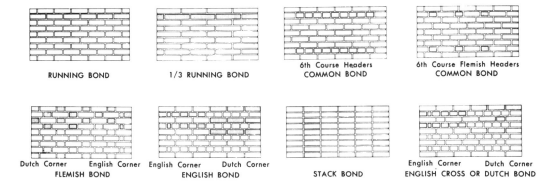

RUNNING BOND

1/3 RUNNING BOND

6th Course Headers
COMMON BOND

6th Course Flemish Headers
COMMON BOND

Dutch Corner        English Corner
FLEMISH BOND

English Corner        Dutch Corner
ENGLISH BOND

STACK BOND

English Corner        Dutch Corner
ENGLISH CROSS OR DUTCH BOND

bond is also created by centering the headers on the stretchers and aligning all the vertical joints of the stretchers. If structural bonding is not required, full-length headers can be broken into.

■ *Block or stack bond:* Strictly a pattern bond, because there is no overlapping of the individual units. To create a block or stack bond, the individual units are placed on top of each other, aligning all the vertical joints.

Several different pattern bonds can be achieved with concrete masonry units. Some of the most popular patterns are running bonds, offset bonds, stack bonds, and several ashlar patterns. (See Figure 9–5.) The type of pattern used is dependent upon the desired effects, availability of the different size units, and how much money can be spent on the wall's construction.

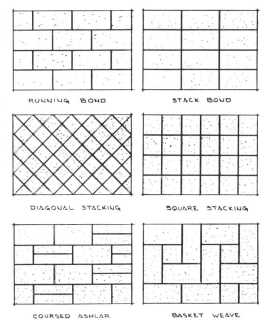

FIGURE 9–5  Popular pattern bonds

## MORTAR JOINTS

The mortar serves three primary functions: (1) it bonds the masonry units together, (2) it compensates for the dimensional variations in the masonry units, and (3) it is used to produce certain architectural effects. The mortar joints can be classified as either troweled or tooled. A troweled joint is created by cutting the extruded mortar flush with the masonry units. A tooled joint is made either by compressing the mortar into the joint with a special forming tool or by raking a portion of the mortar from the joint. Three of the most popular types of tooled joints are concave, V-shaped, and raked joints. (See Figure 9–6.) The concave and V-shaped mortar joints are very effective in resisting rain penetration, and both types of mortar joint are recommended in areas that experience heavy rains and high winds.

The raked joint is made by using a special square-shaped tool to remove the face of the mortar while it is soft. The mortar, however, is not tightly compacted in the joint and offers little resistance against water penetration.

FIGURE 9–6.  Tooled joints

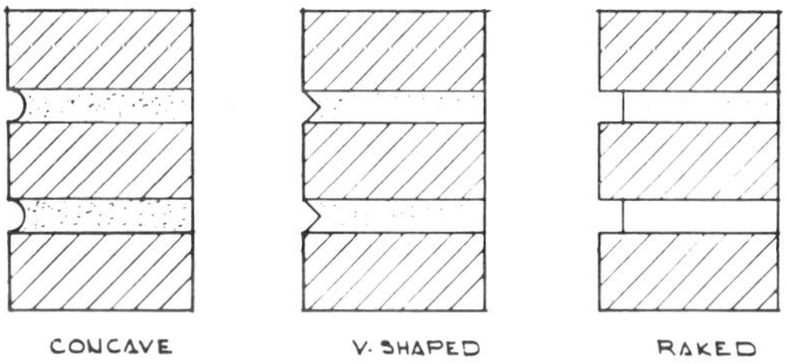

## MORTAR

Mortar is a mixture of portland cement, lime, sand, and water, and is the bonding agent that integrates a masonry wall. Four types of mortar are recognized by most building codes:

- *Type N mortar:* A medium-strength mortar suitable for general use in exposed masonry above grade. It is especially recommended in areas subject to severe exposure. The mortar is a mixture of 1 part portland cement, ½ part hydrated lime or lime putty, and 4½ parts sand by volume.
- *Type S mortar:* Has reasonably high compressive strength, and is recommended for use in reinforced masonry and where high resistance to lateral force is required. The mortar is a mixture of 1 part portland cement, ½ part hydrated lime or lime putty, and 4½ parts of sand by volume.
- *Type M mortar:* Also has high compressive strength, and is specifically recommended for unreinforced masonry below grade and in contact with the earth. The mortar is a mixture of 1 part portland cement, ¼ part hydrated lime or lime putty, and 3 parts sand by volume.

- *Type O mortar:* A low-strength mortar suitable for general interior use in non–load-bearing masonry. In most cases Type O mortar should not be used in areas subject to freezing temperatures. It is a mixture of 1 part portland cement, 2 parts hydrated lime or lime putty, and 9 parts sand by volume.

Mortar should be mixed with the maximum amount of water possible, yet still allow for workability from the bricklayer's standpoint. If the mortar stiffens owing to the loss of water by evaporation, additional water can be added and the mortar remixed. But no mortar should be used that has been mixed for more than two hours.

One of the biggest mistakes made in preparing mortar is oversanding, usually caused by inaccurate measurement of the sand with shovels. Shovels vary in size and the amount of sand that can be placed on a shovel varies with the moisture content of the sand. To insure the finest quality of mortar, the sand should be measured by the cubic foot and in a container that can accurately measure the amount of sand placed in the mixer.

## FLASHING

Flashing serves two primary purposes: it prevents moisture from penetrating a wall, and it directs to the outside any moisture that might penetrate it. Depending on the function and location, it can be classified as either external or internal flashing. (See Figure 9–7.)

*External flashing* is used to direct water away from the intersection of a relatively flat surface and a vertical masonry wall. External flashing usually consists of two parts: base flashing and cap or counterflashing. Base flashing covers the flat surface and turns up the vertical masonry wall. The cap flashing is built into the masonry wall and is turned down over the base flashing.

*Internal flashing* is usually built into

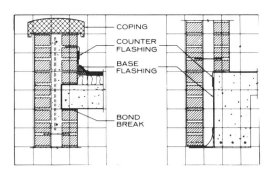

FIGURE 9-7.  External and internal flashing

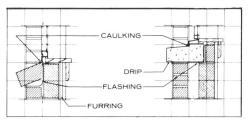

FIGURE 9-8.  Sill flashing, side view

and concealed by masonry walls, and collects and diverts moisture to the outside. To drain moisture through the masonry walls, weep holes must be provided. These holes can be constructed by placing sash cords in the mortar joints, or using plastic soda straws or copper tubing in the bed joint, or omitting a vertical head joint of mortar. Weep holes are usually placed on 16- or 24-inch centers.

In most cases a masonry wall is flashed at the wall base, window sill, and at opening heads. Sill flashing should extend beyond the jamb line on both sides and turn up a minimum of 1 inch into the wall. (See Figure 9–8.) If the sill does not slope away from the wall, a drip notch should be provided in the sill to prevent moisture from running over the sill and down the wall, eventually staining the masonry. Flashing should also be provided above all openings, except those protected by a covering.

For flashing, many different types of materials are used, the commonest being copper, zinc, aluminum, lead, plastics, and bituminous material.

Copper flashing is durable, is an excellent moisture barrier, and is not affected by the caustic alkalis in mortars. But it may stain adjacent masonry, and it is expensive. One of the biggest disadvantages of zinc as a flashing material is its tendency to corrode when placed in fresh mortar. The corrosion, however, forms a film around the zinc flashing and readily bonds with the mortar. Aluminum and lead should not be used as flashing in a masonry wall, because the unhardened mortar attacks and destroys the metal. Plastic is one of the most widely used flashing materials; it is relatively inexpensive and is highly resistant to corrosion. Bituminous flashing consists of fabrics saturated with bitumen. If it can be placed without the coating breaking it is effective; but in most cases the bituminous type is not considered to be as permanent as a good metal flashing.

## DIFFERENTIAL MOVEMENT

All building materials will expand or contract with changes in temperature and some materials will make dimensional changes with changes in moisture content. Because such dimensional changes can cause stresses that result in cracks, expansion joints and control joints are sometimes placed in the masonry wall.

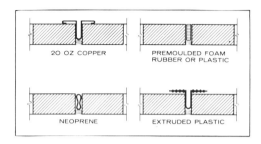

FIGURE 9-9.  Expansion joints

In clay masonry, expansion joints are constructed in a number of ways. Some of the materials successfully used as expansion-joint fillers are copper, premolded foam rubber or plastic, neoprene, and extruded plastic. (See Figure 9–9.) Fiberboard and similar materials are not highly compressible—and once they have been compressed they do not return to their original shape and size—so they should not be used as expansion-joint fillers.

Before control joints are positioned; the entire wall assembly should be analyzed to determine locations of potential movement. In most cases, however, control joints are located at offsets and at junctions of walls in L-, T-, or U-shaped buildings. In cavity and solid walls, expansion joints are sometimes placed at or near an exterior corner.

Control joints in concrete masonry walls are constructed in various ways. Some of the acceptable techniques are:

■ Using a full- and a half-length block to form a continuous joint (See Figure 9–10)
■ Using offset jamb blocks with a noncorroding wall tie (See Figure 9–11)

FIGURE 9–10.  Full- and half-length blocks

FIGURE 9–11.   Offset jamb block

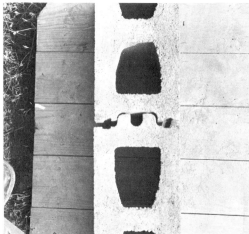

FIGURE 9–12.   Tongue-and-groove control-joint
block

FIGURE 9–13.   Construction of a control joint

■ Using some control-joint blocks provided with tongue-and-groove-shaped
ends to provide lateral support (See Figure 9–12)

■ Inserting building paper in the end core
of the block and filling the core with
mortar (See Figure 9–13); (the building
paper prevents the mortar from bonding
the two units together and the mortar
adds lateral strength to the wall).

## CONCRETE MASONRY CONSTRUCTION TECHNIQUES

Before the wall assembly can be erected,
corner poles or corner leads should be
placed at each corner. A corner lead is a
portion of a wall, located at a corner, that
is racked back on successive courses. The
construction of the leads is extremely im-

portant, for if the bricks are improperly laid the individual courses of brick may not match up.

A corner pole is a wooden or metal pole that is marked for the different courses of brick. It serves the same function as a lead, but eliminates the necessity of building up individual leads. The corner pole is placed vertically and attached to the building, but it is free of the masonry wall that is to be built.

Once the corner po⌐ ⌐ has been positioned or the lead built, a mason's line (usually a nylon line) is attached, to act as a guide in placing the bricks. To maintain a straight course of bricks, the line should be pulled tight, with the same degree of tautness for each course. If the line is stretched over a long distance, its own weight may cause it to sag. To eliminate the sag and to prevent the line from being blown out of proper alignment, a trig is used. A trig is a short piece of line looped over the mason's line and fastened to the top edge of a trig brick that has previously been laid to the proper elevation.

Once the mason's line has been placed, it should not be touched, even when the mortar and the brick are spread. If the line is interfered with, the wall could be laid out of alignment.

The mortar is spread onto the brick with a trowel. This requires a throwing and wrist-snapping action to maintain even distribution of the mortar. The mortar should be spread for only a few bricks at a time so that it will not dry out too fast, but remain soft and plastic. After the mortar has been deposited it should be riffled (grooved) down the center or down the edge. It is usually considered better practice to riffle the mortar back from the edge of the brick,

for if it is riffled down the center, fins sometimes protrude from the brick and into the cavity. A riffle down the center can also lead to improper bonding.

With the mason's line as a guide, the bricks are carefully positioned. A brick should be placed back from the line by about the line's thickness, with the brick's top edge placed even with the bottom edge of the line. When the brick is shoved into place, mortar should be squeezed out of the head and bed joint. Then, to prevent staining the wall, the excess mortar is cut off. When a course of bricks is started from each end, a closure block is used to finish the course. Morter is applied to the ends of the closure brick, and on the ends of the brick already in place. Once the mortar joint becomes thumbprint-hard, the joint should be tooled.

If high-absorption bricks are used, they should be thoroughly wetted before they are placed. If they are not, they will absorb the moisture from the mortar and create a weak mortar bond. A full stream of water should be played on the bricks until the water drips from them; they should be surface-dried before they are placed. If the bricks are placed with surface water on them, they will "float" on the mortar bed.

## Caulking

As the masonry units are placed around the perimeter of exterior door and window frames, a space of no less than $1/4$ inch and no more than $3/8$ inch should be left between the frame and the brick. After all the bricks have been placed, the space between the bricks and the door and window frame should be cleaned and caulked. The caulking compound should be elastic, waterproof, and forced to a uniform depth of ¾ inch. When the caulking compound dries, its outside will be a thin yet tough membrane, but it will remain plastic beneath the membrane. The caulking should not be affected by heat, humidity, or long exposures to extreme temperatures.

## Cleaning New Masonry

In some cases the masonry wall requires a final cleaning and washing down. First, large particles of mortar should be removed with wood paddles and scrapers. Then the wall is saturated with clean water to remove any loose mortar and dirt. A solution of hydrochloric acid or a solution of triso-dium phosphate, water, and household detergent should then be used to scrub down the walls. After this, the wall is washed again with clean pressurized water.

## Chimneys and Fireplaces

There are two basic types of fireplaces—single-faced and multi-faced. The single-faced fireplace has only one that is open, while the multifaced may have two open faces that are adjacent; two open faces that are opposite; or three open faces, two long, and one short. The sizes of faces are shown in Table 9–2.

A typical fireplace is constructed of seven component parts: ash pit, hearth,

**TABLE 9–2   FIREPLACE AND CHIMNEY SIZES**

| Fireplace Type | | Opening Height (h) | Hearth Size (w × d) | Flue Size |
|---|---|---|---|---|
| Single Face | | 30″ | 26″ × 16″ | 8″ × 12″ |
| | | 33″ | 36″ × 16″ | 12″ × 12″ |
| | | 36″ | 36″ × 18″ | 12″ × 16″ |
| Two Face (adjacent) | | 30″ | 34″ × 20″ | 12″ × 16″ |
| | | 36″ | 30″ × 30″ | |
| | | 42″ | 42″ × 24″ | 16″ × 20″ |
| Two Face (opposite) | | 30″ | 34″ × 28″ | 16″ × 16″ |
| | | 36″ | 38″ × 28″ | 16″ × 20″ |
| | | 42″ | 30″ × 24″ | 16″ × 20″ |
| Three Face (2 long, 1 short) | | 24″ | 34″ × 24″ | 16″ × 16″ |
| | | 30″ | 38″ × 28″ | 16″ × 20″ |
| | | 36″ | 38″ × 28″ | 20″ × 24″ |

Reprinted with permission of Brick Institute of America.

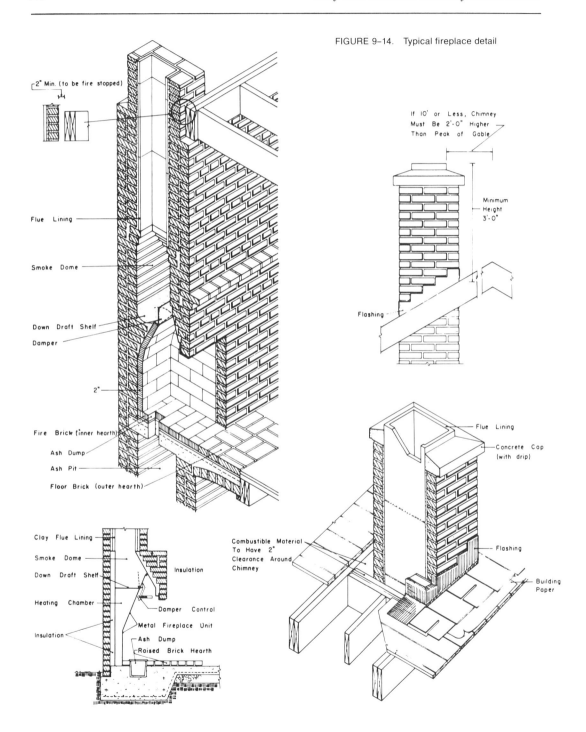

FIGURE 9–14.   Typical fireplace detail

firebox, damper, smoke shelf, smoke dome, and flue. (See Figure 9–14.)

The inner hearth is the area where the fire is built; the outer hearth extends into the room and protects the floor from flying sparks. Because of the intense heat, the inner hearth is usually constructed of firebrick.

Ashes are swept from the hearth through a pivoted metal trap door and into the ash pit. A cleanout door is provided in the basement, or on the backside of the chimney, so that ashes may be removed.

The firebox or combustion chamber is the area where the fire is actually built. It should be lined with firebrick laid with thin joints of fired-clay mortar.

The damper is a device that is placed in the throat of the fireplace to control the flow of air. Its primary purpose is to close off the flue when the fireplace is not in use, thus reducing the possibility of air infiltration. A damper should be large enough to allow both an unrestricted flow of air and a proper draft on all sides of the firebox. It should also be properly located to prevent deflection of smoke into the room.

The smoke shelf, sometimes called the downdraft shelf, is used to deflect any downdraft that might occur in the chimney. It is usually designed as a flat shelf, but it can be built in a bowl shape. The latter is considered the best design, because it is more effective in deflecting the air in the flue. The smoke shelf should be located directly under the bottom of the flue.

The flue is the open area in the chimney that allows for the passage of air and gases. A clay flue liner of the proper size is recommended for the proper function of a residential chimney. The proper flue sizes are shown in Table 9–2.

The chimney should be supported on a reinforced concrete footing and should project a minimum of 3 feet above the ridge line, or be at least 2 feet higher than any portion of the roof within 10 feet. The thickness of the chimney should not be less than 8 inches; but if the wall is less than 8 inches thick, the flue should be separated from the chimney wall. The flue liner should be an approved fired-clay liner with a minimum thickness of $5/8$ inch. If two flues are used in the same chimney, the joints in the adjacent flues should be staggered a minimum of 7 inches.

The flue should be allowed to project 4 inches above the top of the chimney. To provide drainage and to direct air currents, the top of the chimney should be flat or concave. The recommended practice is to have a cap at the top of the chimney, which forms a drip, as shown in Figure 9–14, allowing the walls to remain dry and clean.

## CONCRETE MASONRY CONSTRUCTION TECHNIQUES

The first procedure in concrete masonry is to spread a full mortar bed and then furrow it with a trowel. The masonry units can then be placed with the thicker end of the face shells up, to provide a larger mortar-bedding area. As the units are placed, the vertical joints are mortared to produce a well-filled vertical mortar joint. As the first course is laid, it should be constantly checked for proper grade, and plumbed by tapping the individual units with the trowel handle.

After the first course has been laid, the corners of the wall are built, usually four or five courses higher than the center of the wall. As the corners are laid they should be checked with a level to make sure they are properly aligned, level, and plumbed correctly. (See Figure 9–15.) After the corner leads have been built, a mason's line can be stretched between corners, and the blocks laid to the line. (See Figure 9–16.) As each unit is laid, the extruded mortar is cut with the trowel and thrown back on the mortar board, to be reworked into the fresh mortar. Some masons, however, prefer to use

FIGURE 9–15. (right)   Checking the corner

FIGURE 9–16.   Laying a block to the mason's line

extruded mortar cuts to butter the ends of the vertical face shell of the block just laid. The mortar should not be spread too far ahead of the actual placement of the blocks, or it will stiffen and lose its plasticity. Any adjustments to the blocks should be made while the mortar is soft and plastic; if they are made after the mortar has stiffened, the bond will be broken. When the closure block is laid, all edges of the opening and all edges of the closure block should be buttered with mortar. The closure block is then carefully lowered into place.

## Attaching Plates

Plates are attached to a concrete masonry wall by ⅜-inch-diameter, 18-inch-long anchor bolts spaced on 4-foot centers. The bolts are positioned in cores of the top courses and are supported by mortar placed in the cores. To prevent mortar from dropping through the cores, metal lath that will support the mortar is placed at the base of the cores. Once the mortar has hardened, the plates can be placed and attached to the wall.

## Intersecting Walls

Intersecting load-bearing walls should not be tied together in a masonry bond, except at outside corners. But tie bars are used to provide lateral support to the bearing wall. The tie bars are ¼ inch thick, 1¼ inches wide, and 28 inches long and have two right-angle bends. The bends are placed in mortar-filled cores of the masonry units. To provide adequate lateral support, the tie bars are usually placed on every other course, but never more than 4 feet apart.

If the intersecting wall is not a load-bearing wall, the block walls can be tied to each other with ¼-inch mesh galvanized hardware cloth. The hardware cloth is also placed on alternate courses.

## Setting Door and Window Frames

Door and window frames can be set and braced and the wall built around them, or jamb blocks can be used to build an opening. The offset of the jamb blocks permits the rough opening to be finished before the window and door frames are set. After the rough opening has been completed, the frames can be installed. (See Figure 9–17.) This latter technique is usually preferred.

FIGURE 9–17. Installation of frames

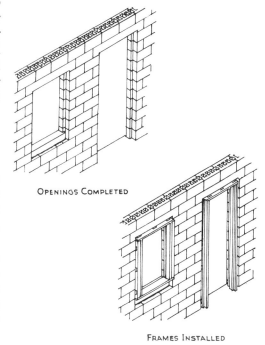

OPENINGS COMPLETED

FRAMES INSTALLED

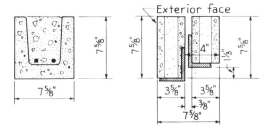

FIGURE 9–18. Lintel block and steel-angle lintels

## Lintels and Sills

One-piece lintels, split lintels, one-piece lintels with stirrups, lintel blocks, and steel angle lintels are used to span the opening over windows and doors. (See Figure 9–18.) The distances these lintels can span are shown in Tables 9–3, 9–4, and 9–5.

Regardless of the type of lintels used, they should be placed on noncorroding metal plates. The plates allow the lintels to move and the control joints to function properly. The lintel is placed in a full bed of mortar, which is allowed to stiffen slightly and is then raked from the joint to a depth of ¾ inch. Caulking compound is then forced into the joint.

Precast concrete sills are often used in concrete masonry construction. The sills are sloped to shed water and are constructed with a drip ledge to prevent water from running down the face of the masonry units. (See Figure 9–19.) The sills should be placed in a full bed of mortar. The joints at the ends of the sill should be filled with mortar or caulking compound.

**TABLE 9–3    LINTELS WITH WALL LOAD ONLY**

| Height in. | Width in. | Clear span of lintel ft. | No. Bars | Size of bars |
|---|---|---|---|---|
| $5\frac{3}{4}$ | $7\frac{5}{8}$ | Up to 7 | 2 | ⅜-in. round deformed |
| $5\frac{3}{4}$ | $7\frac{5}{8}$ | 7 to 8 | 2 | ⅝-in round deformed |
| $7\frac{5}{8}$ | $7\frac{5}{8}$ | Up to 8 | 2 | ⅜-in round deformed |
| $7\frac{5}{8}$ | $7\frac{5}{8}$ | 8 to 9 | 2 | ½-in round deformed |
| $7\frac{5}{8}$ | $7\frac{5}{8}$ | 9 to 10 | 2 | ⅝-in round deformed |

Reprinted with permission of Portland Cement Association.

**TABLE 9–4    SPLIT LINTELS WITH WALL LOAD ONLY**

| Height in. | Width in. | Clear span of lintel ft. | No. bars | Size of bars |
|---|---|---|---|---|
| $5\frac{3}{4}$ | $3\frac{5}{8}$ | Up to 7 | 1 | ⅜-in round deformed |
| $5\frac{3}{4}$ | $3\frac{5}{8}$ | 7 to 8 | 1 | ⅝-in round deformed |
| $7\frac{5}{8}$ | $3\frac{5}{8}$ | Up to 8 | 1 | ⅜-in round deformed |
| $7\frac{5}{8}$ | $3\frac{5}{8}$ | 8 to 9 | 1 | ½-in round deformed |
| $7\frac{5}{8}$ | $3\frac{5}{8}$ | 9 to 10 | 1 | ⅝-in round deformed |

Reprinted with permission of Portland Cement Association.

**TABLE 9–5    LINTELS WITH WALL AND FLOOR LOADS[1]**

| Size of lintel | | | Reinforcement | | Web reinforcement No. 6 gauge wire stirrups. Spacings from end of lintel—both ends the same |
|---|---|---|---|---|---|
| Height in | Width in | Clear span of lintel ft | Top | Bottom | |
| $7\frac{5}{8}$ | $7\frac{5}{8}$ | 3 | None | 2—$\frac{1}{2}$-in round | No stirrups required |
| $7\frac{5}{8}$ | $7\frac{5}{8}$ | 4 | None | 2—$\frac{3}{4}$-in round | 3 stirrups, Sp.: 2, 3, 3 in. |
| $7\frac{5}{8}$ | $7\frac{5}{8}$ | 5 | 2—$\frac{3}{8}$-in round | 2—$\frac{7}{8}$-in round | 5 stirrups, Sp.: 2, 3, 3, 3, 3 in. |
| $7\frac{5}{8}$ | $7\frac{5}{8}$ | 6 | 2—$\frac{1}{2}$-in round | 2—$\frac{7}{8}$-in round | 6 stirrups, Sp.: 2, 3, 3, 3, 3, 3 in. |
| $7\frac{5}{8}$ | $7\frac{5}{8}$ | 7 | 2—1-in round | 2—1-in round | 9 stirrups, Sp.: 2, 2, 3, 3, 3, 3, 3, 3, 3 in. |

[1]Floor load assumed to be 85 lb per sq ft with 20-ft span

Reprinted with permission of Portland Cement Association.

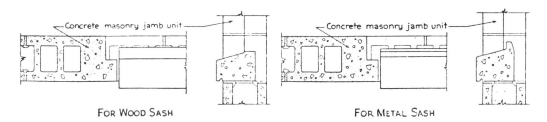

FOR WOOD SASH                    FOR METAL SASH

FIGURE 9–19.   Precast concrete sills

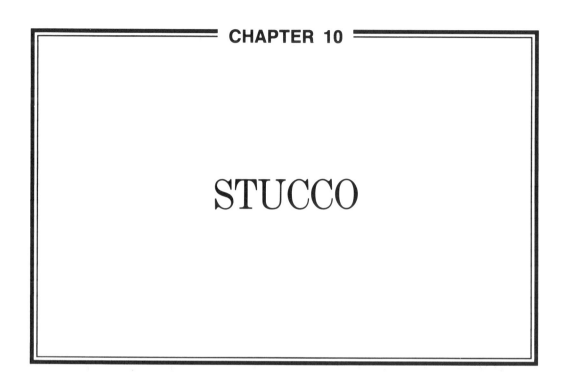

# CHAPTER 10

# STUCCO

STUCCO IS A POPULAR EXTERIOR FACING material that has been successfully used in all areas of the country. For residential construction, it offers an attractive, durable surface with minimum maintenance. For nonresidential construction, it also provides a durable facing material easily adapted to any architectural treatment. Stucco, a mixture of portland cement, masonry cement, lime, sand, and water, is an extremely durable, fire- and weather-resistant facing material. In effect, stucco is a thin concrete slab that is built up with tools and techniques associated with plastering.

The material can also be formed in complex shapes, planes, and surfaces without any appreciable deteriorating effects.

## PORTLAND CEMENT

Portland cement as a base material for stucco should comply with all required specifications. The approved cementitious materials used as a basic ingredient in stucco are: portland cement Types I, II, III, IA, IIA, IIIA, IS, and IS-A.

*Type I* is a general purpose cement that can be used where special properties of the other types are not required. It should be limited, however, to areas where the stucco cannot be exposed to sulfate attack from soil or water, or to severe temperature rise caused by the heat generated by hydration. (Hydration is the heat-generating chemical reaction between cement and water.)

*Type II* portland cement has an improved resistance to sulfate attack, and generally generates less heat and at a slower rate than Type I cement.

*Type III* portland cement provides high strengths at an early period, developing most of its strength in the first few days after application. This type is often desirable to use in freezing temperatures because it eliminates some of the hazards associated with cold-weather construction, and speeds the application of the various coats of stucco.

*Types IA, IIA, and IIIA* cements are the same as Types I, II, and III cement, except that they contain air-entraining materials which improve the concrete's ability to resist the freeze-thaw action. The air-entraining materials also produce small, well-distributed air bubbles throughout stucco, which add to the plasticity of the mix.

*Types IS and IS-A* cements are manufactured by intergrinding granulated blast-furnace slag with the portland cement clinkers. The primary difference between the two is that IS-A contains air-entraining agents and IS does not.

## PLASTICIZERS

Plasticizing agents are added to the other ingredients to improve the workability of stucco. The two most common plasticizers are Type IS finishing hydrated lime and asbestos fibers. Type IS hydrated lime requires no presoaking and can be mixed with the portland cement before water is added. If asbestos fibers are used, they should be clean and free of substances that might reduce the strength of the facing material.

## WATER

The mixing and curing water should be clean potable water, free from oils, acids, alkalis, salts, or organic substances that could retard hydration or cause corrosion of the metal reinforcement. Water is an essential element in the mixing of stucco; not only does it increase the workability of the mix, but it also promotes hydration and carbonation.

Hydration hardens portland cement, while carbonation is the chemical process that hardens lime.

## AGGREGATES

The aggregate most often used in stucco mix is sand, although expanded shale, clay, slate, and slag are sometimes used. Any aggregate used should be clean, well graded, and free of any foreign material that might prevent the cement paste from bonding to the aggregate. Because the aggregate constitutes most of the mix volume, it should meet rigid requirements. A cubic foot of mix contains approximately 0.97 cubic foot of aggregate and 0.02 to 0.03 cubic feet of cementitious material.

The aggregates that are used should be well graduated, with particles ranging from coarse (maximum size $\frac{1}{8}$ inch) to fine. When a well-graduated aggregate is used, the smaller aggregates fill the voids created by the larger aggregates, thus tending to minimize voids. If the aggregates are poorly graduated, the mix will have large voids that require too much cement paste, a problem that often causes cracking. Poorly graduated sand also leads to excessive mixing water, which dilutes the paste and results in a poor-quality portland cement stucco. Too much water in a mix also

Too much water in a mix also causes excessive shrinkage and cracking.

## WATERPROOF BUILDING PAPER OR FELT

Waterproof building paper or felt is often placed over a frame wall before the metal reinforcement is installed. The backing paper is used to segregate the stucco and the supporting construction, and it helps to embed the stucco on the lath. The backing paper is usually 15-pound asphalt-saturated felt, and should be free from holes and breaks. To eliminate the need to install building paper or felt, some metal lath has paper interwoven with the wires. (See Figure 10–1.) If stucco is applied by hand on a horizontal surface and is protected from direct wetting, backing paper is not usually required.

FIGURE 10–1. Paper-backed wire-fabric lath

## FLASHING

Flashing is used both to keep water from penetrating the facing material and to expel water once it has entered. If water is allowed to stand behind stucco, it can cause the reinforcement to rust; freezing water can expand and heave the stucco out of alignment.

Because of the possibility of water penetration, only high-grade rust-resisting materials should be used.

## METAL REINFORCEMENT

The most common metal reinforcement used to strengthen stucco is expanded metal lath, wire lath, expanded stucco mesh, stucco netting, and stucco mesh. The openings in the reinforcement should be large enough to force the scratch coat through the lath, so the lath is completely embedded in the stucco. This technique will prevent the corrosion of the lath.

Expanded metal lath and expanded stucco mesh are manufactured from light-gauge steel sheets that have been cut and expanded to form a diamond-shaped pattern. (See Figure 10–2.) When expanded metal lath is placed over a vertical surface, it should not weigh less than 2.5 pounds per square yard; for horizontal surfaces it should not be less than 3.4 pound per square yard.

Woven wire fabric (stucco netting) is a popular lath that resembles chicken wire. (See Figure 10–3.) It is manufactured in 1-inch hexes with 18-gauge wire, 1½-inch holes of 17-gauge wire, and in 2¼-inch hexagonals of 16-gauge wire. The reinforcement is available in 36-, 48-, and 60-inch-wide rolls and is nailed to the studs or sheathing with furring nails, which are used to project the lath ¼ inch away from the backing paper.

FIGURE 10–2. Expanded metal lath and expanded stucco mesh

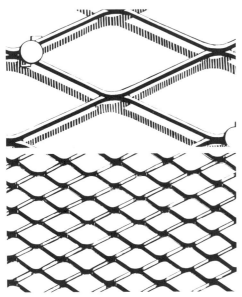

FIGURE 10–3. Stucco netting

## SUPPORTING CONSTRUCTION

Stucco can be placed over clay and concrete masonry, cast-in-place concrete, old stucco or plaster, and properly prepared wood and steel frames. The bond between the supporting construction and the stucco is provided by mechanical keying or by suction. Mechanical keying results from the interlocking of the stucco mix with the lath; suction is developed when the cement paste of the mix is drawn into the small opening of the base. Suction bonding is associated with concrete and masonry, while mechanical keying is associated with metal reinforcement.

In order for stucco to form a good bond with metal reinforcement, the reinforcement should be placed a minimum of ¼ inch away from the supporting construction and the openings in the lath should not exceed 4 square inches in area. The reinforcement should form a continuous net over the supporting construction. The ends and sides of the lath are lapped a minimum of 2 inches, and wired securely. The laps should occur at intermediate framing members and should be staggered.

### Masonry

The bonding of stucco to a masonry surface is achieved by mechanical keying and/or suction. If the surface texture of the masonry is rough enough, a mechanical key can be formed; but if the surface is relatively smooth, the stucco is held in place by suction. To find out if a surface will provide a good suction bond, it should be sprayed with water. If tiny balls of water form, the surface is not porous enough to provide a good suction bond. But if the water is absorbed too quickly, the moisture in the stucco may be extracted too rapidly, creating a facing material that would be difficult to work with. To eliminate this hazard, the wall can be lightly sprayed with water before the stucco is applied.

Concrete masonry usually has a rough texture and provides a good mechanical key and strong suction that will make a reliable bond. Before stucco is applied to a concrete masonry wall, it should be inspected for dust, dirt, oil, grease, paint, or any other substance that might prevent a good bond. If there are any foreign substances on the wall, they should either be removed or the wall covered with waterproof building paper and metal reinforcement.

Clay masonry walls also provide a suitable surface for the application of stucco, but the same precautions required for concrete masonry also apply to clay masonry.

### Cast-in-Place Concrete

Before stucco can be installed over cast-in-place concrete, the concrete should be inspected and any smooth spots roughened by wire-brushing, sandblasting, chipping, or acid etching. The smooth spots are usually created by form oil or the use of smooth formwork.

A solution of 1 part muriatic acid and 6 parts of water should be used to roughen the concrete with acid. Before applying the acid, the wall must be washed with clean water so the acid will act only on the surface. The acid is then applied to the concrete wall in as many applications as necessary. Once the desired finish has been achieved, the wall is again washed with clean water and allowed to dry so that suction may be restored.

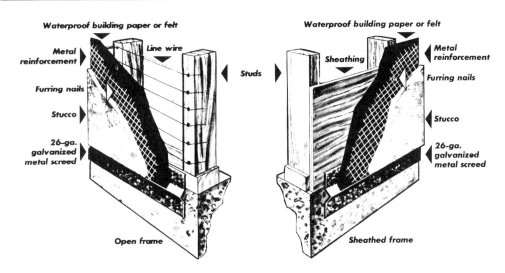

FIGURE 10–4.   Open-frame construction and frame construction with sheathing

## Wood-Frame Construction

There are two categories: open-frame construction and frame construction with sheathing. (See Figure 10-4.)

In open-frame construction a No. 18 gauge or heavier wire is stretched across the outside faces of the studs to serve as a backing for the building paper or felt. The line wire is spaced on 6-inch centers; to be tightened, the wire must be secured on every fifth stud. Then, when the wire is slightly lowered or raised and nailed or stapled to the studs, it is stretched taut.

Once the line wires have been placed, building paper or felt can be applied without fear of the paper sagging. The building paper should be placed with 4-inch end laps and 2-inch side laps, and secured to the

wall frame with large-headed roofing nails. Metal reinforcement is then placed over the building paper or felt and secured to the vertical framing members of the wall frame. When metal reinforcement is installed, it should be started one full framing member away from the corner, and wrapped around the corner.

If sheathing has been placed over the wall frame, the line wires are omitted and the building paper or felt is applied directly to the sheathing. Metal reinforcement is then placed with the long dimension perpendicular to the studs, and is furred out ¼ inch from the paper. Galvanized or rust-resistant nails should be used to secure the reinforcement. Aluminum nails should not be used, however, because of their reaction with freshly placed portland cement.

## STUCCO APPLICATION

Stucco is usually applied in three coats: first the scratch coat; second the brown

coat; and third the finish coat. But if stucco is placed over clay or concrete ma-

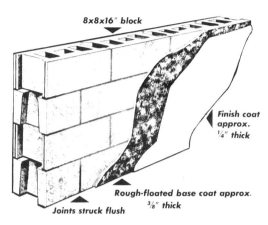

**8x8x16" block**

**Finish coat approx. ¼" thick**

**Rough-floated base coat approx. ³/₈" thick**

**Joints struck flush**

FIGURE 10–5. Application of stucco over masonry construction

sonry, only two coats are required. (See Figure 10–5.) Under no circumstances should the stucco be applied in freezing temperatures or to surfaces covered with frost. A temperature of 50-degrees F or higher should be maintained during application and for 48 hours after the finish coat has been applied.

Stucco can be applied by hand or machine; each method affords certain advantages and disadvantages. Hand application has been successfully used for centuries, and involves only the use of traditional tools. Machine application involves the use of expensive equipment, but the application process is faster and results in a more uniform texture.

## Hand Application

The first or scratch coat of stucco should be applied with sufficient pressure to completely embed it in the lath. It should be approximately ½ inch thick and cover at least 90 percent of the metal reinforcement. Then it is horizontally scored or scratched so that it will provide a good bond with the brown coat.

After the scratch coat has had time to set, usually 4 or 5 hours for sheathed frame construction, the second or brown coat is applied. If open-frame construction is used, the scratch coat should be allowed to set for 48 hours. This time allows the scratch coat to increase in strength, thus providing a rigid base for the application of the brown coat. If the scratch coat dries out, it should be evenly dampened before the brown coat is applied.

When the brown coat is applied, it should be struck to a thickness of ³/₈ inch. To maintain an even thickness of stucco, screeds are often placed over the scratch coat. A stucco screed is nothing more than the placement of stucco to a desired elevation. Stucco can then be placed between the stucco screeds and struck to the proper elevation. When the brown coat is applied, it should cover an entire wall without stopping because, if stops are made, the joints might project through the finish coat. Stops should be made only at corners, wall openings, and abrupt changes in the surface area.

Once the brown coat has been applied, it should be moist-cured for a minimum of 48 hours and then allowed to dry for 5 days. Then before the finish coat is applied, the brown coat must be uniformly dampened.

The third coat is then troweled over the brown coat to a thickness of ⅛ inch. For certain architectural effects, the finish coat is frequently pigmented with decorative colors. To avoid damaging the finish coat, moist-curing should not be started until the finish coat has set for 24 hours. Then a light mist (not a saturating spray) of water should be applied to the surface area.

A variety of finish coats can be applied. Some of the more popular surface finishes include rock dash, a stippled-troweled

FIGURE 10–6.  Rock-dash finish

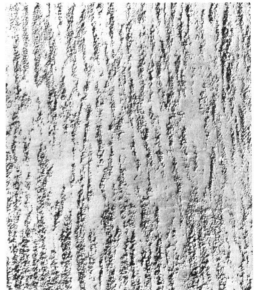

FIGURE 10–7.  Stippled-troweled finish

finish, and a hand-floated finish. (See Figures 10–6, 10–7, and 10–8.) If a rock dash finish is desired, marble chips or rocks are

thrown into a soft bedding coat of stucco. The aggregates can be thrown by hand or by an electrically or mechanically powered blower gun.

FIGURE 10–8.  Hand-floated finish

## Machine Application

Because of the speed of application, stucco is often sprayed from a machine nozzle. The nozzle should be held approximately 12 inches from the wall and moved with a steady and even stroke. It should also maintain a consistent angle with the wall, because if the angle varies there will be excessive buildup of stucco. When it is necessary to finish areas around window and door bucks (rough frames), the nozzle should be held about 2 or 3 inches from the area to be sprayed and pointed away from the area where overspray is to be avoided.

If stucco is sprayed over masonry or

cast-in-place concrete, the surface should be cleaned, dried, and finished with two coats. The first coat should be sprayed on in a thin consistency and allowed to set for 2 or 3 hours. A thicker spray is then placed over the first, until the desired texture is achieved. To form a plumb corner and assure an even thickness of stucco, a straight-edge can be placed at the corner of the wall and the stucco sprayed to it. Once the stucco has been sprayed in place, the plaster can be screeded to the proper elevation.

## CONTROL JOINTS

Differential settlement and building movement can often cause cracks to occur in stucco. It is difficult to eliminate cracks, but they can be controlled by the proper use of metal control joints. (See Figure 10–9.) The control joints are used to localize movement and minimize stresses that might occur. They should be positioned so that they divide the walls and panels into rectangular panels that do not exceed 20 feet either vertically or horizontally. When control joints are properly used, they divide a wall into various sections and can create an architectural effect that is quite pleasing.

FIGURE 10–9. Accordian-pleat control joint

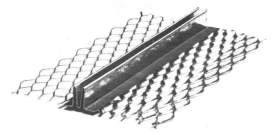

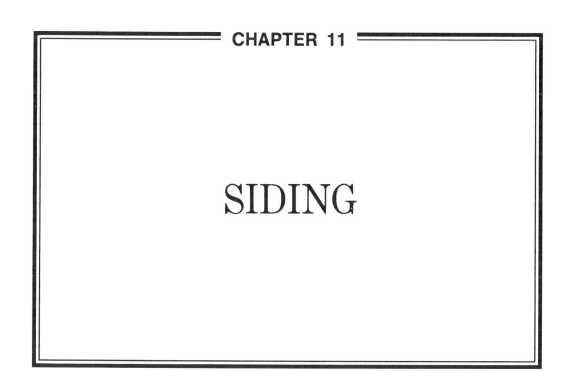

# CHAPTER 11

# SIDING

THERE ARE MANY DIFFERENT TYPES OF siding, but some of the more common types used in light construction are: mineral-fiber, aluminum and vinyl, plywood, hard-board lap and panel, and bevel. All these sidings should have the ability to resist the effects of rain, snow, heat, and wind; and they should hold a finish.

## HARDBOARD SIDING

Hardboard siding is made from clear selected wood fibers pressed under heat. Natural and synthetic binders are used to bind the fibers together. The formation of the fibers accounts for varying characteristics in density, appearance, thickness, and finishing properties.

Hardboard siding is available in a wide variety of patterns and finishes, made possible by two basic coating systems—flat stock printing and laminating. Flat stock-printing applies liquid coatings directly to the hardboard in such patterns as simulated wood grain, mosaic, and marble. Laminating involves the application of a plastic film, such as vinyl or acrylic, over the hardboard.

Some of the major advantages of hardboard siding are economy, resistance to moisture, acoustical or sound-control quality, and workability. This siding can be readily worked with ordinary carpenter's tools and will not split, chip, or check. Its high density and acoustical quality make hardboard siding an excellent sidewall covering. It is also highly resistant to moisture and to all the extremes of the weather.

Hardboard siding may be placed over sheathed or unsheathed walls with studs placed not more than 16 inches on center. To prevent air infiltration, building paper should be applied directly to studs or over wood sheathing. If the walls are not insulated, a vapor barrier should be used to prevent damaging condensation from occurring within the walls.

### Application of Flat-Panel Hardboard Siding

Flat-panel hardboard siding must first be positioned and plumbed, starting at either an outside or inside corner. When the panel is positioned, all the edges and joints should be placed opposite a framing member. The panel is then nailed in place, with noncorrosive nails spaced on 6-inch centers on all panel edges, and 12 inches on center on intermediate supports. If the panels are applied directly to the studs, 6d nails should be used; but if it is placed over sheating, 8d nails should be used. Color-matched stainless steel nails should be used if the nails are exposed. These should be driven flush with the siding, and not set below the surface of the siding.

If it is necessary to cut the panels, a fine crosscut hand-saw or a power saw with a combination blade should be used. If the siding is prefinished and a table saw or portable electric saw is used to cut it, the marking and cutting should be done on the back side. If a handsaw is used, however, the marking and cutting should be done on the face side.

If it is necessary to make a butt joint, a $\frac{1}{16}$-inch gap should be left between the panels. Caulking is sometimes applied to

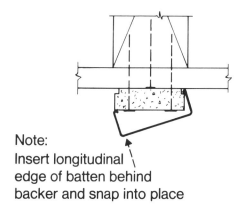

Note:
Insert longitudinal
edge of batten behind
backer and snap into place

FIGURE 11–1.   Snap-on batten cover

the joint before the installing of batten strips, to increase weather resistance. A batten is then used to cover the joint.

There are several different types of battens: each is designed to cover joints and nailing patterns. In some cases a batten backer is nailed in place and a snap-on batten cover is installed over the backer. (See Figure 11–1.) Snap-on batten covers eliminate exposed nail heads and add to the weather resistance of the joint.

Some hardboard flat-panel sidings do not use battens, but have grooves on 2-inch to 8-inch centers and shiplapped edges. When the panels are joined, a $\frac{1}{16}$-inch gap should be left between them. The space allows for the possible expansion and contraction of individual units. Because of possible dimensional changes in paneling, panels should not be forced or sprung into place.

Inside and outside corners can be finished by inside corner posts or wood corner boards. (See Figure 11–2.) If a wood inside corner is used, a $\frac{1}{16}$-inch gap should be left between the two intersecting pieces of paneling. If batten covers have been used over the vertical panel edges, they should also be used on the inside and outside corners. When the inside corner is finished, a $\frac{1}{4}$-inch gap should be left between the two battens. If optional battens are used on the outside corner, a $\frac{1}{8}$-inch gap should be left between the two battens.

## Application of Hardboard Lap Siding

Before the installation of hardboard lap siding, a $\frac{3}{8}$-inch $\times$ $1\frac{3}{8}$-inch wood starter

FIGURE 11–2.   Inside corner post and outside corner boards

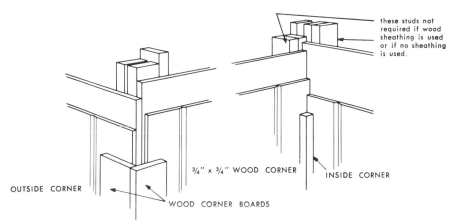

strip is placed at the base of the frame wall. (See Figure 11–3.) The starter strip should be placed level, and installed not closer than 8 inches to the finished grade. Next, the first course of lap siding can be placed; it must be level, and the bottom edge must be at least $\frac{1}{8}$ inch below the starter strip. To level the siding, chalklines are usually placed on the building paper or panel sheathing. The siding is secured with 8d noncorrosive nails placed 1 inch down from the top edge for 8-inch-wide lap siding, and $1\frac{3}{4}$ inches down from the top edge of 12-inch-wide siding.

The butt end of the siding should be placed in a butt-joint connector that has been snugly positioned and securely fastened. (See Figure 11–4.) The joint should fall over a stud but, if sheathing is used, butt joints may occur between studs.

If it is necessary to field-cut lap siding, the cuts should be sealed with edge sealer. After the first course of siding has been placed, additional courses are installed

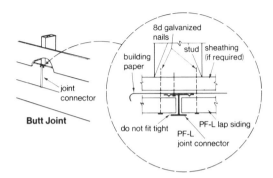

FIGURE 11–4.   Butt joint connector

with a minimum 1-inch lap. If face nailing is required, color-matched stainless steel nails are used. The nails should be nailed flush with the paneling, and not set below its surface.

A $\frac{1}{16}$-inch gap should be left between the ends of the siding that butt into a door or window casing. Once the siding has been installed, all openings between siding and casings should be caulked.

Inside and outside corners can be finished with metal accessories. (See Figure 11–5.) If a metal inside corner is used, it is installed before the installation of the siding, but the outside is finished after the siding has been placed. When a continuous inside corner is used, a $\frac{1}{16}$-inch gap should be left between the siding and the corner.

FIGURE 11–3.   Starter grip

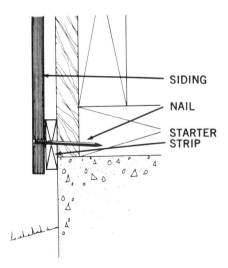

FIGURE 11–5.   Metal inside and outside corners

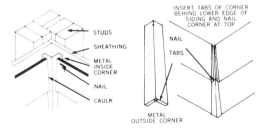

The gap allows for the possible expansion of the siding. In place of metal corners, a wood inside corner post and wood corner boards can be used. Both the corner post and the corner boards are installed before the siding is put in. When the siding is placed, a $\frac{1}{16}$-inch gap should be left between it and the corner member.

## ALUMINUM AND VINYL SIDING

Aluminum and vinyl siding provide virtually maintenance-free exterior sidewall coverings. Solid vinyl siding is made from polyvinyl chloride, a rigid vinyl easily adapted for building applications. In addition to being durable, the material is noncombustible and will not conduct electricity. Most vinyl siding is approximately 0.045 inch thick and has a $\frac{3}{8}$-inch-thick insulation board backing. The most popular siding is extruded or molded in two basic patterns: beveled siding with 8 inches to the weather, siding molded to give the look of 4 inches to the weather. (See Figure 11–6.)

Aluminum siding also simulates 4-inch or 8-inch beveled siding and is available in a smooth or textured finish. Most aluminum horizontal siding has interlocking horizontal edges, a butt or shadow leg, and a drip bead in the butt to direct the flow of water from the panel surface. Breather holes are usually placed in the shadow leg of every panel, permitting the wall to breathe and condensation and water vapor to escape. Elongated nail holes are placed in the nailing hem of the siding to allow for possible expansion and contraction.

## Installation

Before the actual installation of the siding, a starter strip should be installed at the base of the wall frame. (See Figure 11–7.) It should be level, and secured to studs or solid sheathing with corrosion-resistant nails. The nails should be flush with the siding, but not driven so tight that the siding binds. If aluminum siding is used, the nails should be aluminum, but for vinyl siding the nails can be of aluminum or other corrosion-resistant metal. They should be at least $1\frac{1}{2}$ inches long, and penetrate a sound nailing base.

After the starter strip has been put in,

FIGURE 11–6. Vinyl siding

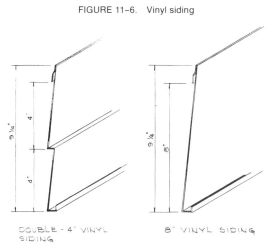

DOUBLE - 4" VINYL SIDING

8" VINYL SIDING

FIGURE 11–7. Starter grip

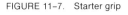
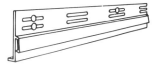

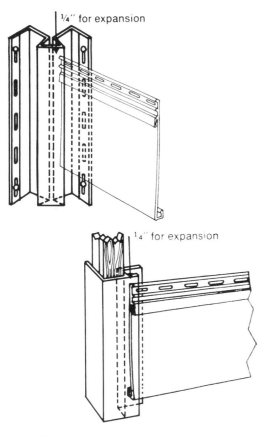

FIGURE 11–8.   Inside corner post and outside
corner post

is placed over the cap and down the sides of the windows and doors. (See Figure 11–10.) The J channel is used to trim door and window frames and to finish off cut ends of siding panels. When it is installed, a ¼-inch tab should be cut and bent down over the J channel on each side of the window or door to keep water from running down behind it.

The first piece of siding is placed in the starter strip, and securely locked in. If a backer board is used, it should be dropped into place behind the siding, which is fastened to the wall by nails spaced not over 16 inches on center.

If it is necessary to trim around windows, undersill trim should be first fitted under all of them. (See Figure 11–11.) Next, the siding can be cut to the appropriate shape with a sharp knife, snips, or power saw equipped with an abrasive cutting blade. In order for the siding to be securely fastened under the window it is sometimes necessary to perforate its trimmed edge at 8-inch intervals. The top

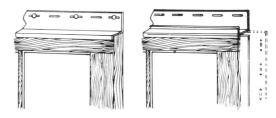

FIGURE 11–9. (left)   Cap molding, and 11–10 (right)
"J" channel

FIGURE 11–11.   Undersill trim

the inside corner post should be positioned, plumbed, and nailed. (See Figure 11–8A.) The corner post is first nailed at the top and the other nails are then placed on 12-inch centers. The inside corner post is used to terminate the siding, and must be installed before the siding. The outside corner post is also used to terminate siding. (See Figure 11–8B.)

Cap molding is installed over each window and door for flashing. (See Figure 11–9.) If furring is necessary (for alignment with adjacent panels), a piece of J channel

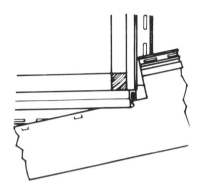

FIGURE 11–12. Placement of siding under a window

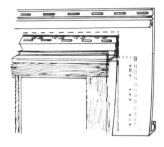

FIGURE 11–13. Placement of siding in a "J" channel

edge of the panel can then be placed into the undersill trim and the bottom edge locked into the adjacent piece of siding or starter strip. (See Figure 11–12.) When the siding is fitted over a window or door, the material is cut out from the bottom of the panel and the siding is placed in the J channel. (See Figure 11–13.) A ¼-inch space for expansion should be left between the siding and the trim.

## PLYWOOD SiDING

Plywood siding is available in many grades and textures, each designed to meet a specific need and structural requirement. Some of the more popular types of plywood siding are reverse board and batten, rough-sawn and kerfed, channel-groove, brushed, fine-line, and striated. (See Figure 11–14.)

*Reverse board and batten plywood siding* has deep, wide grooves cut into the surface. The grooves are cut approximately ¼ inch deep and to 1½ inches wide. To join individual panels and provide a continuous pattern, the edges are shiplapped.

*Rough-sawn and kerfed plywood siding* has a rough-textured surface with narrow grooves placed 4 inches on center. So they will form a continuous pattern, the long edges are shiplapped.

*Channel-groove siding* is similar to reverse board and batten, the main difference being in the smaller size of the latter. Channel grooves are usually ¹⁄₁₆ inch deep, ³⁄₈ inch wide, and are placed on 2-inch or 4-inch centers.

*Brushed siding* is often referred to as a relief-grain surface; it is used to accent the natural grain pattern.

*Fine-line siding* has fine grooves cut into the surface. They are placed on ¼-inch centers, and are about ¹⁄₃₂ inch wide.

*Striated siding* uses closely spaced grooves and fine striations to produce a vertical pattern. The striations cover joints, nail heads, and surface defects.

### Care of Plywood

When plywood siding is delivered to the job site, it should be handled with care and stored in a cool, dry place. If it has to be stored outdoors, it should be stacked to

## REVERSE BOARD AND BATTEN
(303 SIDING)

Deep, wide grooves cut into brushed, rough sawn, coarse or scratch sanded or natural textured surfaces. Grooves 1/4″ deep, 1″ to 1½″ wide, spaced 8″, 12″ or 16″ o.c. with panel thickness of 5/8″. Also available in 3/8″ or 1/2″ thickness panels with 3/32″ deep grooves. Provides deep, sharp shadow lines. Long edges shiplapped for continuous pattern. Finish with exterior pigmented stain or leave natural. Available in redwood, cedar, Douglas fir, lauan, southern pine and other species.

## CHANNEL GROOVE
(303 SIDING)

Shallow grooves typically 1/16″ deep, 3/8″ wide, cut into faces of 3/8″ thick panels, 4″ o.c. Other groove spacings available. Shiplapped for continuous patterns. Available in similar surface patterns and textures as Texture 1-11. Finish with exterior pigmented stain. Available in redwood, Douglas fir, cedar, lauan, southern pine and other species.

## BRUSHED
(303 SIDING)

Brushed or relief-grain surfaces accent the natural grain pattern to create striking textured surfaces. Available in 3/8″ or 5/8″ panels. For finishing, follow individual manufacturer's recommendations. Available in redwood, Douglas fir, cedar, Sitka spruce, lauan and white fir.

## ROUGH SAWN AND KERFED
(303 SIDING)

Rough sawn surface with narrow grooves providing a distinctive effect. Long edges shiplapped for continuous pattern. Grooves are typically 4″ o.c. Also available with grooves in multiples of 2″ o.c. May be variable. Especially suited for exterior pigmented stain. Available in Douglas fir, cedar, redwood, lauan, southern pine and other species.

## FINE-LINE
(303 SIDING)

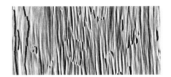

Fine grooves cut into the surface to provide a distinctive striped effect; reduces surface checking and provides additional durability of finish. Finish with exterior pigmented stain or acrylic latex emulsion finishes. Available factory-primed. Shallow grooves about 1/4″ o.c., 1/32″ wide. Also available combined with Texture 1-11 or channel grooving spaced 2″, 4″ or 8″ and reverse board and batten. Available in several species.

## STRIATED
(303 SIDING)

Fine striations of random width, closely spaced grooves forming a vertical pattern. The striations conceal nailheads, checking and grain raise, and also conceal butt joints. Finish with exterior acrylic latex paint system or pigmented stain.

FIGURE 11–14.  Plywood siding

provide good air circulation and covered to prevent possible wetting.

The end grain of siding readily absorbs moisture much faster than side grain, so to minimize possible moisture damage the edges of the plywood should be sealed. If the siding is to be painted, the edge can be sealed with a coat of exterior house primer, but if it is to be finished with a stain, a water-repellent preservative can be applied to the edges. In most cases it is easier to seal the edges while the panels are stacked. If it is necessary to cut a panel, the cut edges should also be sealed.

## Application

If the edges of plywood siding are ship-lapped or if the joints are covered with battens, no building paper is required. The siding can be placed directly to the studs, or over wood sheathing. Plain (ungrooved) panels of ½-inch or ⅝-inch thickness can be installed over studs placed on 24-inch centers. If the panels are grooved, however, the studs should be placed on maximum centers of 16 inches.

The nails used to secure the paneling to the wall frame should be hot-dip galvanized or aluminum nails, long enough to penetrate the studs 1 inch or more if sheathing is used. If sheathing is not used, the nails should penetrate a minimum of 1½ inches in the studs. In most cases 6d nails are recommended for ⅜-inch or ½-inch panels. For thicker panels, 8d nails are recommended. These should be nailed every 6 inches along the edges of the panel, and every 12 inches at intermediate supports. To allow for possible expansion, a ¹⁄₁₆-inch space should be left between all ends and edges of the panels.

It is not necessary to caulk shiplapped

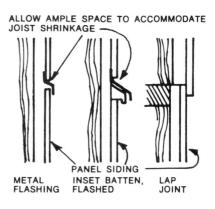

ALLOW AMPLE SPACE TO ACCOMMODATE JOIST SHRINKAGE

METAL FLASHING · PANEL SIDING · INSET BATTEN, FLASHED · LAP JOINT

FIGURE 11–15. Horizontal-joint treatment

joints or joints backed by building paper. But all joints at inside and outside wall corners should be caulked, using any polyurethane, thiokol, or silicone caulks.

Outside corners may be treated either by placing 1-inch × 2-inch and 1-inch × 3-inch corner boards over the siding, or by butting the siding against 1-inch × 2-inch and 1-inch × 3-inch corner boards. If it is necessary to create a horizontal joint, one of three techniques may be used: lap joint, inset batten flashed, or metal flashing. (See Figure 11–15.)

## BEVEL SIDING

Bevel siding is produced by resawing kiln-dried surfaced lumber on a bevel to produce two pieces thicker on one edge than on the other. Several different types of wood are used for bevel siding, but because of its natural resistance to decay, Western red cedar is often preferred. It is highly resistant to weather checks, and isn't likely to cup and pull loose from fastenings. The heartwood of Western red cedar varies in

color from deep reddish-brown to a light brown. The texture is fine with even, straight, narrow summer rings, and it is completely free of pitch.

There are five basic grades of bevel siding: Clear V.G. All Heart, A, B, Rustic, and C. With the exception of siding intended for rough side use, bevel siding is graded from the surfaced side.

- *Clear V.G. All Heart bevel siding:* Used where the highest quality of siding is needed. This is manufactured from 100 percent heartwood and is free of any imperfections.

- *Grade A bevel siding:* Usually has a mixed grain; intended for use where good appearance is desired. In some cases the siding might have some minor growth irregularities.

- *Grade B bevel siding:* May contain minor imperfections and occasional cutouts, but when painted makes an economical and good-quality sidewall covering.

- *Rustin bevel siding:* May also be mixed grain, but is usable full length and adds a rustic charm and appearance as a sidewall covering. The siding is graded for the resawn side and may be surfaced or rough.

- *Grade C bevel siding:* An economical siding suitable for temporary construction, for buildings with minimum shelter requirements, and for various industrial uses.

The size of bevel siding varies from widths of 3½ inches to 11½ inches, with a top thickness of $\frac{3}{16}$ inch, and bottom thicknesses from $\frac{15}{32}$ to ¾ inch. Also, the butt edge of the siding can be plain or rabbeted.

## Nails

The nails used to secure bevel siding have a definite bearing on the appearance and performance of the finished work. To provide the finest-quality construction, the nails should not rust or cause discoloring, should

have good holding power, should be easy to install, must not pop out or pull through, and should be relatively unobtrusive in the finished wall.

For exterior installation the nails should be corrosion-resistant—for example, high-tensile-strength aluminum, stainless steel, or galvanized—hot-dipped. To prevent the siding from splitting during the nailing process, a blunt or medium diamond point should be used. For good holding power a rink-threaded shank is usually recommended, although smooth-shanked nails can be used. The head diameter of the nails ranges in size from $\frac{13}{64}$ to $\frac{17}{64}$ inch. The nail heads should be flat casing or slightly countersunk (sinker-head). A sinker-head nail is usually used for face-nailing. The size of the nail varies, depending upon type and thickness of the siding; but usually it is considered necessary for the nails to penetrate at least 1½ inches of the studs and wood sheathing combined.

## Application

Before the bevel siding is installed, a chalkline should be positioned along the foundation wall. This line will serve as a guide in placing the first course and is an important procedure, for it determines the uniformity of all succeeding courses. Once the chalkline has been marked, the butt edge of the siding is placed along the chalkline and the siding is nailed to the bottom plate. The nails should be located at points below each stud.

Once the first course has been placed, additional courses can be placed with a 1-inch head lap. For proper alignment of the siding, a chalkline can be placed 1 inch from the top of the bevel siding. The siding is secured to the studs, with only one nail per stud. The nails should be driven so they miss the upper edge of the previously placed piece of siding by ⅛ inch. The ⅛-inch space allows for possible expansion and contraction of the individual units. When the nails are driven they should be tapped in place until the head is flush with the siding surface. The siding should be secured snugly but not tightly to the wall frame. As the individual pieces of siding are placed, the vertical butt joints should be staggered along the sidewall; all joints must fall on studs.

Outside corners can be either mitered or fitted with metal corners. If the corner is mitered, the saw blade should be set at approximately 45° and the siding cut from butt end to tapered edge. The cut should start at the butt end and angle toward the tapered edge to a point that is equal to the

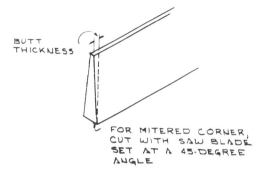

FIGURE 11–16. Cutting an outside corner

butt thickness. (See Figure 11–16.) If metal corners are used, the siding can be allowed to project past the corner during application. It is later trimmed and fitted with a metal corner.

Inside corners are usually finished by installing a wood corner post in the inside corner and butting the siding to it. The corner post is square and usually about 1⅛ by 1⅛ inch in size.

## MINERAL-FIBER SIDING

Mineral-fiber siding is manufactured from two inorganic materials—asbestos fiber and portland cement. Because of the composition of the siding, it will not burn or rot or be eaten by termites or insects or destroyed by the ravages of weather and time.

Mineral-fiber siding units are available in sizes from 9 to 16 inches in width and from 24 to 48 inches in length. The units are usually rectangular in shape and have straight or wavy butt edges. The units are usually ³⁄₁₆ inch thick, but the thickness can vary, depending upon the overall size of the units. The individual units are also available in a wide variety of colors and textures.

Each individual siding unit is manufac-

tured with correctly sized and located holes to receive exposed face nails. The holes are properly aligned above the bottom edges of the units in order to establish the recommended amount of top lap. (See Figure 11–17.) For standard application, nails are

FIGURE 11–17. Placement of individual siding units

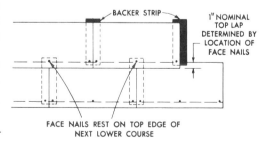

placed through the holes of the siding and allowed to rest on the top edge of the next-lower course. This technique allows for the correct amount of top-lap exposure and unit alignment.

## Nails

The nails used to secure mineral-fiber siding should be permanently and effectively corrosion-resistant and nonstaining. Approved face nails are usually furnished with the purchase of the siding. They should be long enough to penetrate and hold securely in a sound nailing base, on both new and residing work. Some of the recommended nail types are aluminum screw thread, aluminum spiral thread, and ring barb.

## Cutting

Mineral-fiber siding can be cut by one of the following methods:

■ *Shingle cutters:* Portable mineral-fiber siding cutting machines are used for cutting, notching, and punching all sizes of mineral-fiber siding units.

■ *Score and snap:* A mineral-fiber siding unit can be cut by first deeply scoring the unit, preferably on both sides. The unit is then placed along a firm edge and the unit is snapped off.

■ *Saw:* A power saw or handsaw can be used to cut mineral-fiber siding. For handsawing, coarse-toothed crosscut saw is used. If a power saw is used, it should be equipped with a tungsten carbide blade or a flexible abrasive saw disc.

## Application

Before the siding units are placed, a ¼-inch × 1½-inch cant strip should be nailed along the bottom edge of the sheathing. The cant strip should be level and overhang the top of the foundation to seal the joint between the top of the foundation and the sill. The cant strip is used to provide a solid backing for the siding and to create a pitch to the first course of shingles.

Chalklines can then be snapped around the building to locate the top edge of the different courses of siding. The lines should be spaced to provide the required exposure for the type of mineral-fiber siding being used.

The placement of the shingles is started at either the left- or right-hand corner of the wall with a full-size unit. With the chalkline as a guide, the first unit is placed, leveled, and secured with the proper nails, driven snug but not too tight. Before the last nail is placed in the right- or left-hand corner, a joint strip must be centered at the joint between the siding units. The lower edge of the joint strip should overlap the cant strip or head of the lower course. Next, other individual units are placed, with their top edges aligned with the chalkline. The last unit placed in a course should not be less than 6 inches wide. If a space is smaller than 6 inches, a few inches should be cut from some of the previously placed units. previously placed units.

The second course is started with either a half-unit or a two-thirds unit. A half-unit is used for a 24-inch or 32-inch shingle, and a two-thirds unit is used for a 48-inch shingle. The first unit in the second course is started by aligning the head edge with the chalkline. The lower edge is allowed to overlap the head of the next-lower course by the correct distance to provide the necessary top lap between courses. The top lap can be established by inserting a nail in a face-nail hole and allowing the nail to rest on the top edge of the next-lower course. The piece of siding is then nailed in place. The course is then continued with full-size units, and joint strips are placed at each vertical joint.

Depending on the type of sheathing, mineral-fiber siding may be applied using one of four application techniques: the wood nailing strip method, the shingle backer method, the channel method, and the direct application.

### Wood nailing strip method

This technique can be used over any wood lumber, plywood, or any thickness of nonlumber plywood or lumber sheathing. The wood nailing strips are $3/8$ inch $\times$ $3\frac{1}{2}$ inch and are placed to overlay the head of a lower course. (See Figure 11–18.) The overlay should equal the desired top lap less $1/4$ inch. When the strip is placed, it firmly clamps the head of the underlying course to the wall. The wood nailing strip should be secured to the underlying studs with two 6d nails placed at each stud location. As the wood strips are placed, the joints should be broken over a stud. Additional courses of siding are placed with their butt edges placed over the wood strip and with the butt edge of the siding extending down past the butt edge of the wood strip by not more than $1/4$ inch. The siding is then nailed to the nailing strips with $1\frac{1}{8}$-inch screw face nails.

### Shingle backer method

This method of applying mineral-fiber siding requires the siding to be placed over wood lumber or plywood sheathing. For this technique, a piece of $5/16$-inch rigid insulation backs each course of mineral-fiber siding. The rigid insulation should not be less than $5/16$ inch thick and should completely underlay the width and full length of each course. The length of the shingle backer can vary, but the width should be about $1/4$ inch less than the full width of the siding. The siding is placed with its head edge flush with the head edge of the shingle

FIGURE 11–18. Wood nailing strips

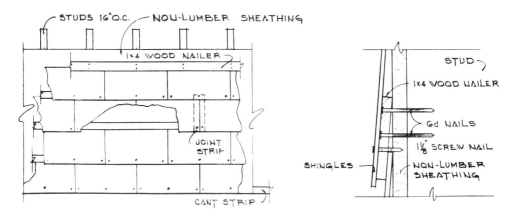

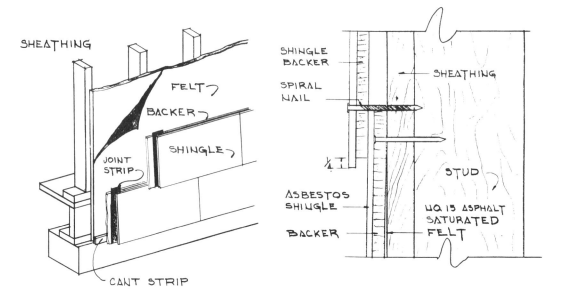

FIGURE 11–19. Shingle-backer method

backer, and its butt edge extended past the butt edge of the underlying shingle backer. (See Figure 11–19.) As the backer strips are placed, the joints on succeeding courses should be broken and the ends of the shingle units and shingle backer must not occur at the same location. The shingle backer and the mineral-fiber siding are secured to the wall frame with spiral- or screw-thread face nails hammered into the underlying sheathing. Before the shingles are installed, the backer is held in position by 1¼-inch galvanized roofing nails placed 2 or 3 inches from the top edge of the backer.

### Channel Method

With this method, 1³⁄₁₆-inch channels are used to support the siding. (See Figure 11–20.) The channels are made of noncorroding and nonstaining metal and are often prepainted to match the siding. They are secured to the wall frame at 16-inch cen-

ters, using 10 to 12 hot-dipped galvanized large-head roofing nails. The nails are driven at an angle to the wall frame, and should be long enough to penetrate and hold securely in the nailing base.

### Direct application method

Mineral-fiber siding can also be applied directly to the sheathing without shingle backers, wood strips, or channel molding. If direct application is used, 1⅛-inch-long face nails are usually satisfactory. In some cases, however, shorter nails should be used, because the face nails should not be allowed to protrude beyond the inside face of the sheathing.

## Flashing

To eliminate the possibility of water and air infiltration, all openings and corners should be properly flashed. To flash win-

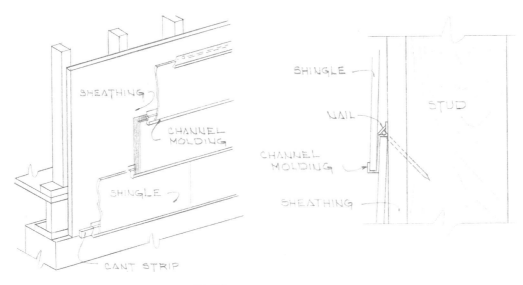

FIGURE 11-20.   Channel method

dows and doors, a drip cap must be placed above the opening. Metal flashing is then placed over the drip cap and held in place by the siding. (See Figure 11–21.)

Inside and outside corners are flashed by placement of a 12-inch wide strip of underlayment down the center of the corner. The outside corners can then be finished by metal corners, wood corner boards, or by lapping the adjoining courses of siding. Inside corners are usually butted, although metal molding or wood corner strips can be used.

To further weather-seal the structure, a nonshrinking caulking compound should be used for all the joints. The compound is available in white or matching colors, and

is placed where the siding butts wooden trim, masonry, or other projections.

FIGURE 11–21.   Metal-flashing placement

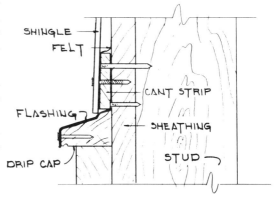

## WOOD SHAKES AND SHINGLES

Wood shakes and shingles, as sidewall covering, provide a durable and decorative material that blends a structure into the environment. The primary difference be-

### TABLE 11-1   APPROXIMATE COVERAGE OF ONE SQUARE OF SIDEWALL SHAKES

| Shake Type, Length, and Thickness | No. of Courses per Bundle | No. of Bundles per Square | Approximate coverage (in sq ft) of one square, when shakes are applied with ½" spacing, at following weather exposures (in inches): | | | | | | | | |
|---|---|---|---|---|---|---|---|---|---|---|---|
| | | | 5½ | 6½ | 7 | 7½ | 8½ | 10 | 11½ | 14 | 16 |
| 18" × ½" to ¾" Resawn | 9/9[1] | 5[2] | 55[3] | 65 | 70 | 75[4] | 85[5] | 100[6] | | | |
| 18" × ¾" to 1¼" Resawn | 9/9[1] | 5[2] | 55[3] | 65 | 70 | 75[4] | 85[5] | 100[6] | | | |
| 24" × ⅜" Handsplit | 9/9[1] | 5 | | 65 | 70 | 75[7] | 85 | 100[8] | 115[9] | | |
| 24" × ½" to ¾" Resawn | 9/9[1] | 5 | | 65 | 70 | 75[3] | 85 | 100[10] | 115[9] | | |
| 24" × ⅜" to 1¼" Resawn | 9/9[1] | 5 | | 65 | 70 | 75[3] | 85 | 100[10] | 115[9] | | |
| 24" × ½" to ⅝" Tapersplit | 9/9[1] | 5 | | 65 | 70 | 75[3] | 85 | 100[10] | 115[9] | | |
| 18" × ⅜" True-Edge Straight-Split | 14[11] Straight | 4 | | | | | | | | 100 | 112[12] |
| 18" × ⅜" Straight-Split | 19[11] Straight | 5 | 65[3] | 75 | 80 | 90[10] | 100[9] | | | | |
| 24" × ⅜" Straight-Split | 16[11] Straight | 5 | | 65 | 70 | 75[3] | 85 | 100[10] | 115[9] | | |
| 15" Starter-Finish Course | 9/9[1] | 5 | Use supplementary with shakes applied not over 10" weather exposure. | | | | | | | | |

[1] Packed in 18"-wide frames.

[2] 5 bundles will cover 100 sq ft roof area when used as starter-finish course at 10" weather exposure; 6 bundles will cover 100 sq ft wall area when used at 8½" weather exposure; 7 bundles will cover 100 sq ft roof area when used at 7½" weather exposure; see footnote[13].

[3] Maximum recommended weather exposure for three-ply roof construction.

[4] Maximum recommended weather exposure for two-ply roof construction; 7 bundles will cover 100 sq ft roof area when applied at 7½" weather exposure; see footnote[13].

[5] Maximum recommended weather exposure for side wall construction; 6 bundles will cover 100 sq ft when applied at 8½" weather exposure; see footnote[13].

[6] Maximum recommended weather exposure for starter-finish course application; 5 bundles will cover 100 sq ft when applied at 10" weather exposure; see footnote[13].

[7] Maximum recommended weather exposure for application on roof pitches between 4-in-12 and 8-in-12.

[8] Maximum recommended weather exposure for application on roof pitches of 8-in-12 and steeper.

[9] Maximum recommended weather exposure for single-coursed wall construction.

[10] Maximum recommended weather exposure for two-ply roof construction.

[11] Packed in 20" wide frames.

[12] Maximum recommended weather exposure for double-coursed wall construction.

[13] All coverage based on ½" spacing between shakes.

Reprinted with permission of Red Cedar Shingle & Handsplit Shake Bureau.

tween wood shakes and wood shingles is that wood shakes are split and wood shingles are sawed.

Wood shakes and shingles can be constructed from several different species of wood, but Western red cedar is usually used, for it is a highly durable wood. It has natural preservative oils and multiple grooves that act as channels to shed water from the exposed wall. It is a fine-grained wood, completely devoid of pitch or resin and decay-resistant under most conditions.

Wood shakes and shingles are available in three different lengths: 16, 18 and 24 inches; and with respective butt thicknesses of 0.40, 045, and 0.50 inch. The individual units are packaged in bundles and sold by the square. The number of bundles per square depends on the size of the shake and the weather exposure. (See Table 11–1.)

## Application

Two basic techniques are used to install wood shakes and shingles—the single-coursed method and the double-coursed method. When the single coursed technique is used, the weather exposures are slightly less than half of the shingle length. The correct exposure for the shingles are: 16-inch length—7½ inch; 18-inch length—8½ inch; 24-inch length—11½ inch. For the double-coursed technique, two layers of shingles or shakes are applied, one directly over the other. The top course is usually a No. 1 grade, while the lower course is a lower grade. If the double-coursed technique is used, the weather exposures are:

16-inch length—12 inch; 18-inch length—14 inch; and 24-inch length—16 inch.

If wood shakes or shingles are to be placed over solid sheathing, the sheathing should first be covered with 15-pound asphalt-saturated felt. The felt should have 2-inch side laps and 4-inch end laps, and should be fastened to the sheathing with large-headed galvanized nails. But if sheathing has not been used, the asphalt-saturated felt must be nailed directly to the studs and 1- × 4-inch nailing strips nailed over the felt so that the course lines fall 2 inches below the strip centers. If the double-coursed technique is used, the bottom course is laid with two undercoursings and one outer course. The triple-coursed method eliminates the need for a drip cap and it makes the bottom course as thick as the succeeding courses. A straight piece of shiplap is placed along a chalkline marked along the bottom course. With the piece of shiplap as a guide, the undercourse is placed on the upper lip of the straightedge and the outer course is placed on the bottom lip. Each shingle should be secured with two 5d (1¾ inch) small-head, rust-resistant nails driven about 2 inches above the butts and ¾ inch in from each side. The nails should never be driven so hard that its head crushes the wood fibers of the shingle.

Inside corners of wood shakes or shingled walls may be either "woven" in alternate overlaps, or the corner can be closed by butting the shingles to a 1 inch × 1 inch corner post. Outside corners can be finished by alternately overlapping the edges and by mitering the two adjoining shakes.

# CHAPTER 12

# INTERIOR WALL COVERING

THERE ARE SEVERAL DIFFERENT TYPES OF materials that can be used to finish interior walls. Some of the most popular are panel- ing, gypsumboard, plaster, and ceramic tile. Each material has certain advantages and unique functions.

## PANELING

Paneling, one of the most popular means of finishing an interior wall, can be solid lumber, hardboard, or plywood.

*Solid lumber* is an excellent paneling material; not only is it decorative, but it adds structural strength to the wall frame. Paneling is available with a rough-sawn texture or a smooth, planed surface. Rough-sawn boards are usually square-edged, tongued-and-grooved, or have a shiplapped edge. (See Figure 12–1A.) Planed boards usually have machine-molded patterns (see Figure 12–1B), but they can be purchased without a molded edge.

*Hardboard* paneling is made from clean selected wood fibers, pressed under heat. The panels are actually wood, but they are hard, dense, and grainless. The standard panel is 4 feet × 8 feet × ¼ inch, and is available in a wide variety of colors, grains, and textures.

*Plywood*, also a popular type of paneling, is available in plain or processed surfaces. Both hardwood and softwood plywood can be used as paneling, but in most cases hardwood is chosen. The panels can either be prefinished at the factory or finished after they have been installed.

### Application

Before the panels are nailed in place, they should be allowed to condition to room temperature for 48 hours. If the area where they are to be applied has unusual moisture conditions, it should first be thoroughly dried out. To allow the panels to adjust to the area of application, they should be unwrapped and placed on their long edges around the room.

The walls should also be properly conditioned and prepared before the panels are positioned. If they are to be placed over old walls that are not true or over masonry walls, furring strips should first be installed on the walls, placed so they run at right angles to the direction of panel application. These strips are nominal 1 × 2s placed on 16-inch centers; if they are attached to masonry walls, masonry nails must be used.

Vertical furring blocks are used to provide a solid backing at the panel edges. The blocks are spaced 48 inches on center, and for ventilation purposes a ½-inch space should be left between the horizontal furring strips and the vertical furring blocks. If the furring strips are not level, shims can be used.

In new construction the panels can either be nailed directly to the studs or placed over a backerboard. It is recommended, however, that paneling be placed over solid backing.

FIGURE 12–1. Solid-lumber paneling and machine-molded paneling

A          B

## Nailing

If paneling is nailed directly to the studs, 3d (1¼ inch) finish nails should be spaced 6 inches on center along the panel edges. On intermediate framing members, the nails are spaced 12 inches on center. If paneling is nailed to furring strips, 3d finish nails are placed 8 inches on center along the panel edges and 16 inches on center along the horizontal furring strip. When a backerboard is used, or if the paneling is placed directly over an existing wall, 6d (2-inch) finish nails are spaced 4 inches on center along panel edges and 8 inches on center on the intermediate framing members.

After the nails have been placed, they are countersunk ¹⁄₃₂ inch and the hole is filled with putty from a color-coordinated putty stick. To eliminate countersinking, colored nails that match the paneling can be used.

## Adhesives

To eliminate the use of nails, the panels can be secured with adhesive. Before adhesive is applied, the studs, furring strips, or backerboards should be checked to make certain they are clean and free of any dust or dirt.

When the panels are placed over open studs, a ⅛-inch-thick continuous ribbon of adhesive should be applied to the studding. The adhesive applied to the intermediate framing members should be placed in an intermittent ribbon (a 3-inch bead for every 6 inches of open space).

If the paneling is applied over a backerboard, a ⅛-inch-thick continuous ribbon of adhesive should be placed where the panel edge is to be bonded. Intermittent ribbons of mastic are then placed on the backerboard on 16-inch centers.

## Cutting Paneling

Certain precautions should be taken to avoid splintering or chipping paneling. When the panel is cut with a crosscut handsaw or a table saw, the panel face should be kept up. But if a portable or sabre saw is used, the panel face should be placed down. A ripsaw should *not* be used to cut paneling.

## Installation Procedures

The first panel is placed in position either by using a chalkline as a guide, or by using a level to plumb the panel. A shim is used to maintain a ¼-inch clearance between the floor and paneling. If the corner is irregular, it should first be plumbed; then a compass should be used to scribe the panel. (See Figure 12–2.)

As mentioned earlier, if nails are used to attach the panels they should be color-matched or countersunk below the surface of the paneling.

If an adhesive has been applied, the panel should be gently pressed into position and then pulled away from the wall. The adhesive should be allowed to set for 2 to 5 minutes , after which the panel is repositioned and pressed into place. Many mechanics place a nail in each corner to hold the panel in place while the adhesive is drying.

## Molding Installation

To terminate the paneling at an inside corner, the panels may be butted to each

FIGURE 12–2. Scribing a panel

other, or a piece of molding can be used to cover the intersection. At an outside corner, the paneling may be mitered, or a piece of molding can be placed over the intersection. At the intersection of wall and ceiling, a piece of cove molding can be used to cover the intersection. If the wall is wainscoted, or if the paneling does not extend the full height of the wall, cap molding can be used to trim the paneling. Paneling can be finished at the edge of a window or door jamb with a rabbeted casing, a plain casing, a cap molding, or a base shoe. At the intersection of wall and floor, a piece of base is used to cover the intersection.

## Application of Solid-Lumber Paneling

Solid-lumber paneling should be placed over a firm nailing surface. If it is installed over a frame wall, nominal 2 × 4 blocks should be placed between the studs. To increase the bearing surface, the blocks are often turned so the larger surface backs up the panels. The spacing of the blocks depends on the size of the paneling. If 1-inch paneling is used, the blocks are spaced on 48-inch centers, and if ½-inch paneling is used, the blocks should be spaced on 24-inch centers.

The first piece of paneling should be plumbed with a level and then nailed into place. The nails should be long enough to penetrate at least 1 inch into the framing member. For nominal 1-inch paneling, 2-inch or 6d finishing nails are recommended. If the paneling is tongue-and-groove and narrower than 6 inches, it should be blind-nailed. The nails should be placed at the base of the tongue and should be nailed into the cross-blocking. To avoid damaging the paneling, the nails are not driven home, but are countersunk to allow for flush application of the next piece of paneling. If the tongue-and-groove panels are wider than 6 inches they are blind-nailed and face-nailed. If face-nailed, they must be countersunk, and the holes later filled with a matching putty.

Lapped paneling less than 6 inches in width is face-nailed with one nail to be placed in the center of the paneling. If this is wider than 6 inches, two nails should be used to secure each piece of paneling to the wall frame. Nails should be placed so that they will not penetrate the underlying lap.

When board-and-batten paneling is used, the boards are first placed with a ½-inch gap between them. Each board should be secured to the wall frame with one 8d nail placed in the center of the board. Battens are then placed over the opening be-

FIGURE 12–3.  Placement of battens

BOARD
ON
BOARD

FIGURE 12–4.  Board-on-board application

tween the boards and secured with one 10d finishing nail. (See Figure 12–3.) The shank of the nail should pass between the edges of the two underboards.

If board-on-board paneling is used, the underboard is secured with one 8d finishing nail placed in its center, and spaced to the top boards to project over the edges a minimum of 1 inch. The top board is fastened to the wall frame with two nails placed to miss the underboards. (See Figure 12–4.)

## GYPSUMBOARD

Gypsumboard is excellent as an interior wall and ceiling finish material, for it is noncombustible, fire-resistant, and reduces sound transmission between rooms. The large panels have a core of gypsum sandwiched between gray liner paper on the back side and a special paper covering on the front side. The front covering of paper allows for easy painting and decorating.

Gypsum wallboard is available in standard thicknesses, lengths, and widths. The standard thicknesses are ¼, ⅜, ½, and ⅝ inch. The ¼-inch panels should be applied over a solid surface, and in most cases are used for the rehabilitation of old walls and ceiling surfaces. The ⅜-inch panels are usually preferred for the outer layer in multi-layer construction. The ½-inch-thick panels are used for direct application to vertical and horizontal framing members. The ⅝-inch-thick panels are also used in single-layer construction, but the added

### Paneling Applied over Plaster

If paneling is to be placed over an existing plaster wall, the wall should first be strapped with 1 × 3s placed on 24-inch centers. The first piece of strapping should be placed at ceiling height and secured with nails long enough to penetrate the framing members. If there are any low spots in the strapping, they should be shimmed so that they come to the proper elevation.

thickness insures additional fire resistance and greater damping of sound. The standard width of gypsum wallboard panels is 4 feet; the length, however, is available in 8, 10, 12, and 14 feet.

The edges of gypsum wallboard are also standard being rounded, beveled, tapered, or square-edged.

### Types of Gypsum Wallboard

Each type of gypsum wallboard is designed for a unique and specific function.

*Insulating gypsum wallboard* has a sheet of aluminum foil bonded to the back of the panel. The foil acts as a vapor barrier and thermal insulation, and is placed next to the framing members, reducing the outward flow of heat in the winter and the inward flow in the summer. This particular type of panel should not be used as a base for ceramic tile.

*Predecorated gypsum wallboard* has a sheet of decorative vinyl or paper bonded to it. The decorative covering is available in several different colors and surface patterns (stipples, linens, wood grains, textiles, and character woods).

*Gypsum backing board* is a low-cost panel used as a base layer in multiple construction. The backing board gives extra fire resistance, sound isolation, and strength.

*Water-resistant gypsum backing board* has a gypsum core and a specially treated face paper that is water-repellent. Because of its resistance to moisture, it is used as a base for tile in baths and showers.

## Wall and Ceiling Framing

To achieve quality construction in the application of gypsum wallboard, the framing members must be in proper alignment. Neither top plate nor studs should be out of alignment, and the studs must not be twisted. To assure an adequate bearing surface, no framing or furring member should vary more than ¼ inch from the plane of the faces of the adjacent framing or furring

member. If the studs are bowed or warped, they can be straightened by means of a partial cut made through the stud. A wedge should then be driven into the saw kerf until the stud is in proper alignment. To reinforce the stud and keep it aligned, a 1 × 4 scab should be nailed over it. To straighten slightly crooked or bowed ceiling joists, stringers can be placed perpendicular to the joist. (See Figure 12–5.) If a joist is excessively bowed or crooked, it should not be used; but if it has a slight crown, it should be placed with the crown up. Natural deflection and the use of stringers will produce a level surface for the application of gypsumboard.

In addition to being placed in proper alignment, the framing members should not exceed the maximum spacing requirements shown in Table 12–1. But if they do, furring strips should be used to provide a rigid base.

When furring strips are used in ceiling construction, they should have a minimum cross-section of 1½ inches by 1½ inches. If screws are used to attach the panels, the furring members can be ¾ inches by 2½ inches. When furring members are placed

FIGURE 12–5.  Straightening bowed ceiling joints

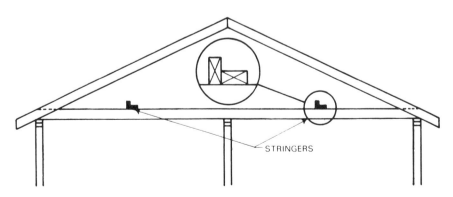

STRINGERS

## TABLE 12 –1   MAXIMUM FRAMING SPACING FOR APPLICATION OF GYPSUMBOARD

Framing spacing for two-ply, ³⁄₈-inch gypsumboard wall may be 24 inch o.c. if adhesive is used between plies.

| Gypsumboard Thickness | | | | Maximum Spacing | Two Layers | |
| Base Layer | Face Layer | Location | Application | One Layer Only | Fasteners Only | Adhesive Between Layers |
|---|---|---|---|---|---|---|
| $\frac{3}{8}''$ | — | Ceilings | Horizontal | 16″ o.c. | 16″ o.c. | 16″ o.c. |
| $\frac{3}{8}''$ | $\frac{3}{8}''$ | Ceilings | Horizontal | NA | 16″ o.c. | 16″ o.c. |
| $\frac{3}{8}''$ | $\frac{3}{8}''$ | Ceilings | Vertical | NA | NR | 16″ o.c. |
| $\frac{1}{2}''$ | — | Ceilings | Horizontal | 24″ o.c. | 24″ o.c. | 24″ o.c. |
| $\frac{1}{2}''$ | — | Ceilings | Vertical | 16″ o.c. | 16″ o.c. | 16″ o.c. |
| $\frac{1}{2}''$ | $\frac{3}{8}''$ | Ceilings | Horizontal | NA | 16″ o.c. | 24″ o.c. |
| $\frac{1}{2}''$ | $\frac{3}{8}''$ | Ceilings | Vertical | NA | NR | 24″ o.c. |
| $\frac{1}{2}''$ | $\frac{1}{2}''$ | Ceilings | Horizontal | NA | 24″ o.c. | 24″ o.c. |
| $\frac{1}{2}''$ | $\frac{1}{2}''$ | Ceilings | Vertical | NA | 16″ o.c. | 24″ o.c. |
| $\frac{5}{8}''$ | — | Ceilings | Horizontal | 24″ o.c. | 24″ o.c. | 24″ o.c. |
| $\frac{5}{8}''$ | — | Ceilings | Vertical | 16″ o.c. | 16″ o.c. | 24″ o.c. |
| $\frac{5}{8}''$ | $\frac{3}{8}''$ | Ceilings | Horizontal | NA | 16″ o.c. | 24″ o.c. |
| $\frac{5}{8}''$ | $\frac{3}{8}''$ | Ceilings | Vertical | NA | NR | 24″ o.c. |
| $\frac{5}{8}''$ | $\frac{1}{2}''$ or $\frac{5}{8}''$ | Ceilings | Horizontal | NA | 24″ o.c. | 24″ o.c. |
| $\frac{5}{8}''$ | $\frac{1}{2}''$ or $\frac{5}{8}''$ | Ceilings | Vertical | NA | 16″ o.c. | 24″ o.c. |
| $\frac{1}{4}''$ | — | Walls | Vertical | NR | 16″ o.c. | 16″ o.c. |
| $\frac{1}{4}''$ | $\frac{3}{8}''$ | Walls | NR | NA | NR | NR |
| $\frac{1}{4}''$ | $\frac{1}{2}''$ or $\frac{5}{8}''$ | Walls | Horizontal or Vertical | NA | 16″ o.c. | 16″ o.c. |
| $\frac{3}{8}''$ | — | Walls | Horizontal or Vertical | 16″ o.c. | 16″ o.c. | 24″ o.c. |
| $\frac{3}{8}''$ | $\frac{3}{8}''$ or $\frac{1}{2}''$ or $\frac{5}{8}''$ | Walls | Horizontal or Vertical | NA | 16″ o.c. | 24″ o.c. |
| $\frac{1}{2}''$ or $\frac{5}{8}''$ | — | Walls | Horizontal or Vertical | 24″ o.c. | 24″ o.c. | 24″ o.c. |
| $\frac{1}{2}''$ or $\frac{5}{8}''$ | $\frac{3}{8}''$ or $\frac{1}{2}''$ or $\frac{5}{8}''$ | Walls | Horizontal or Vertical | NA | 24″ o.c. | 24″ o.c. |

NA—Not Applicable

NR—Not Recommended

Reprinted with permission of Gypsum Association.

over solid surfaces, they should not be less than ¾ inch thick and 1½ inches wide.

The framing lumber should also be of a good grade, and at the time of panel installation the moisture content of framing members should not exceed 15 percent. To prevent dimensional changes in these, the area that is to be paneled should be properly ventilated beforehand and kept at a temperature between 55° and 70°.

## Fasteners and Adhesives

Nails, screws, staples, and special adhesives are used to secure gypsumboard to framing members, furring members, or to an underlying layer of gypsumboard.

*Nails* are available in several different sizes and shapes, the two most common types being the cement-coated and the annular ring nail. The latter is preferable, for it has about 20 percent greater holding power than a cement-coated nail of the same length. The nail shanks should be long enough to sufficiently penetrate the underlying framing members. If a smooth-shank nail is used, the shank should penetrate the framing member $\frac{7}{8}$ inch, but an annular ring nail must penetrate the framing member only $\frac{3}{4}$ inch. The sizes of the nail head are also important, for if they are either too small or too large they might cut into the face paper. The nail heads are usually at least $\frac{1}{4}$ inch, and not more than $\frac{5}{16}$ inch in diameter.

*Screws* are also a popular means of attaching gypsumboard to framing members. The screws have recessed Phillips heads and are installed with a drywall power screwdriver. There are three basic types of drywall screws—type W for wood, type S for sheet metal, and type G for solid gypsum construction. (See Figure 12–6.) Type W screws, available in several lengths, are used to secure $\frac{3}{8}$-, $\frac{1}{2}$-, and $\frac{5}{8}$-inch single-layer panels to wood framing. Type S gypsum drywall screws, also available in several lengths, are designed to secure gypsum drywall to metal studs or furring. Most type G screws are $1\frac{1}{2}$ inches in length and are used in multilayer adhesive-laminated gypsum-to-gypsum partitions. But drywall screws are not recommended for double-

layer $\frac{3}{8}$-inch panels, because they lack the desired holding strength.

*Staples* are used only to attach the base sheet in multi-ply construction. The staples are constructed from .16 guage galvanized wire, and the crown should be $\frac{7}{16}$ inch

FIGURE 12–6. Selector guide for screws

### Selector Guide for USG Brand Screws

| Fastening Application | Fastener Used |
| --- | --- |
| **GYPSUM PANELS TO STANDARD METAL FRAMING (1)** | |
| $\frac{1}{2}$″ single-layer panels to standard studs, runners, channels | $\frac{7}{8}$″ Type S Bugle Head |
| $\frac{5}{8}$″ single-layer panels to standard studs, runners, channels | 1″ Type S Bugle Head |
| $\frac{1}{2}$″ double-layer panels to standard studs, runners, channels | $1\frac{5}{16}$″ Type S Bugle Head |
| $\frac{5}{8}$″ double-layer panels to standard studs, runners, channels | $1\frac{5}{8}$″ Type S Bugle Head |
| 1″ coreboard to metal angle runners in solid partitions | $1\frac{1}{4}$″ Type S Bugle Head |
| $\frac{1}{2}$″ panels through coreboard to metal angle runners in solid partitions | $1\frac{5}{8}$″ Type S Bugle Head |
| $\frac{5}{8}$″ panels through coreboard to metal angle runners in solid partitions | $2\frac{1}{4}$″ Type S Bugle Head |
| **GYPSUM PANELS TO 12-GA. (MAX.) METAL FRAMING** | |
| $\frac{1}{2}$″ and $\frac{5}{8}$″ panels and gypsum sheathing to 20-ga. studs and runners | 1″ Type S-12 Bugle Head |
| USG Self-Furring Metal Lath through gypsum sheathing to 20-ga. studs and runners | $1\frac{1}{4}$″ Type S-12 Bugle Head |
| $\frac{1}{2}$″ and $\frac{5}{8}$″ double-layer gypsum panels to 20-ga. studs and runners | $1\frac{5}{8}$″ Type S-12 Bugle Head |
| Multi-layer gypsum panels to 20-ga. studs and runners | $1\frac{7}{8}$″ Type S-12 Bugle Head |
| **WOOD TRIM TO INTERIOR METAL FRAMING** | |
| Wood trim over single-layer panels to standard studs, runners | $1\frac{5}{8}$″ Type S Trim Head |
| Wood trim over double-layer panels to standard studs, runners | $2\frac{1}{4}$″ Type S Trim Head |

wide. The length of the legs varies, depending on the thickness of the gypsum board. If it is ⅜-inch thick, the legs should be 1 inch in length. The legs should be 1⅛ inches for ½-inch-thick gypsumboard, and 1¼ inches for ⅝-inch-thick gypsumboard.

FIGURE 12–6. *(cont.)*

**Selector Guide for USG Brand Screws**

| Fastening Application | Fastener Used |
|---|---|
| **METAL STUDS TO DOOR FRAMES, RUNNERS** | |
| Standard metal studs to runners | ⅜" Type S Pan Head Also available with Hex Washer Head  |
| Standard metal studs to door frame jamb anchor clips 20-ga. studs to runner Other metal-to-metal attachment (12-ga. max.) | ⅜" Type S-12 Pan Head  |
| Standard metal studs to door frame jamb anchor clips (heavier shank assures entry in clips of hard steel) | ½" Type S-12 Pan Head  |
| Strut studs to door frame clips, rails, other attachments in ULTRAWALL partitions | ½" Type S-16 Pan Head Cadmium Plated 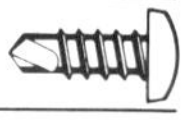 |
| **TRIM AND ACCESSORIES TO METAL FRAMING** | |
| Door hinges and trim to door frame Aluminum trim to metal framing (screw matches hardware and trim) | 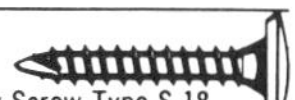 ⅞" Finishing Screw Type S-18 Oval Head Cadmium Plated |
| Metal base splice plates through panels and runner | 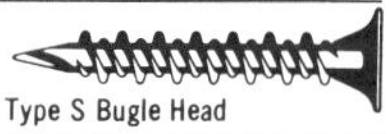 1¼" Type S Bugle Head |
| Batten strips to standard metal studs in demountable partitions | 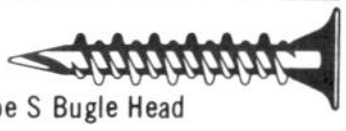 1⅛" Type S Bugle Head |
| Aluminum trim to interior metal framing in Demountable and ULTRAWALL partitions | 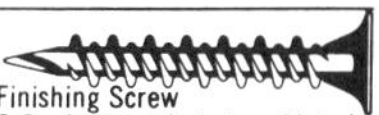 1¼" Finishing Screw Type S Bugle Head Cadmium Plated |
| **GYPSUM PANELS TO WOOD FRAMING** | |
| ⅜", ½" and ⅝" single-layer panels to wood framing | 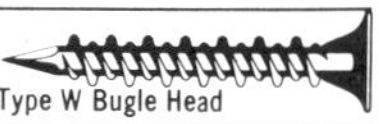 1¼" Type W Bugle Head |
| **RC-1 RESILIENT CHANNEL TO WOOD FRAMING** | |
| Screw attachment required for ceilings, recommended for partitions | 1¼" Type W, ⅞" or 1" Type S Bugle Head (see details above) |
| For fire-rated construction | 1¼" Type S Bugle Head (see details above) |
| **GYPSUM PANELS TO GYPSUM PANELS** | |
| Multi-layer adhesively laminated gypsum-to-gypsum partitions (not recommended for double-layer ⅜" panels) | 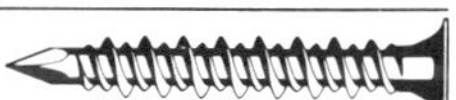 1½" Type G Bugle Head |

Notes: (1) Includes USG Standard Metal Studs, Metal Runners, Metal Angle Runners, Metal Furring Channels, RC-1 Resilient Channels. If channel resiliency makes screw penetration difficult, use screws ⅛" longer than shown to attach panels to RC-1 channels. For 20-ga. Metal Studs and Runners, always use Type S-12 screws. For steel applications not shown, select a screw length which is at least ⅜" longer than total thickness of materials to be fastened. USG Brand Screws are manufactured under U.S. Patent Nos. 2,871,752; 3,056,234; 3,125,923; 3,207,023; 3,221,588; 3,204,442; 3,260,100.

*Adhesives* are used to attach gypsumboard to framing members, furring strips, masonry, concrete, or underlying gypsum panels. The three adhesives most often used are stud, laminating, and modified contact adhesives. They are used to reduce up to 75 percent of the number of fasteners; and are stronger than conventional nail application, for they provide as much as 100 percent more tensile strength, and 50 percent more shear strength. Adhesives are not affected by moisture changes, and their use results in fewer loose panels.

When adhesive is placed on framing members, it should be applied in a ⅜-inch bead and should provide a 1/16-inch thickness of adhesive over the entire support. Where two gypsum panels meet, zigzag bead should be applied. On the intermediate supports, a straight bead is used. In multiply construction, strips of adhesive can be applied to the base ply on the wall, or the adhesive can be spread on the back of the face panel. To improve resistance to sound, dabs of adhesive can be spotted on the back of the face panel.

## Cutting Gypsum Wallboard

To properly cut gypsum wallboard, a T-square is placed in the proper location as a guide, and a wallboard knife is used to score through the face paper and into the core. Pressure is then applied to the panel to snap the board, after which the knife is used to cut the back paper. The rough edges are then smoothed with a rasp, coarse sandpaper, or metal lath wrapped around a block of wood. If it is necessary to cut around electrical outlets, light receptacles, or ducts, the locations should be properly measured and transferred to the back

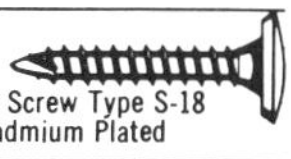

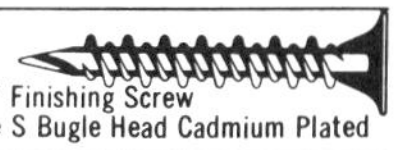

side of the gypsum wallboard. After the opening is outlined, it can be cut with a keyhole saw.

## Single-Ply Application (Wood Frame)

Gypsum wallboards are nailed first to the ceiling frame and then to the wall frame. When the panels are installed, they should be firmly pressed against the supporting framing members and nailed in place, starting from the center and working toward the edges. When the nails are driven, the last blow of the hammer should create a slight dimple in the wallboard. (See Figure 12–7.)

The spacing of the nails can conform to either the single-nailing or the double-nailing technique. For the single-nailing technique, the nails are spaced 7 inches on center for ceilings and 8 inches on center for the walls. (See Figure 12–8A.) The nails

FIGURE 12–7.  Correct driving of nails

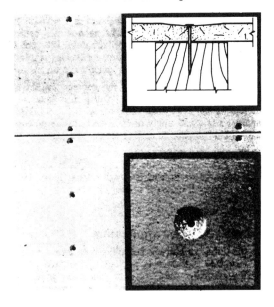

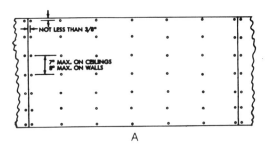

A

**Double Nailing**

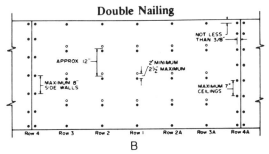

B

FIGURE 12–8.   Single-nailing and double-nailing techniques

around the perimeter of the wallboard should be placed a minimum of ⅜ inch from the edge.

For the double-nailing technique, a row of nails is placed 12 inches on center; additional rows of nails are then located 2 inches from those previously placed. (See Figure 12–8B.)

The driven nails should draw the panels firmly against the supporting member. The panels should be properly cut so that they easily fit into their spaces. If they are forced into an area, the framing member and panel cannot contact properly. But loose nails also can be responsible for loose boards and nail pops. To prevent loose nails: (1) mark the location of the supporting members before placing nails; (2) drive the nails perpendicular to the framing members; (3) pull all nails that miss the framing members.

Face-paper fractures can also cause nail popping and loose boards. This defect results from both improper nailing practices and poorly designed nailheads. If the nails are bent, they can crunch the gypsum core and fracture the face paper, or the core can be crushed by overdriving. In either case a new nail should be placed 2 inches from the fracture. Face-paper fracture can also occur because of misaligned or twisted supporting members. Such framing irregularities prevent the panels from coming in contact with the supporting members, and hammer contact ruptures the paper.

If screws are used to attached the gypsumboard panels, they should be spaced 12 inches on center on ceilings, and 16 inches on center on walls in which the framing members are spaced 16 inches on center. If the framing members are spaced 24 inches on center, the screws should be placed on 12-inch centers for both wall and ceilings.

The gypsumboard panels can also be placed by using an adhesive nail-on application. When the panels are glued and nailed to the wall, nails are placed only around the perimeter; but when ceiling panels are positioned, field fasteners are placed on 24-inch centers. When the panels are nailed into place, the adhesive should spread to an average width of 1 inch. In some cases it is not desirable to have fasteners placed in the field, so temporary braces are put under the gypsumboard for a minimum of 24 hours. To avoid locating fasteners at edge joints, the panels should first be prebowed and braced to the supporting members. The bow keeps the panel in constant contact with the beads of adhesives.

## Corner Cracking

To eliminate the possibility of corner cracking and nail popping, a technique called "floating angle construction" is used. This particular technique omits the last row of fasteners on the wall and ceiling. (See Figure 12–9A, B, and C.) If the joists

FIGURE 12–9. Horizontal *(a)* and vertical *(b)* ceiling applications

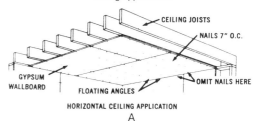

HORIZONTAL CEILING APPLICATION
A

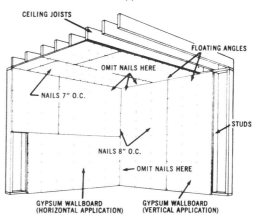

B

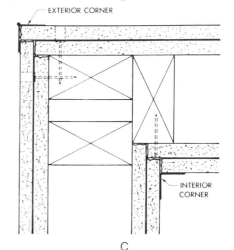

C

**METAL ACCESSORIES**

Cornerbead — (Numbers indicate width of flanges
—i.e.—118 is 1-1/8 in. W.
Flange)

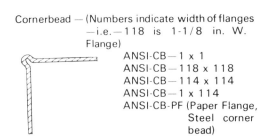

    ANSI-CB—1 x 1
    ANSI-CB—118 x 118
    ANSI-CB—114 x 114
    ANSI-CB—1 x 114
    ANSI-CB-PF (Paper Flange,
        Steel corner
        bead)

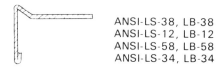

    ANSI-LS-38, LB-38
    ANSI-LS-12, LB-12
    ANSI-LS-58, LB-58
    ANSI-LS-34, LB-34

"U" Bead — (Numbers indicate thickness of
board to be used) "S" indicates
square nose, "B" Bull or Round
Nose)

"LK" Bead — (For use with Kerfed jamb) ("S"
indicates square nose, "B" Bull
or Round Nose)

    ANSI-US-38, UB-38
    ANSI-US-12, UB-12
    ANSI-US-58, UB-58
    ANSI-US-34, UB-34

    ANSI-LK-S
    ANSI-LK-B

"L" Bead — (Number indicate thickness of board
to be used) ("S" indicates square
nose, "B" Bull or Round Nose)

"LC" Bead — (Numbers indicate thickness of
board to be used)

    ANSI-LC-38
    ANSI-LC-12
    ANSI-LC-58
    ANSI-LC-34

FIGURE 12–10.   Metal accessories

are perpendicular to the intersection of ceiling and wall, the nails are started 7 inches from the intersection. But if the joists are parallel to the intersection of wall and ceiling, the nails are started at the intersection. The wall nails should be started 8 inches from the ceiling. This floating angle joint created by the omission of fasteners is intended to minimize damage that might be caused by structural stresses.

## Accessories

Metal accessories are often used to protect exposed gypsum wallboard corners and edges. (See Figure 12–10.) A cornerbead is applied to protect outside corners from damage, while metal flanged beads can be used to protect the wallboard where inconspicuous edge protection is needed. The edges must be crimped so the metal accessories can be attached.

## Multi-Ply Application (Wood Frame)

Multi-ply application involves the use and application of two layers of gypsum wallboard. The base layer is usually wallboard, backing board, or sound-deadening board attached to the framing members with nails, staples, or screws. The spacing requirements for each fastener are shown in Table 12–2. When the face ply is attached,

### TABLE 12–2 BASE-PLY FASTENER SPACING ON WOOD FRAMING

| | Nail Spacing | | Screw Spacing | | Staple Spacing | |
|---|---|---|---|---|---|---|
| Location | Laminated Face Ply | Nailed Face Ply | Laminated Face Ply[2] | Screwed Face Ply | Laminated Face Ply | Nailed or Screwed Face Ply |
| Walls | 8-inch o.c. | 16-inch o.c. | 16-inch o.c. | 24-inch o.c. | 7-inch o.c. | 16-inch o.c. |
| Ceilings | 7-inch o.c. | 16-inch o.c. | 16-inch o.c. | 24-inch o.c. | 7-inch o.c. | 16-inch o.c. |

[1] Fastener size and spacing for applying sound-deadening boards varies for different fire and sound-rated constructions. The manufacturer's recommendations should be followed.

[2] 12-inch o.c. for both ceilings and walls when supports are spaced 24 inches o.c.

Reprinted with permission of Gypsum Association.

its joints should offset the joints of the base ply by at least 10 inches. The face ply can be attached by mechanical fasteners or adhesives.

When mechanical fasteners are used, the maximum spacing and minimum penetration recommended for screws and nails should be the same as for single-ply construction. To prevent corner cracking, the inside corners should be allowed to float, with only the base ply nailed. (See Figure 12–11.)

Temporary braces or fasteners are usually used if an adhesive is used to bond the

FIGURE 12–11. Floating inside corner

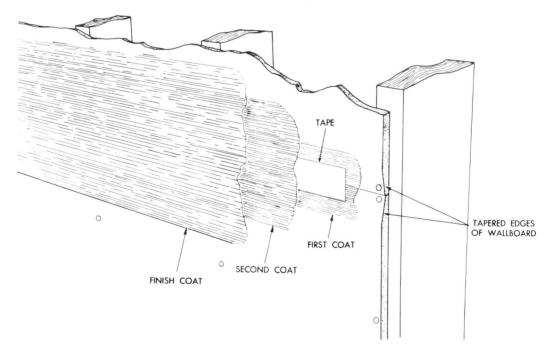

TAPE

TAPERED EDGES OF WALLBOARD

FIRST COAT

SECOND COAT

FINISH COAT

face ply to the base ply. The temporary fasteners are usually double-headed nails that can be removed once the adhesive has set. To properly secure the face ply, the temporary fasteners are usually spaced on 24-inch centers.

If a fire-rated assembly is required, permanent fasteners must be used in the field. Their number and spacing depends on the tested assembly construction.

## Taping and Finishing

After the gypsum panels have been placed, it is necessary to tape and finish the joints. The tape reinforces them, reducing the possibility of cracking. A recommended minimum of three coats of joint compound should be placed over the tape to give the area a smooth finish. The first coat embeds the tape, and the other two are used to feather the edges.

These are the recommended steps for joint treatment:

■ Butter the intersections of the panels with joint compound. (See Figure 12–12.) Place the compound in even

FIGURE 12–12.   Buttering the joint

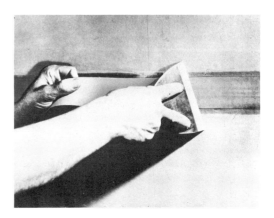

FIGURE 12–13.   Placement of reinforcing tape

thicknesses, avoiding heavy fills that might cause shrinkage and cracking.

■ Press reinforcing tape into the freshly placed joint compound. (See Figure 12–13.) Hold the knife at an approximate angle of 45°, with enough pressure exerted to extract any excess compound. To bond the tape to the wallboard, a minimum of $\frac{1}{32}$ inch of compound should be left under the edge of the embedding tape. Before the joint compound can dry, place a skim coat over the tape to prevent the edges wrinkling or curling.

FIGURE 12–14.   Feathering the edges of the second coat

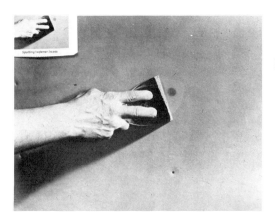

FIGURE 12–15. Treating the nail heads

■ Place a second application over the embedding and covering coat, feathering the edges 2 inches beyond the edges of the first coat. (See Figure 12–14.)

■ After the second coat has dried, apply a third, finishing coat, feathering the edges 2 inches beyond the edges of the second coat.

■ Apply joint compound over the nail heads, using just enough pressure to bring the compound level with the wallboard. (See Figure 12–15.) Three coats

FIGURE 12–16. Finishing an inside corner

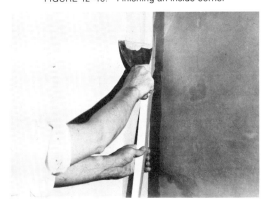

of joint compound are required to properly finish the nail heads.

■ Finish the inside corners by first folding the tape along the center crease, then buttering the corner with joint compound and applying the tape. (See Figure 12–16.) Second and third coats are applied over the corner in the same manner as for flat joints.

■ Once the fastener heads have been spotted and the joints properly treated, lightly sand the entire surface.

The joints can also be finished by mechanical applicators. An automatic taping machine can be used to tape and fill joints, and an automatic treating machine places an even coat of compound over the previously placed tape.

In order to assure quality construction, there are certain precautionary steps that should be followed in applying and finishing gypsum wallboard. If the weather is hot and dry, the doors and windows should be closed; the humidity can be raised by sprinkling the subfloor with water. If the weather is wet and humid, each coat of compound should be allowed to dry thoroughly before additional coats are applied. In cold weather, an even temperature of 55° F or above should be maintained during and after the application of the joint compound. Regardless of weather conditions, the compound should be kept free of contamination and residue.

## Predecorated Wallboard

If predecorated wallboard is used, the joints should first be finished with a quick-setting compound. (See Figure 12–17.) After the compound has set, the area between

FIGURE 12–17  (above)   Finishing the joints of predecorated wallboard, and 12–18 (below) Pasting the flaps down

the two vinyl flaps should be coated with wheat paste. The flaps are then pasted down, the right-hand flap first and the left-hand flap next. (See Figure 12–18.) (When the left-hand flap is pasted down, it will project over the right-hand flap.) A straightedge is then put over the intersection of the two flaps, and a wallboard knife is used to cut through both flaps. (See Figure 12–19.) The two strips that have been cut off are then peeled from the wall. (See Figure 12–20.) The strip cut from the

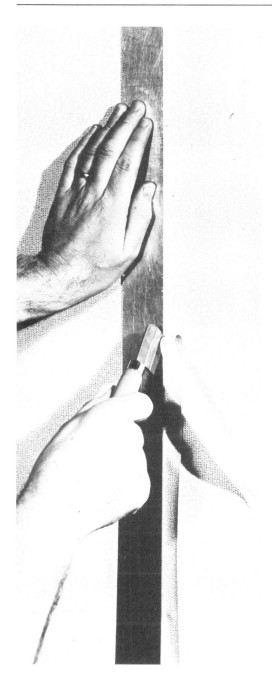

FIGURE 12–19.  Cutting the wallboard flaps

FIGURE 12–20.  Peeling the cut-off strips

right-hand flap must be pulled from under the left flap. The intersection of vinyl is then smoothed flat and rolled, to eliminate bubbles and excess paste. After the joint has been rolled, the area should be sponged and allowed to dry.

## PLASTER

Plaster is a durable wall-finish material that is made from gypsum, one of the common minerals of the earth. Gypsum is sold in powder form; when it is mixed with water and an aggregate, a fire-resistant product is produced. Once the plaster has been mixed to the proper consistency, it is applied to a plaster base.

### Plaster Bases

There are two basic classifications of plaster bases—those that are attached to framing members, and masonry bases. In light construction the plaster base is most often gypsum lath or metal lath. Gypsum lath is either plain or perforated and in most cases is $3/8$ inch thick, 16 inches wide, and 48 inches long. (See Table 12–3.) If the spacing of the framing members exceeds 16 inches, $1/2$-inch lath must be used. The lath is placed with its long dimension perpendicular to the framing members and may be nailed, stapled, screwed, or clipped to the framing members. It is secured to each framing member with nails, staples, or screws. The heads of the fasteners should be placed slightly below the face paper, which should not be broken. In addition to holding the lath in place, the fasteners are used to meet necessary fire rating requirements.

Once the lath has been set in place, corner beads should be placed on all external corners. These beads are made from 26-gauge galvanized steel and provide plaster protection, true and straight lines at corners, and grounds for plastering. The inte-

**TABLE 12–3    GYPSUM LATH**

| Type | Thickness (Inches) | Width (Inches) | Length (Inches) | Use |
|---|---|---|---|---|
| PLAIN | $3/8$ | 16 | 48 or 96 | Application to wood or metal framing, by nails, staples, screws, or clips. |
|  | $1/2$ | 16 | 48 | |
| PERFORATED | $3/8$ | 16 | 48 or 96 | Same as above, except not used for ceiling attachment, where the only attachment is by clips at edges of lath or where insulation is placed on the ceiling. |
|  | $1/2$ | 16 | 48 | |
| INSULATING | $3/8$ | 16 | 48 or 96 as requested to 12 feet | Same as for plain and where a vapor barrier is required. |
|  | $3/8$ or $1/2$ | 24 | | |
| LONG LENGTH | $1/2$ | 24 | as requested to 12 feet | Primarily used in 2-inch solid gypsum lath and plaster partitions. |

Reprinted with permission of Gypsum Association.

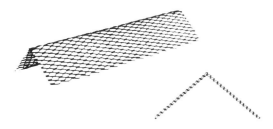

FIGURE 12–21.  Conerite

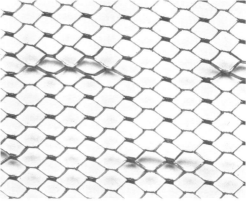

FIGURE 12–23A.  Diamond mesh

rior corners are reinforced with cornerite. (See Figure 12–21.) Cornerite is made from a strip of copper alloy, diamond mesh lath. However, if clips are used to fasten lath to framing members, cornerite is not recommended. To provide an even thickness of plaster, grounds are placed around all openings and wherever baseboards and moldings are to be used. (See Figure 12–22.) The grounds, usually constructed from 1 × 2s, serve as a leveling surface during the application of plaster, and later as a nailing base. If gypsum lath is used, the grounds should be set to provide a minimum plaster thickness of ½ inch; and if metal lath is used, the grounds should be set to provide a minimum plaster thickness of ⅝ inch. The size of the grounds allows for the proper basecoat thickness, plus allowing 1/16 inch for the thickness of the finish coat.

Metal is a popular lath, because it forms a mechanical bond with the plaster. The gypsum laths provide a suction bond with the plaster. The two most popular types of metal lath are diamond mesh and riblath. Diamond mesh is a general all-purpose lath that can have as many as 11,000 meshes per yard. (See Figure 12–23A.) Riblath has a herringbone mesh pattern with U-shaped ribs running lengthwise of the sheet. (See Figure 12–23B.)

FIGURE 12–23B.  Riblath

FIGURE 12–22.  Placement of the grounds

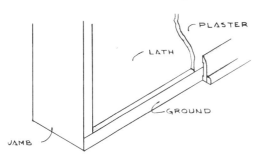

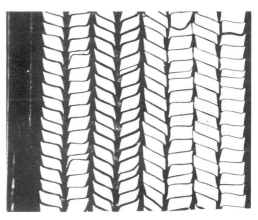

## General Requirements

Before the plaster is placed on the lath, certain precautionary steps should be taken. The area to be plastered should be kept at a constant temperature for seven days before application, and also during the plastering phase and until the plaster is dry. The temperature should not be less than 55° F and evenly distributed.

Ventilation should be provided to allow the plaster to dry properly. Windows can be slightly opened or, if natural ventilation is absent, fans can be used to circulate the air.

## Gypsum Basecoat Plasters

Gypsum basecoat plaster is placed over lath to provide resistance against structural movements, as well as a true and level surface of the finish coat of plaster. There are three basic types of basecoat plasters used in light construction:

- *Gypsum neat plaster:* A mixture of calcined gypsum plaster, sand, perlite or vermiculite aggregate, and water. This particular type of plaster is often referred to as a gypsum cement plaster or hardwall.

- *Gypsum ready-mixed plaster:* Mixed at the mill; requires only the addition of clean water. The plaster is a mixture of calcined gypsum plaster and an aggregate, usually sand or perlite.

- *Gypsum wood-fibered plaster:* Mixed at the mill; requires only the addition of clean water. But instead of using sand as an aggregate, wood fibers are used to give the plaster more bulk and coverage. This particular type of plaster is used

where good plastering sand is not available and where greater base strength is desired.

The basecoat of plaster can be placed over lath by hand (see Figure 12–24A) or by machine (see Figure 12–24B), using either a three- or a two-coat application. In three-coat application, a scratch coat of plaster is placed over the lath and cross-raked. (See Figure 12–25.) A brown coat is then troweled over the scratch coat and allowed to partially dry. (See Figure 12–26.) The brown coat should be brought out to the grounds, darbied, and allowed to partially dry; then it should be topped by the finish coat. Three-coat application is required:

- When plaster is placed over a metal lath

- When gypsum lath is placed on horizontal framing members that are put on centers greater than 16 inches

- Over ⅜-inch perforated gypsum ceiling lath

- When gypsum ceiling lath is supported by clips

Two-coat application is similar to three-coat application except that the basecoat is not cross-raked, and the brown coat is doubled back over the scratch coat minutes after its application. Two-coat application is usually used over gypsum lath, although the three-coat method is stronger because the scratch coat is allowed to partially dry, thus drawing out the excess water from the brown coat.

## Finish-Coat Plasters

After the basecoat of plaster has been placed and allowed to dry partially, a finish

FIGURE 12–24A (above) Application of basecoat plaster by hand, and 12–24B (right) Application of basecoat plaster by machine

coat of plaster is applied. There are six different types of finish coat plasters that can be used; are classified according to texture, color, and hardness:

■ *Gypsum-lime putty-trowel finish:* One of the most popular; sometimes referred to as "white-coat" finish. This finish consists of a mixture of lime putty and gypsum gauging plaster; it has high plasticity and provides a hard finish at a

FIGURE 12-25.  Scratch coat of plaster

FIGURE 12-26.  Brown coat of plaster

relatively low cost. But it must be carefully mixed according to manufacturer's specifications; otherwise check-cracking, crazing, bond failure, and lack of hardness will occur.

■ *Keene's cement-lime putty-trowel finish:* A mixture of lime putty and Keene's cement. This finish can be extremely hard and can be troweled to a smooth, uniform surface. It should be placed only over a strong basecoat, and the surface should be occasionally troweled until a complete set has been achieved.

■ *Keene's   cement-lime   and   sand-float*

*finish:* A mixture of sand, Keene's cement, and dry hydrated lime. The finish can be placed over all plaster bases and, by varying the amount of sand, different surface textures can be achieved. It is also much less likely to crack than smooth-trowel finishes, and can also be mixed with a color before its application. plication.

■ *Prepared gypsum trowel finishes; prepared gypsum-sand float finishes:* Ready-mixed plasters that require only the addition of water. These finishes possess basically the same characteristics as other trowel and float finishes.

■ *Acoustical plasters:* Ready-mixed plasters that are used to absorb sound. The primary advantages of this finish are noise-reduction qualities, fire resistance, noncombustibility, and ability to follow complex shapes.

Trowel finishes are achieved by applying the finish onto a dry or nearly dry basecoat. The dry basecoat then draws the moisture from the finish coat, resulting in a tight bond between the two layers of plaster. A second finish coat is applied once the initial bond is complete. The last finish coat is applied by moistening an area with a wet brush and troweling over the area to produce a smooth, dense surface. A floated finish is similar to a trowel finish, except that the first finish coat is floated to the basecoat with a wooden float, which is immediately followed by floating the area with a rubber float. (See Figure 12–27.)

Once the plaster has been placed, it should be allowed to dry properly. If it dries too fast before it sets, or if it dries too slowly, its strength may be impaired.

FIGURE 12–27.  Floated finish

## Veneer Plaster

Veneer plaster is another type that is gaining in popularity and acceptance. An apparent reason for its popularity is the speed with which it can be applied and its rapid setting time. Other advantages are durability, variety, versatility, and economy.

Veneer plaster is never mixed with anything except potable water and may be applied by hand (see Figure 12–28A) or machine (see Figure 12–28B). When done

FIGURE 12–28.   Veneer plaster applied by hand (below) and by machine (bottom)

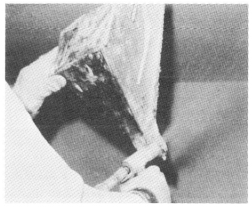

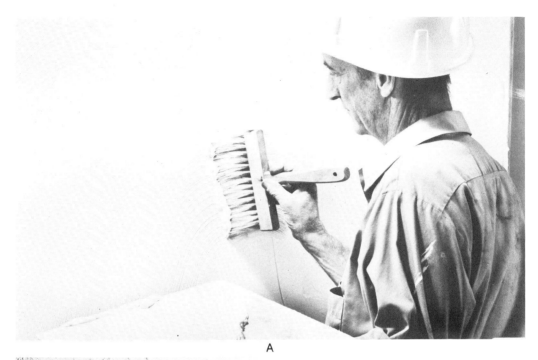

A

B

FIGURE 12–29    (above and left) Textured finish

by hand, a thin coat is applied to an approximate depth of $\frac{1}{16}$ inch. When the plaster is firm, a second coat is doubled back over the first coat, bringing the total thickness of the two coats to a minimum of $\frac{3}{32}$ inch. Before the plaster sets, a trowel is used to give the surface a smooth finish. If a textured finish is desired, deep swirls can be shaped with sponges, wood floats, or brushes. (See Figure 12–29A, B.)

If the plaster is applied with a machine, it should be sprayed in two even passes and troweled. Finally, a hard finish coat should be applied. If a textured finish is desired, the troweling operation is omitted and the final coat is sprayed on.

In some cases, electric radiant heating cables are embedded in veneer plaster. To properly install the system, the veneer base

FIGURE 12–30.   Placement of radiant-heating cables

A

B

FIGURE 12–31.   Application of plaster over heating cables

must be attached to the ceiling and the joints reinforced with glass fiber or paper tape. The heating cables are then attached to the ceiling. (See Figure 12–30.) The cables should be installed flat. In areas that require a splice, a notch should be cut in the veneer base to recess the splices. Once the cables are secured, a full coat of plaster is applied to cover the cables. (See Figure 12–31A, B.) If the plaster is applied in one coat, sand is sometimes added to give it density and increase the bonding properties. After application, the plaster should be leveled with a darby or a feather edge; but the heating cables should not be used for grounds. Once the fill coat has set, a finish coat can be troweled, floated, or textured as desired.

## CERAMIC TILE

Ceramic tile is very popular as an interior wall finish in kitchen and bath areas. It is available in a variety of colors, shapes, and textures, which can complement or enhance any decor. The glazed wall-tile units, made from potter's clay, are available in individual units; or the tile can be purchased in pregrouted sheets of $4\frac{1}{4} \times 4\frac{1}{4}$ inches,

6 × 4¼ inches, or 8½ × 4¼ inches with thin-set or conventional trim already attached. Because the tile has an impervious glaze, it is dent-proof, stainproof, and has less than ½ percent absorption.

The tile units can be bonded to the wall with conventional portland cement mortar (thick-bed) or thin-bed bonding materials (dry-set mortars, latex–portland cement mortar, epoxy mortar, epoxy adhesive, furan mortar, and organic adhesives).

## Conventional Portland Cement Mortar

Conventional portland cement mortar is a mixture of portland cement and hydrated lime and water in proportions of 1:5:½ to 1:7:1. The mortar is reinforced with metal lath or mesh, and can be placed over open framing members or over a solid backing. Before the lath is placed over wood studs or furring strips, a layer of building felt or polyethylene should be placed over the studs. This layer keeps the framing members from coming in direct contact with any moisture. If the mortar is placed over open framing members, a scratch coat and a float coat of mortar are required; but if mortar and lath are placed over a solid backing, such as gypsumboard, one coat is required. When the mortar is placed over wood studs or furring, the mortar base should not exceed 1½ inches. With two coats, the scratch coat should be allowed to cure for a minimum of 24 hours and thoroughly dampened before the float coat is applied.

## Dry-Set Mortar or Latex Portland Cement Mortar

Dry-set mortar is a mixture of portland ce-

ment and sand additives which impart water retentivity; latex portland cement mortar is a mixture of portland cement, sand, and special latex additives. Both of these mortars are classified as thin-set, and can be used over any smooth, clean, and true surface. The mortar is used in one layer and may be troweled as thin as ³⁄₃₂ inch. But if the surface is not level or if the mortar bed exceeds ¼ inch in thickness, a leveling coat is necessary.

## Organic Adhesives (Mastics)

Organic adhesives are classified as either Type I or Type II. Type I adhesives are used in areas that require prolonged water resistance, while Type II adhesives are used in areas that receive intermittent wetting. Before the adhesive is troweled, the surface of the wall should be sealed with a sealer recommended by an adhesive manufacturer. To regulate the amount of adhesive applied, the adhesive is spread with a notched trowel. The depth of the ribbons of adhesives should not exceed ¹⁄₁₆ inch.

## Application

Before ceramic tile is installed, a level base must first be established. A chalkline or a line scribed adjacent to a level can be used, but most mechanics prefer to tack a board to the wall and use it as a base line. Some mechanics also prefer to use a vertical line as a guide. With a notched trowel, an emulsion-type adhesive or mortar is then troweled on the wall.

The individual tile units or pregrouted sheets of tile can then be placed, with the level base as a guide. As each tile or sheet is placed, it should be pressed firmly into the adhesive or mortar and then gently tapped

A

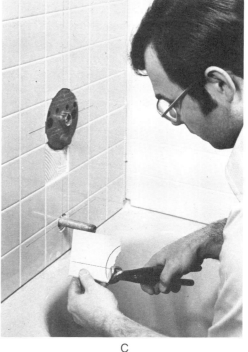

C

FIGURE 12–32A (upper left), B (lower left), and C (above) Application of ceramic tile

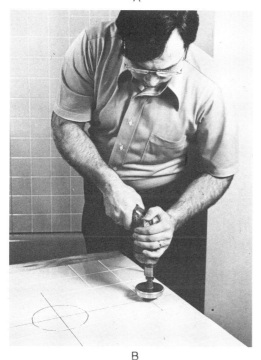

B

with a rubber mallet. If it is necessary to cut the tile, the glazed face can be scored and the tile unit snapped into two pieces. (See Figure 12–32A.) A power hole cutter can be used to cut openings for pipes (See Figure 12–32B), or nippers can be used to make the tile conform to the diameter of the pipe. (See Figure 12–32C.) Once the tile has been placed and allowed to set, it can be grouted.

## Grouts

There are many different types of grouting materials, each designed to meet the requirements of the different kinds of tile and exposures. Portland cement is the base for most grouts and is modified to provide whiteness, hardness, flexibility, and water retentivity.

Some of the different types of grout are:

■ *Sand and portland cement grout:* A mixture of 1 part portland cement to 1 part fine graded sand.

■ *Dry-wall grout:* A mixture of portland cement with sand and additives, providing water retentivity.

■ *Latex grout:* A mixture of portland cement and special latex additives.

■ *Epoxy grout:* Made from epoxy resin and a hardener to resist stain.

When the grout is placed, it should be troweled diagonally across the joints, to force grout into each joint. Then excess grout is removed from the surface of the tile with a burlap or rough-textured cloth. The tile surface should also be washed and sponged to remove any loose grout particles. If pregrouted sheets are used, grout should be applied between sheets with an air-gun or a standard hand-gun. After the grout is placed and the tile cleaned, any voids are filled with caulking compound.

# FLOORING

MANY DIFFERENT TYPES OF MATERIALS CAN be used as finish flooring, but some of the more common are wood flooring, resilient flooring, carpet, and ceramic tile. *Wood*

*Wood* flooring is not as popular as it once was, but many floors are still covered with hardwood and softwood strip flooring, random-width planks, unit blocks, and laminated blocks.

*Resilient* flooring has been popular for many years and is still a leading floor finish material. Some basic types of resilient floors are sheet vinyl, vinyl-asbestos tile, plain linoleum, and asphalt tile.

*Carpet* has increased in popularity in recent years, for carpets have proven to be extremely practical and economical.

*Ceramic tile* is also a popular floor finish material used in areas that are exposed to intermediate wetting.

## WOOD FLOORS

Wood floors are commonly constructed from hardwoods, such as oak, maple, beech, birch, and pecan. Oak, however, is the most popular species of hardwood flooring and constitutes over 90 percent of the hardwood flooring in the United States. The different types of hardwood flooring are broadly classified as strip, plank, and block.

*Strip flooring,* made from thin strips of wood of varying thicknesses and widths, is the most popular type of wood flooring. (See Table 13–1.) The preferred pattern size is $\frac{25}{32}$ inch thick and 2¼ inches in width. One edge of the strip flooring has a tongue and the other edge a groove; the ends are sometimes similarly matched.

*Plank flooring* is constructed from random-width pieces of flooring that are usually tongued-and-grooved and can have either square edges or beveled edges if an early hand-hewn plank effect is desired. The boards can also have simulated wood plugs glued into them for effect.

*Block flooring* is manufactured in two basic types: the unit block and the laminated block. The unit block is constructed from several short lengths of standard strip flooring joined together edgewise to form a square unit. (See Figure 13–1A.) The laminated block is constructed from three or more plies of veneer bonded together. (See Figure 13–1B.) To increase the strength of the unit, the grain directions of the different plies are placed at right angles to each other. Regardless of whether the block flooring is unit block or laminated block, most of the units are tongued on two adjoining or opposing edges and grooved on the other two.

### Storage and Delivery of Wood Flooring

Before wood flooring is laid, it must have time to equalize its moisture content with the installation area. The flooring should be delivered to the job site 48 hours before its installation, but it should not be unloaded if it is raining or snowing. In cold weather the building should be heated to

FIGURE 13–1. Unit and laminated blocks

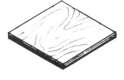

226

A                    B

**TABLE 13–1.   STANDARD SIZES, COUNTS, AND WEIGHTS OF STRIP FLOORING**

"Nominal" is the size designation used by the *trade*, but it is not always the actual size. Sometimes the actual thickness of hardwood flooring is $\frac{1}{32}$-inch less than the so-called nominal size. "Actual" is the *mill* size for thickness and face width, excluding tongue width. "Counted" size determines the board feet in a shipment. Pieces less than 1 inch in thickness are considered to be 1 inch.

| Nominal | Actual | Counted | Weights M Ft. |
|---|---|---|---|
| **OAK** | | | |
| **Tongue-and-Grooved, End Matched** | | | |
| $\frac{25}{32} \times 3\frac{1}{4}$ in | $\frac{3}{4} \times 3\frac{1}{4}$ in | $1 \times 4$   in | 2210 lbs |
| $\frac{25}{32} \times 2\frac{1}{4}$ in | $\frac{3}{4} \times 2\frac{1}{4}$ in | $1 \times 3$   in | 2020 lbs |
| $\frac{25}{32} \times 2$   in | $\frac{3}{4} \times 2$   in | $1 \times 2\frac{3}{4}$ in | 1920 lbs |
| $\frac{25}{32} \times 1\frac{1}{2}$ in | $\frac{3}{4} \times 1\frac{1}{2}$ in | $1 \times 2\frac{1}{4}$ in | 1820 lbs |
| $\frac{3}{8} \times 2$   in | $\frac{11}{32} \times 2$   in | $1 \times 2\frac{1}{2}$ in | 1000 lbs |
| $\frac{3}{8} \times 1\frac{1}{2}$ in | $\frac{11}{32} \times 1\frac{1}{2}$ in | $1 \times 2$   in | 1000 lbs |
| $\frac{1}{2} \times 2$   in | $\frac{15}{32} \times 2$   in | $1 \times 2\frac{1}{2}$ in | 1350 lbs |
| $\frac{1}{2} \times 1\frac{1}{2}$ in | $\frac{15}{32} \times 1\frac{1}{2}$ in | $1 \times 2$   in | 1300 lbs |
| **Square Edge** | | | |
| $\frac{5}{16} \times 2$   in | $\frac{5}{16} \times 2$   in | face count | 1200 lbs |
| $\frac{5}{16} \times 1\frac{1}{2}$ in | $\frac{5}{16} \times 1\frac{1}{2}$ in | face count | 1200 lbs |
| **BEECH, BIRCH, HARD MAPLE AND PECAN** | | | |
| **Tongued-and-Grooved, End Matched** | | | |
| $\frac{25}{32} \times 3\frac{1}{4}$ in | $\frac{3}{4} \times 3\frac{1}{4}$ in | $1 \times 4$   in | 2210 lbs |
| $\frac{25}{32} \times 2\frac{1}{4}$ in | $\frac{3}{4} \times 2\frac{1}{4}$ in | $1 \times 3$   in | 2020 lbs |
| $\frac{25}{32} \times 2$   in | $\frac{3}{4} \times 2$   in | $1 \times 2\frac{3}{4}$ in | 1920 lbs |
| $\frac{25}{32} \times 1\frac{1}{2}$ in | $\frac{3}{4} \times 1\frac{1}{2}$ in | $1 \times 2\frac{1}{4}$ in | 1820 lbs |
| $\frac{3}{8} \times 2$   in | $\frac{11}{32} \times 2$   in | $1 \times 2\frac{1}{2}$ in | 1000 lbs |
| $\frac{3}{8} \times 1\frac{1}{2}$ in | $\frac{11}{32} \times 1\frac{1}{2}$ in | $1 \times 2$   in | 1000 lbs |
| $\frac{1}{2} \times 2$   in | $\frac{15}{32} \times 2$   in | $1 \times 2\frac{1}{2}$ in | 1350 lbs |
| $\frac{1}{2} \times 1\frac{1}{2}$ in | $\frac{15}{32} \times 1\frac{1}{2}$ in | $1 \times 2$   in | 1300 lbs |
| **Special Thickness (T and G, End Matched)** | | | |
| $\frac{17}{16} \times 3\frac{1}{4}$ in | $\frac{33}{32} \times 3\frac{1}{4}$ in | $\frac{5}{4} \times 4$   in | 2400 lbs |
| $\frac{17}{16} \times 2\frac{1}{4}$ in | $\frac{33}{32} \times 2\frac{1}{4}$ in | $\frac{5}{4} \times 3$   in | 2250 lbs |
| $\frac{17}{16} \times 2$   in | $\frac{33}{32} \times 2$   in | $\frac{5}{4} \times 2\frac{3}{4}$ in | 2250 lbs |
| **Jointed Flooring—i.e., Square Edge** | | | |
| $\frac{25}{32} \times 2\frac{1}{2}$ in | $\frac{3}{4} \times 2\frac{1}{2}$ in | $1 \times 3\frac{1}{4}$ in | 2160 lbs |
| $\frac{25}{32} \times 3\frac{1}{4}$ in | $\frac{3}{4} \times 3\frac{1}{4}$ in | $1 \times 4$   in | 2300 lbs |
| $\frac{25}{32} \times 3\frac{1}{2}$ in | $\frac{3}{4} \times 3\frac{1}{2}$ in | $1 \times 4\frac{1}{4}$ in | 2400 lbs |
| $\frac{17}{16} \times 2\frac{1}{2}$ in | $\frac{33}{32} \times 2\frac{1}{2}$ in | $\frac{5}{4} \times 3\frac{1}{4}$ in | 2500 lbs |
| $\frac{17}{16} \times 3\frac{1}{2}$ in | $\frac{33}{32} \times 3\frac{1}{2}$ in | $\frac{5}{4} \times 4\frac{1}{4}$ in | 2600 lbs |

Reprinted with permission of National Oak Flooring Manufacturers Association.

70° F before the flooring is delivered, and after delivery a constant temperature must be maintained. If the atmosphere is damp, the flooring should be protected by a tarpaulin or polyethelene sheet, and stored in a well-ventilated area, not in a cold, damp building. It should not be placed on concrete slabs or on floors that are less than 18 inches above the ground. Crawl spaces should also be properly ventilated, and the ground underneath covered with a layer of polyethylene or rolled roofing.

## The Subfloor

The subfloor is used to stiffen the floor frame and provide a level base for the finish floor. The two most popular types of subfloor are boards and plywood.

If boards are used as a subfloor, they should not be less than 1 inch thick nor more than 6 inches wide. Boards wider than 6 inches are subject to dimensional changes due to absorption of moisture. Square-edged, shiplapped, or tongued-and-grooved boards can be used as a subfloor, but square-edged boards spaced ¼ inch apart are usually recommended. These should be nailed to each joist with two 8d nails. The ends of each board should rest on a joist, and also be secured to that joist with two 8d nails.

If plywood is used as a subfloor, it should be installed with the outer plies at right angles to the joist. The plywood is nailed to the joist with 8d or 10d nails, which are spaced 6 inches along the panel edges, and 10 and 6 inches on intermediate supports. A ¹/₁₆-inch space should be left between all panel end joints, and a ⅛-inch space should be left between all panel edge joints. Table 13–2 gives minimum plywood thicknesses and maximum joist spacing.

Green lumber should *never* be used as a subfloor, because its moisture would be transferred to the finish floor, causing it to cup or buckle and later develop cracks.

Once the subfloor has been laid, it is common practice to place a layer of 16-pound asphalt-saturated felt over it. The felt reduces air infiltration, prevents moisture from rising through the floor, and acts to dampen sound.

## Installation of Strip Flooring

Strip floors should be laid perpendicular to the joists; they usually look best when they are laid lengthwise to the longest dimension of a room. The first piece of strip flooring is laid parallel to the wall, with a ½-inch space left between the wall and flooring strip. This space allows for expansion of the flooring without buckling. The strip is also placed with the grooved edge nearest the wall.

The flooring is then face-nailed, with the nails placed so that they may later be covered by the baseboard and shoe molding. The nails are either 7d or 8d screw-type or cut nails, spaced 10 or 12 inches apart. Screw-type nails are usually preferred because of their greater holding power.

When the second course of strip flooring is placed, the butt joints should be well separated from the joints of the first course. To secure the second course, a nail should be driven where the tongue intersects the shoulder, at an approximate angle of 45°. The nail should not be completely driven down—if it is, the edge of the flooring may be damaged. A nail set should be used to finish driving the nail. (See Figure 13–2.)

Several different types of nailing machines are also available to nail strip flooring.

## Installation of Strip Flooring over a Concrete Slab

Hardwood strip floors can also be placed over concrete without harming the flooring.

## TABLE 13–2 PLYWOOD SUBFLOORING[1,2,3]

For direct application of T&G wood strip and block flooring and lightweight concrete or for a separate underlayment layer; plywood continuous over two or more spans, face grain across supports

| Panel Identification Index | Plywood Thickness (inch) | Maximum Span[4] (inches) | Nail Size & Type | Nail Spacing (inches) | |
|---|---|---|---|---|---|
| | | | | Panel Edges | Intermediate |
| 30/12 | $5/8$[5] | 12[6] | 8d common | 6 | 10 |
| 32/16 | $1/2, 5/8$ | 16[7] | 8d common[8] | 6 | 10 |
| 36/16 | $3/4$[5] | 16[7] | 8d common | 6 | 10 |
| 42/20 | $5/8, 3/4, 7/8$ | 20[9] | 8d common | 6 | 10 |
| 48/24 | $3/4, 7/8$ | 24 | 8d common | 6 | 10 |
| $1\frac{1}{8}''$ Groups 1 & 2 | $1\frac{1}{8}$ | 48 | 10d common | 6 | 6 |
| $1\frac{1}{4}''$ Groups 3 & 4 | $1\frac{1}{4}$ | 48 | 10d common | 6 | 6 |

[1] These values apply for C-D INT-APA, STRUCTURAL I and II C-D INT-APA, C-C EXT-APA and STRUCTURAL I and II C-C EXT-APA grades only.

[2] In some nonresidential buildings, special conditions may impose heavy concentrated loads and heavy traffic requiring subfloor construction in excess of these minimums.

[3] Edges shall be tongue-and-groove or supported with blocking for square-edge wood flooring, unless separate underlayment layer is installed, a minimum of $1\frac{1}{2}''$ of lightweight concrete applied over the plywood, or finish floor is $25/32''$ wood strip. Minimum thickness of this underlayment layer should be $1/4''$ for subfloors up to $48/24''$, and $3/8''$ for thicker panels on spans longer than 24″.

[4] Spans limited to values shown because of possible effect of concentrated loads. Allowable uniform loads vary, but at indicated maximum spans, floor panels carrying Identification Index numbers will support uniform live loads of more than 160 psf.

[5] Check dealer for availability in your area.

[6] May be 16″ if $25/32''$ wood strip flooring is installed at right angles to joists.

[7] May be 24″ if $25/32''$ wood strip flooring is installed at right angles to joists.

[8] 6d common nail permitted if plywood is $1/2''$.

[9] May be 24″ if $25/32''$ wood strip flooring is installed at right angles to joists, or minimum $1\frac{1}{2}''$ of lightweight concrete is applied over plywood.

Reprinted with permission of American Plywood Association.

The technique has been field-tested by the National Oak Flooring Manufacturer's Association, and has proved to be highly satisfactory. The technique involves the use of a moisture barrier sandwiched between 1- × 2-inch wood sleepers. (See Figure 13–3.)

FIGURE 13–2. Setting a nail

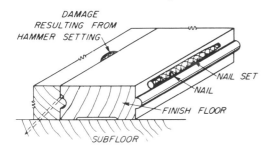

DAMAGE RESULTING FROM HAMMER SETTING
NAIL SET
NAIL
FINISH FLOOR
SUBFLOOR

These are the necessary steps for placing strip flooring over a concrete slab:

- The slab should first be cleaned and primed. Chalklines are then marked 16 inches on center and covered with 2-inch strips of mastic. (See Figure 13–4.) This adhesive should be the kind that bonds wood to concrete.

- Bottom sleepers are then placed over the mastic and secured to the slab with 1½-inch concrete. (See Figure 13–5.) The sleepers should be of 1- × 2-inch wood that has been treated with a wood preservative.

- Next, a layer of 0.004-inch polyethylene should be placed over the sleepers. (See

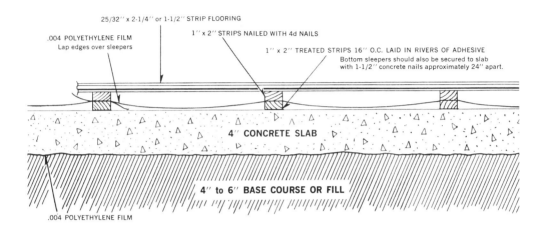

25/32'' x 2-1/4'' or 1-1/2'' STRIP FLOORING

.004 POLYETHYLENE FILM
Lap edges over sleepers

1'' x 2'' STRIPS NAILED WITH 4d NAILS

1'' x 2'' TREATED STRIPS 16'' O.C. LAID IN RIVERS OF ADHESIVE
Bottom sleepers should also be secured to slab
with 1-1/2'' concrete nails approximately 24'' apart.

4'' CONCRETE SLAB

4'' to 6'' BASE COURSE OR FILL

.004 POLYETHYLENE FILM

FIGURE 13-3.   Installation of strip flooring over a concrete slab

Figure 13–6.) The edges of the polyethylene sheets should lap over the edges of the sleepers.

■ Top nailing sleepers are then nailed to the bottom sleepers. (See Figure 13–7.) The two sleepers are nailed together

FIGURE 13-4.   Application of mastic along the chalked lines

FIGURE 13-5.   Placement of bottom sleepers

FIGURE 13–6.   The bottom sleepers are covered with polyethylene

FIGURE 13–7.   The top sleepers are nailed to the bottom sleepers

FIGURE 13–8.   Installation of strip flooring

with 4d nails spaced 16 to 24 inches apart.

■ Next, strip flooring can be installed at right angles to the sleepers. (See Figure 13–8.)

## Installation of Block Flooring

Block flooring can either be nailed to the subfloor, or the flooring may be laid in mastic. The units are nailed to the subfloor where the tongue intersects the shoulder. If the units are laid in mastic, a thin layer is first troweled over the subfloor. A layer of 30-pound asphalt-saturated felt is then placed over the mastic. After the felt is in place, a $\frac{3}{32}$-inch layer of mastic is troweled over the felt. The mastic can be applied either hot or cold, depending on the manufacturer's recommendations.

The individual flooring units can be placed in either a diagonal or a square pattern. For a diagonal pattern, a line should be chalked diagonally across the room and the units laid to the line. The installation procedure is facilitated by laying succeeding blocks in a pyramid fashion.

For a unit-block pattern, allowance must be made for expansion. Rubber expansion inserts are often used for this purpose. A ½-inch expansion gap should also be left between the edge units and walls.

The diagonal pattern can be used in any room, but because this pattern minimizes expansion under high-humidity conditions, it is usually recommended in corridors or in rooms where the length is more than one-and-a-half times the width.

If a square pattern is used, two chalklines are placed perpendicular to each other, and perpendicular to the walls. The center point of the chalklines usually represents the center of the room. The individual flooring units are then placed in a pyramid fashion, with the chalklines as a guide. If the units are placed in mastic, they should be tapped into position with rubber mallets.

## Finishing

After the flooring has been laid, it must be properly sanded. Most manufacturers recommend four sandings, starting with a No. 2 sandpaper and graduating down to No. ½, No. 0, and No. 00.

If a stain is used, it should be brushed on before the application of wood filler or other finishes. The stain is usually brushed on in a lengthwise direction and in strips about 36 inches wide. Any excess stain should be wiped off with a soft cloth.

Wood filler is used on hardwood floors to fill the pores of the wood. This is not necessary, however, if the pores are small, such as those in maple and beech. Filler serves two purposes: it fills the pores to make the surface of the wood smooth; and, if the filler is colored, it makes the pores show more prominently. The filler is applied with a brush, first across the grain and then lightly with it. As soon as the gloss appearance of the filler dulls, the filler should be wiped off with a burlap rag, first across the grain, then by lighter strokes with the grain. In most cases the filler should be allowed to dry for 24 hours before further finish is applied.

Once the surface of the wood flooring has been properly prepared, the finish should be applied. The four basic types of finish used on wood floors are floor seal, varnish, shellac, and urethane.

*Floor seal* is a popular and durable floor finish; instead of just forming a surface coating, it penetrates the wood fibers and wears only as the wood wears. It is manufactured in two types—normal-drying and rapid-drying. Normal-drying floor seal drys in 40 minutes to 1½ hours. For optimum results, most manufacturers recommend two coats for new floors.

*Varnish,* as a floor finish, provides a glossy appearance, is durable and fairly resistant to stains and spots; however, the finish shows scratches. In most cases three coats of varnish are used, but if the finish is placed over wood filler or a wash coat of shellac, two coats can be used.

*Shellac* is used to finish floors because of its ease of application and its quick-drying properties. The biggest disadvantage of shellac is its lack of resistance to moisture; it will spot if moisture is left on the finish for an extended length of time.

*Urethane,* increasingly popular as a floor

finish, is classified as either moisture-curing or oil-modified. A urethane finish is easily maintained, and spills and soils can be readily removed, but the finish in its natural state is extremely glossy. A low-gloss appearance can be achieved by using additives, but these may affect the performance of the finish.

## CARPET

There are several different types of carpet, each designed to meet specific needs and requirements. All carpet falls under three broad classifications, according to the principal method of manufacture: tufted, woven, and knitted.

*Tufting* is one of the newest construction techniques used in carpet manufacturing and accounts for about 90 percent of the total yardage. (See Figure 13–9A.) This method is used to produce shag, plush, pattern, and sculpture carpet. Thousands of needles are threaded with yarn, which is forced through the backing to form loops or tufts.

*Woven carpet* is manufactured on looms that interweave pile yarns and backing yarns in one machine operation. (See Figure 13–9B.)

*Knitted carpets* are very similar to woven carpets in that they are manufactured in one operation. (See Figure 13–9C.) In knitted carpets, usually in a solid color or a tweed, the pile and backing are looped together by means of several sets of needles.

Regardless of its classification and quality, if carpet is improperly installed it will not perform satisfactorily. Before installation, the floors should be cleaned and freed of any loose paint or varnish. If an adhesive is used to attach the carpet, any wax should be removed and a painted or finished floor lightly sanded. Any rough spots should be sanded smooth, and joints and cracks filled with a good-quality patching compound.

Carpet can be installed by means of a tackless strip, or by a direct glue-down installation procedure.

FIGURE 13–9. The three kinds of carpet

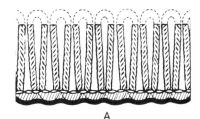

A

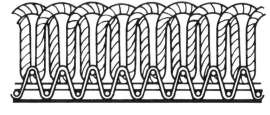

B

C

## Tackless Fastening

Tackless fastening is the most common installation technique used to install carpet. The technique involves the use of wood strips with inserted rows of pins angled toward the wall to hold the carpet in place.

There are three basic types of tackless strips and also three different pin lengths for the strips. One tackless strip is used for nailing or cementing on wood, concrete, tile, or terrazzo; another is a prenailed strip for concrete floors; and the third is a prenailed strip for wood floors.

The three pin lengths are classified as Types C, D, and E. (See Figure 13–10.) A Type C pin is ¼ inch long and used with high pile carpets. A Type D pin is used with low-profile carpets and/or single-back construction. A Type E pin is 7/32 inch in length, and used with double backing and medium or high-pile carpets.

The tackless strips are place a maximum of ¼ inch from the baseboard, but

FIGURE 13–11.   Hooking the carpet over the tackless strip with a knee kicker

FIGURE 13–10.   Tackless strips

**TYPE C-1/4" pin length**—use with high pile carpets with double backing.

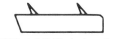

**TYPE D-3/16" pin length**— use with low profile carpets and/or single back construction.

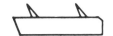

**TYPE E-7/32" pin length**— use with double backing and medium or high pile carpets.

this distance should equal the thickness of the carpet.

After the tackless strips are in place, the underlay cushion can be installed if a bonded-cushion carpet is not used. The cushion should be laid in the longest possible lengths and with a minimum of sections. When the cushion is placed over a concrete slab it should be cemented to the floor; and when placed over a wood subfloor it should be stapled to the floor. The attachment of the cushion to the subfloor keeps the cushion from shifting or buckling.

After the cushion has been installed, the carpet can be laid and allowed to condition at room temperature, which should not be less than 60° F. The carpet should be overcut approximately 1 foot in each direction, and allowed to project up each vertical wall about 6 inches. The following steps should then be executed:

■ Starting from one corner of a room, hook 18 inches of carpet over the tackless strip. (See Figure 13–11.) Using the same corner as a starting point, hook 18 inches of carpet over the tackless strip on the adjacent wall.

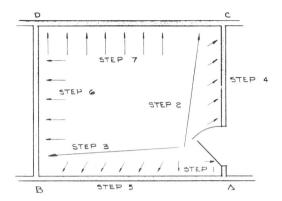

FIGURE 13–12.   Stretching the carpet along wall A–C

■ Using a power stretcher, stretch the carpet along wall A–C (see Figure 13–12), and hook the carpet on the tackless strip near corner C (see Figure 13–13). The carpet should be stretched enough to provide a firm, tight installation. Carpet is usually stretched 1 to 1½ inches for every 10 feet of carpet. The amount of stretch can be gauged by the extension of the carpet up the wall.

■ Next, stretch the carpet along wall AB and hook it to BD at corner B.

■ Then set wall AC in with a knee kicker, which should be placed at an approximate angle of 10° to 15°. A knee kicker is a tool used to stretch carpet. It has rows of protruding teeth on the head; the tail is padded and is struck with the knee. Wall AB is then set in the same manner as wall AC.

■ Stretching from wall AC, wall BD is now stretched at an approximate angle of 15° and is hooked over the tackless strip.

■ Then stretch and hook the remaining wall, DC.

■ When all the walls have been stretched and hooked, trim and tuck the edges. (See Figure 13–14.) The carpet should be trimmed so the edges can be tucked between the wall and the tackless strip. A thin blade, such as a hawkbill knife, is usually used to tuck the carpet. As it is tucked, the carpet should be firmly pressed onto the pins with the side of a hammer head.

If it necessary to join two pieces of carpet, either the heated-seam method or the sewn-seam method can be used. Tapes for *heat seams* have heat-activated thermoplastic adhesive placed on a cotton/fiberglass scrim (fabric). To keep the adhesive from spreading onto the pad, a 4-

FIGURE 13–13 (below)   Hooking the carpet over the tackless strip with a power stretcher, and 13–14 (bottom) Trimming and tucking the edges of the carpet

inch-wide paper back is placed on the back of the tape.

To join two pieces of carpet by the heat-seaming method, the tape must be properly located and the iron moved at a steady rate along the tape. (See Figure 13–15.) The iron should be moved in the direction of the pile lay, and at a rate of 18 to 24 inches per minute. In most cases, the iron should be set at a temperature of approximately 340° F at the base and 240 degrees F on the top of the shield. The temperature of the shield iron should not exceed 240° F because a higher temperature could damage the carpet. As the iron is passed over the seaming tape, only moderate pressure should be applied; too much can push the adhesive off the tape. After the iron has passed over a section of carpet, the carpet should be adjusted before the melt has had a chance to cool. Localized pressure should be avoided until this has occurred. If pressure is applied to the seam before it cools, the carpet may peak along the seam.

FIGURE 13–15.   Joining two pieces of carpet using heat seaming

FIGURE 13–16. Gripper bars

Seams can also be sewed with No. 18 waxed linen thread. There should be a minimum of three stitches per inch, located at least ⅝ inch from the cut edge. To reinforce the stitching, the raw edges at seams are latexed and seam tape is placed beneath the intersection of the two pieces of carpet.

Gripper bars are used to protect the carpet at doorways and island applications. (See Figure 13–16.) These are available in many different sizes and styles, but most are made of aluminum and are folded over to protect the carpet edges.

## Glue-Down Installation

Carpet laid by the glue-down installation procedure is applied without a cushion in areas that are to receive heavy, wheeled traffic.

Before installation, the floor must be properly cleaned, and any faults over ⅛ inch in size should be filled. If curing compounds have been used they should be removed, and if any oil or grease is on the concrete floor, the area should be cleaned with caustic soda or a commercial cleaner.

The concrete should also be checked for alkaline moisture (pH.) This can be measured by first wetting the concrete and then placing pHydrion test paper on the floor. The paper will change color, indicating the floor to be in an acid, neutral, or alkaline condition. If an excessive amount of alkaline moisture is present in the concrete, the surface should be washed with an acid solution, such as 5 percent muriatic acid, then rinsed with clear water and allowed to dry.

In addition, to properly prepare a concrete slab for a glue-down installation, the slab should first be sealed so there will be a better bond between carpet and floor, and the amount of adhesive needed for a job will be reduced. Sealers also help to equalize the porosity of the concrete and to prevent atmospheric conditions from affecting the installation procedure.

If carpet is to be glued down over a wood base, the floor should also be smooth, clean, dry, and free of any foreign elements.

A glue-down installation procedure consists of the following eight steps:

- To provide a straight starting point when more than one piece of carpet is used, chalk a line on the floor.

- Cut the carpet and place it along the chalkline. The carpet should be overcut so that 1 inch extends up the wall.

- Roll back the carpet edge adjacent to the wall to half of its width. Then trowel an adhesive onto the subfloor next to the wall.

- Roll the carpet back onto the adhesive, and remove air bubbles with a cardboard tube or push-broom.

- Cut a second piece of carpet and place it next to the chalkline. When the carpet is in position, trim the two unglued edges with a sharp cutting tool. (See Figure 13–17.) If loop pile is used, the second piece of carpet should be cut along the stitch row, then overlapped over the first piece of carpet. This piece should then be scribed and cut to fit the second piece of carpet. If cut pile carpet is used, the carpet should be cut from the back, using a straight edge and cushion-back cutter. The cushion-back cutter should be adjusted so that it cuts only through the back.

- Then roll the two pieces of carpet back from the intersection of the seams, and trowel adhesive onto the floor. A latex base adhesive is usually used for interior installation on all floors on, above, or below grade for direct glue-down installation of jute-, sponge-, foam-, polypropylene-, and neoprene-backed carpeting.

- Install one section of carpet and remove the bubbles and wrinkles. Next, apply a bead of seam adhesive to the base of the carpet where the two pieces of carpet will be joined.

FIGURE 13–17.   Glue-down carpet installation

FIGURE 13–18.   Fitting two carpet sections together

■ Install the intersecting pieces of carpet, making sure the seams fit snugly. (See Figure 13–18.) If necessary, use a knee kicker to nudge the two pieces of carpet together. (See Figure 13–19.) If there are any wrinkles in the carpet they can be removed with a cardboard tube.

FIGURE 13–19 (right)   Nudging for a good fit

## CERAMIC TILE

Ceramic tile is a popular floor finish especially in areas that receive occasional wettings. This tile is available in a variety of colors, shapes, and textures that can complement or enhance any decor. The floor units are glazed and are available in individual units, ungrouted sheets, or pregrouted sheets. Because the tile has a hard glaze, it is dentproof, stainproof, and has less than $\frac{1}{2}$ percent absorption.

The tile units can be bonded to the floor with either conventional portland cement mortar (thick-bed) or thin-bed bonding materials (dry-set mortars, latex portland cement mortar, epoxy mortar, epoxy adhesives, furan mortar, and organic adhesives).

### Conventional Portland Cement Mortar

Conventional portland cement mortar is a mixture of portland cement, sand, and water. The mortar is reinforced with metal lath or mesh backed by a vaporproof membrane which keeps the framing members from coming in direct contact with any moisture. When the mortar is placed it should be spread $1\frac{1}{4}$ inches thick.

### Dry-Set Mortar or Latex Portland Cement Mortar

Dry-set mortar is a mixture of portland cement with sand and additives that impart water retentivity, while latex portland cement mortar is a mixture of portland cement, sand, and special latex additives. Both of these mortars are classified as thinset, and can be used over any smooth, clean, and true surface. The mortar is used in one layer and is troweled as thin as $\frac{3}{32}$ inch. If the surface is not level, however, or the mortar bed does not exceed $\frac{1}{4}$ inch in thickness, a leveling coat is necessary.

### Organic Adhesives (Mastics)

Organic adhesives are classified as either Type I or Type II. Type I is used in areas that require prolonged water resistance; Type II, in areas that receive intermittent wetting. The adhesive is applied with a notched trowel to a depth of $\frac{1}{16}$ inch. Before this, however, the surface of the subfloor should be sealed with a material recommended by an adhesive manufacturer.

## Application

Before ceramic tile is installed, chalklines should be marked to divide a room into equal quarters. The intersection of the two lines indicates the starting point for the application of the tile; but to avoid joints in the subfloor and also to keep cutting to a minimum, the starting point may have to be adjusted. In some cases it may also be necessary to adjust the starting point so that the sheet can be cut along a grout line. Sheets of tile should be laid along the chalkline to check the location of this line. The last sheet can overlap the previously placed sheets to determine where the tile sheet will be cut (See Figure 13–20.)

After the lines have been chalked and adjusted if necessary, adhesive should be troweled over one-quarter of the floor area. It should *not* be troweled over areas where it is necessary to cut the tile, or where it would take longer than one hour to finish. The first sheet of tile is then placed at the intersection of the two chalked lines, and additional sheets are butted to each previously placed sheet of tile. (See Figure 13–21.) For proper placement, the lower edge of the sheet should be butted to another tile edge, and the body of the sheet then rolled into position. The sheets *should not be slid* into position, for this will cause the mastic to ooze up through the joints.

Before the edge sheets are placed, all the full sheets should be laid. Then the edge sheets can be cut along a grout line and the partial sheet placed along the wall. (See Figure 13–22.) If it is necessary to cut the tile, a full sheet should be laid on top of the last full sheet laid; another sheet is then laid on top of the two sheets of tile, but it is butted to the wall. With the last sheet serving as a straightedge, a line can be scribed

FIGURE 13–20.   Measurement of the border tile

FIGURE 13–21.   Placement of ceramic tile

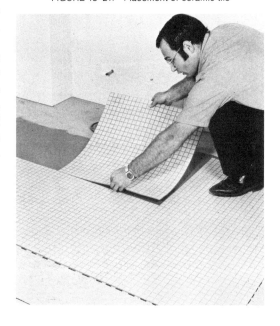

FIGURE 13–22. Placement of a partial sheet

FIGURE 13–23. Scribing ceramic tile

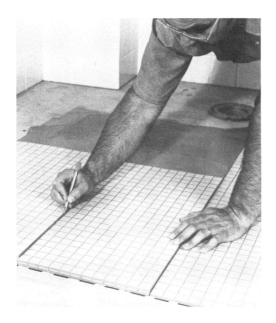

along the sandwiched sheet of tile. (See Figure 13–23.) The marked sheet can then be placed in a tile cutter and cut to the appropriate size. If it is necessary to make contour cuts, nippers are usually used to break and chip the ceramic tile to the correct size.

To insure the proper contact between ceramic tile and floor, and 150-pound carpet-covered roller is rolled over the floor, or a small block of wood is placed over the tile and a mallet is used to set the tile. If a pregrouted system was not used, the tile can be grouted after it has been placed and allowed to set.

## Grout

There many different types of grouting materials, each designed to meet the requirements of the different kinds of tile and exposures. Portland cement is usually the base for most grouts and is modified to provide whiteness, hardness, flexibility, and water retentivity. Some of the different types of grout are:

- *Sand-portland cement grout:* A mixture of 1 part portland cement to 1 part fine graded sand.
- *Dry-wall grout:* A mixture of portland cement with sand and additives that provide water retentivity
- *Latex grout:* A mixture of portland cement and special latex additives.

When the grout is placed, it should be troweled diagonally across the joints, forcing grout into each joint. Then excess grout is removed from the surface of the tile with a burlap or rough-textured cloth. The tile surface should also be washed and sponged to remove any loose particles of grout. If pregrouted sheets are used, grout is applied

between the sheets. Last, any voids are filled with caulking compound. (See Figure 13–24.)

FIGURE 13–24.   Placement of the caulking compound

## RESILIENT FLOORING

There are several different types of resilient flooring, each designed to meet a specific need. In most cases resilient flooring is used in areas that require a dense and nonabsorbent surface. Some of the more common types of resilient flooring are sheet vinyl, vinyl-asbestos tile, linoleum, and asphalt tile.

*Sheet Vinyl* is available in varying thicknesses and widths. Because it is one of the most popular flooring materials, vinyl is manufactured in large quantities and a variety of colors. It has good resistance to wear, grease, and alkalis. Some sheet-vinyl flooring has a foamed vinyl backing, and is used to reduce the noise level and increase comfort.

*Vinyl-asbestos tile* is, as the name implies, available in tile form and is made from a mixture of vinyl resins and asbestos fillers. This particular type of resilient flooring can be used over suspended subfloors, on-grade slabs, and below-grade concrete.

*Linoleum* is available either plain (sometimes called "battleship linoleum") or embossed. Plain linoleum is composed of oxidized linseed oil, cork, wood flour, color-stable mineral pigments, plasticizers, and stabilizers. It is primarily used when a solid-color floor of exceptional wearing quality is desired. Embossed linoleum is available in a wide variety of colors and designs. It is well suited to suspended floors, but should not be placed over concrete subfloors or on or below grade.

Another type of resilient tile flooring is *asphalt tile*, a low-cost material that is a mixture of asphaltic and/or resinous binder, asbestos fibers, pigments, and fillers.

## Subfloors

If resilient flooring is placed over a concrete subfloor that is on or below grade, the concrete should be separated from direct contact with the soil by a suitable vapor barrier. If one is not used, the concrete will absorb and retain moisture that will destroy the bond between the adhesive and subfloor. Other precautionary techniques that should be followed when a resilient floor is placed over concrete are to allow the concrete to dry for several months in a well-ventilated area. If concrete curing and parting compounds were used, they should be removed by grinding (using a concrete or terrazzo grinder), or by sanding and scarifying (using a power-driven wire brush). The compound must be removed because it keeps the adhesive from making a good bond with the slab.

After the curing compound has been removed, a bond test should be conducted to see if an adhesive will bond to the concrete. Test panels are secured to the subfloor, with the same adhesives that will be used in the actual installation. If they hold for two weeks, it can be concluded that the concrete is dry and sufficiently clean for satisfactory installation.

Resilient flooring can also be placed over a combination subfloor-underlayment, or a double-layer construction composed of separate subfloor and underlayment. If a combination subfloor-underlayment is used, plywood laid perpendicular to the joists is usually recommended. If square-edge plywood is used, the longitudinal edges should be supported by lumber blocking. But if $5/8$-inch tongued-and-grooved plywood is used, the blocking may be omitted. The thickness, span, and installation of plywood combination subfloor-underlayment should conform to the finish flooring manufacturer's instructions, but not be less than the minimum thickness and maximum joist spacing shown in Table 13–3. When the subfloor panels are installed, a $1/32$-inch space should be left for all joints, and a $1/8''$ space should be left between the subfloor and all vertical surfaces. Once the subfloor-underlayment has been placed, the floor area should be sanded smooth.

If an underlayment is used in conjunction with a subfloor, it is usually either a

**TABLE 13–3    PLYWOOD COMBINATION
SUBFLOOR-UNDERLAYMENT[1]**

Plywood continuous over two or more spans.

| Species Groups | Maximum Spacing of Joists (inches)[2] | | |
|---|---|---|---|
| | 16 | 20 | 24 |
| 1 | $\frac{1}{2}''$ | $\frac{5}{8}''$ | $\frac{3}{4}''$ |
| 2,3 | $\frac{5}{8}''$ | $\frac{3}{4}''$ | $\frac{7}{8}''$ |
| 4 | $\frac{3}{4}''$ | $\frac{7}{8}''$ | $1''$ |

[1] Applicable to C-C plugged, underlayment with intermediate or exterior glues, and underlaminate grades of plywood. For T&G laminated wood-block flooring, laid with $\frac{1}{2}$ bond pattern, sheathing grades of plywood may be used.

Leave $\frac{1}{16}$ inch spacing between joints. Set nails $\frac{1}{16}$ inch and lightly sand subfloor at joints if resilient flooring is to be applied.

[2] Plywood with approved T&G edges ($\frac{1}{2}''$—5 ply or thicker) shall be used or solid blocking shall be installed under all unsupported edges.

mastic-type or board-type underlayment. The most effective *mastic-type underlayment* contains a binder of latex, asphalt, or polyvinyl-acetate resins in the mix. A mastic that contains cement, gypsum, and sand can be used, but it often breaks down under traffic. Mastic-type underlayment can be troweled to a feather edge in leveling worn or damaged areas.

The three basic types of *board-type underlayments* are hardboard, plywood, and particleboard. In most cases hardboard is used in remodeling projects, and plywood or particleboard is used in new construction. Hardboard, minimizes the building up of old subfloors; but in new construction, it may be necessary to increase the thickness of the subfloor.

If *Particleboard* is used as an underlayment, it should be installed immediately before the resilient flooring is laid. The underlayment should be placed $\frac{3}{8}$ inch from the walls, and the panels should be placed $\frac{1}{4}$ inch apart at all edges and endjoints. The panels should also be placed so they overlap plywood panel joists by 2 inches.

To secure the underlayment to the subfloor, ring-grooved underlayment nails, staples, or a glue-nailing method can be used. If nails are used, they should first be placed in the center of the panel, then worked toward the edges. Along the panel edge the nails should be placed no closer than $\frac{1}{2}$ inch and no further than $\frac{3}{4}$ inch from the edge. Panels thinner than $\frac{3}{8}$ inch require 4d nails spaced 3 inches on center around the panel edges, and 6 inches on center each way throughout the field of the panel. If the panels are $\frac{3}{8}$ inch thick or thicker, 6d nails are spaced 6 inches on center around the panel edges, and 10 inches on center each way throughout the field of the panel.

When staples are used, they are placed $\frac{1}{2}$ inch from the panel edge and on 3-inch centers around the edges. They are also placed 6 inches on center each way throughout the field of the panel. Most of the staples used to secure $\frac{1}{4}$-inch underlayment are at least $\frac{7}{8}$ inch long, 18 gauge, and have a $\frac{3}{16}$-inch crown. For $\frac{3}{8}$-inch underlayment, the minimum length of the staples is $1\frac{1}{8}$ inches, 16 gauge, with a $\frac{3}{8}$-inch crown. Once the staples have been placed, they should not be countersunk more than $\frac{1}{16}$ inch.

One of the best particleboard floor systems incorporates the use of nails and glue as a fastening agency. The glue is a hard-setting casein material that is applied in strips. A 12-inch-wide strip is placed along each panel end; a 3-inch wide strip is placed along each side edge; and a 6-inch-wide strip is placed down the center of each panel. The glue can be spread with a roller, notched trowel, or brush.

When the underlayment is positioned in the glue, nails are placed on 16-inch centers around the panel edges, and 16 inches on center throughout the field. After the panels have been set in place, all gouges, gaps, and chipped edges should be filled and sanded flush with a belt sander.

Plywood, used as an underlayment and placed over subflooring, should also be laid just before installation of the finish flooring. The panel and joints should be staggered with respect to each other, and also staggered with the joints in the subfloor. When the panels are placed, there should be a $\frac{1}{32}$-inch space between all ends and edges. The panels are secured with 3d ring-shank nails for thicknesses of $\frac{1}{2}$ inch or less; 4d for $\frac{5}{8}$ inch and $\frac{3}{4}$ inch; or the panels can be secured with 16-gauge staples at 3 inches on center along panel edges and 6 inches on center throughout the field. If nails are used to secure the underlayment, they should be placed on 6-inch centers along panel edges and 8 inches on center throughout the field. Before the flooring is placed, the nails should be set $\frac{1}{16}$ inch; if staples are used, they should be set $\frac{1}{32}$ inch. (A nail is set when its head is below the surface of the wood.)

## Adhesives

Selection of the correct adhesive is just as important as selection of the right type of resilient flooring. The life and serviceability of the floor depends largely on the type of adhesive used to bond the flooring to the subfloor. It must be strong enough to prevent the flooring from separating from the subfloor, but it must not be so strong as to prevent the flooring from being removed when its useful life is over. The six basic types of adhesives used in the application of resilient flooring are linoleum paste, asphalt, asphalt-rubber, waterproof resin, latex, and epoxy.

*Linoleum paste* can be used to bond all sheet flooring and vinyl tile with a backing to a suspended wood or concrete subfloor. But it should not be used either below grade or over a suspended subfloor that is not properly ventilated; or for solid vinyl, asphalt, and vinyl-asbestos tile.

*Asphalt* is available in two basic types—cutback asphalt and asphalt emulsion. Both types of adhesives are used to bond asphalt tile and vinyl-asbestos tile over all subfloors. *Asphalt rubber* is also used in conjunction with asphalt tile and vinyl-asbestos tile over all subfloors. Once the asphalt adhesive has been troweled to the subfloor, it should be allowed to set for 30 to 60 minutes.

*Waterproof resin* adhesives can be used with linoleum and vinyl coverings that are placed over a suspended subfloor. But they should not be used on or below grade, or for asphalt or vinyl-asbestos tile.

*Latex* and *epoxy* adhesives can be used with resilient vinyl flooring and can be placed over all subfloors. When latex is used, the flooring should be installed immediately, but epoxy adhesive remains effective for as long as $2\frac{1}{2}$ hours without being covered. The disadvantages of an epoxy adhesive are that it must be mixed on the job and it is difficult to spread.

## Installation of Resilient Floor Tile

Once the underlayment has been properly placed and cleaned, the center of the room must be located. To do this, two perpendicular lines should be chalked, each line located at the center of the wall it intersects.

After the lines have been chalked, a trial layout of the tiles is placed along the center lines. If the distance between the last tile and the wall is 2 inches or less, or more than 8 inches, the center line should be moved 4½ inches closer to the wall. This makes it unnecessary to use small strips of border tile.

Before the adhesive is spread, it is important to check the floor to make sure it is free of foreign materials and is thoroughly dry. The temperature of the room should not be less than 70° F for 24 hours before, during, and after application of the resilient tile.

A notched trowel is used to spread the adhesive over a quarter of the floor area. The notches in the trowel regulate the amount of adhesive that will be placed on the floor. If too much adhesive is used, it will creep through the tile joints; but if too little is used the tile will not bond properly. After the adhesive has been troweled to the underlayment, it should be allowed to set before the tiles are placed. The necessary time lapse varies, but when the adhesive feels tacky but doesn't stick to the fingers, it has set.

Following this, the first piece of tile can be placed next to the two intersecting lines. Tile should never be slid into place—this causes the adhesive to ooze up between the cracks.

When the main floor area has been covered, the border tiles can be cut and positioned. To assure a good bond between underlayment and floor, the finished floor should be rolled in both directions.

Vinyl cove base is increasing in popularity because it adds a neat appearance to the resilient flooring and covers the intersection between flooring and wall. The base is available in 2½-, 4-, and 6-inch heights. It

A

FIGURE 13-25. Steps in installing a vinyl cove base for an outside corner: A (above), heat-treating the base; B (below), chilling the base

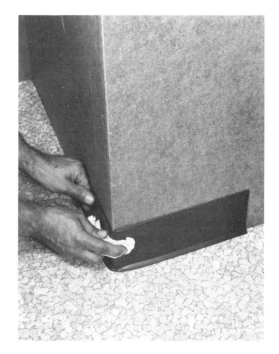

is attached to the wall with conventional adhesives. When an outside corner must be turned, the base is heated so it will bend easily. (See Figure 13–25A.) The toe of the base is then raised and the base is formed to the outside corner. When it is placed in position against the wall, a cold, damp cloth is used to chill the base and make it conform to the outside corner. (See Figure 13–25B.) The adhesive is then applied, and the base is placed against the wall.

## Installation of Resilient Sheet Flooring.

Sheet flooring can be installed with or without seams; when seams are used, they should be placed in inconspicuous areas, and out of the path of heavy traffic. When a "loose-lay" technique is used, the seams are omitted as is adhesive. The flooring is cut to fit around the room's perimeter and fixtures, then is dropped into place.

Sheet flooring is usually cut by one of three methods: knifing, scribing, or seam cutting. In the *knifing* method, the flooring is overcut and placed into position, and then the excess material gradually cut away. Knifing is frequently used, but results in some waste.

*Scribing* is often used to transfer the shape of a vertical wall to the sheet flooring. (See Figure 13–26.) With this method, the flooring is placed as close to the wall as possible and a pair of dividers is positioned to transfer the design. One leg of the divider should be placed at the base of the vertical wall and the other leg on the sheet floor. Then, when the dividers are drawn along the base of the wall, its shape is transferred to the flooring.

*Seam cutting* can be done either manually or by a machine, but both techniques

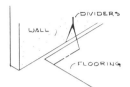

FIGURE 13–26. Scribing resilient sheet flooring with dividers

utilize the same principle. Before the seam can be cut, the flooring must be lapped; then a seam can be cut by slicing through the overlapped edges of adjoining sections. Next, the overlying and underlying scraps can be removed, and the flooring is positioned into the adhesive, which has already been spread. The seam is then rolled so the material will be well-seated in the adhesive.

Once the flooring has been properly cut, the adhesive is applied with a notched trowel, spreader, or brush. To facilitate the spreading of an adhesive, the sheet flooring is folded back, by either a tubing or lapping technique. (See Figure 13–27.) For the tubing technique, the flooring is folded back

FIGURE 13–27. Tubing and lapping techniques

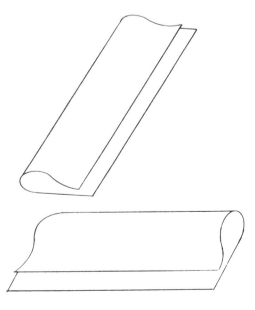

along its length; for the lapping technique, the flooring is folded back along its width. Then the adhesive can be placed over half the floor area, and the folded portion of the flooring placed in the adhesive. The other half of the flooring is then folded back, and the operation is repeated. When the second half of the flooring has been secured, the seams should be hand-rolled with a heavy roller.

# CHAPTER 14

# ACOUSTICAL

# CEILING SYSTEMS

ACOUSTICAL CEILING SYSTEMS ARE USED TO decorate and complement a room's decor. In addition, they also reduce the noise level, provide protection against fire, cover exposed pipes and ducts, and in some cases are used to lower a ceiling that is too high.

The two basic types of acoustical ceiling systems are the suspended ceiling and the tile system. The *suspended* ceiling consists of a simple metal grid framework suspended on wires from above. The ceiling panels are dropped into and supported by this framework. The ceiling panels are usually 2 × 2 feet or 2 × 4 feet and made from cellulose or mineral fibers.

Individual *tiles* are usually 12 × 12 inches and are also manufactured from cellulose or mineral fibers. The tiles are designed with tongue-and-groove edges that allow them to fit together snugly. In most cases they are installed either by being stapled to furring strips or by being cemented directly to a level, sound ceiling.

Both ceiling panels and ceiling tiles have a wide range of attractive designs, including embossed white, two-tone effects, and smooth-surfaced decorator styles; there is also a design featuring small perforations or fissures to enhance the noise-dampening properties.

## INSTALLATION CONDITIONS

Before installation, the individual tiles and panels should be delivered to the job site so they can adjust to room temperature and achieve a stabilized moisture content. (The ceiling should be installed under the same temperature conditions it will have when the job is completed.)

If the ceiling tile is to be secured with adhesive, this should not be applied if the room temperature is above 100° F or below 50° F. The surface to which the tile will be cemented must be dry, and free of oil, residue, and foreign matter. If any moisture is present, the tile will not properly adhere to the surface. A simple test for moisture is to place ¼ teaspoon of calcium chloride crystals in a small jar and tape the jar tightly to the ceiling for 8 hours. If there is any moisture present, the crystals will stick together or drops of water will form on the inside of the jar.

A painted ceiling should also be checked before acoustical tiles are cemented in place, because some painted surfaces do not bond properly with tile adhesives.

## ACOUSTICAL TILE CEILINGS

In most cases, acoustical tile ceilings are stapled to furring strips placed 12 inches on center. The first furring strip is placed adjacent to the wall; the second furring strip is equal to the width of the border; and the third and remaining strips are placed on 12-inch centers.

To determine the width of the border tile or the location of the second furring strip, the width of the room should be measured in feet and inches. The dimension of feet is then dropped and 12 inches is added to the remaining inch figure. The width of the border tile is then determined by dividing the total number of inches by 2. For example, if a room is 10 feet 3 inches wide, add 12 inches to the 3 inches for a total of 15 inches. Then divide the 15 inches by 2, for a border tile width of 7½ inches.

The furring strips should be placed perpendicular to the joist and secured to it with two 6d nails. To insure a level surface, the furring strips should be periodically checked with a straight edge, and if necessary leveled with shims. (See Figure 14–1.)

At this point a chalkline should be placed down the center of the second furring strip. The border tiles are then cut ¼ inch less than the distance from the chalkline to the vertical wall. When the first row of border tiles is cut, the tongue edge should be cut off. This cut should be made with the face of the tile up.

The installation of the individual tile units is started in the corner of the room. The first tile is placed in position, with the chalkline as a guide. The two flanged edges should always point to the area that will be covered by an adjacent tile. The flange is

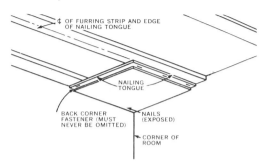

FIGURE 14–2. Fastening the individual tiles

then stapled to the furring strip and the edges adjacent to the wall are face-nailed. (See Figure 14–2.) Once the first tile has been set, additional border tiles can be placed on each side. A full tile is then placed between and adjacent to the border tiles. The procedure of installing two border tiles and filling in with full-size tiles is repeated until the ceiling is completed. The edges can be aligned by placing a 2 × 4 block on the flange edge of the tile, and tapping it with a hammer.

If a broken-joint effect is desired, alternate rows are started with a half-tile. This technique is relatively simple, since the tiles need to be aligned in only one direction. The technique should not be used with a design that carries through from one tile to the next.

If the ceiling tiles are cemented to the existing ceiling, it should first be inspected for dust, soot, dirt, peeling paint, and moisture content. If the ceiling is dirty, it should be washed. If it is concrete, it should be primed with either a commercial size or a field site mixture. One mixture consists of water and glue, the other is a combination of 50 percent shellac and 50 percent alcohol. The primers should be placed in thin layers and allowed to dry before the tiles are cemented in place.

FIGURE 14–1. Shimming furring strips

When the ceiling has been properly prepared, a chalkline should be marked parallel to one wall, to be used as a guide in the
placement of the border tiles. The size of
the border tile and the location of the
chalkline are calculated in the same manner as for ceiling tiles placed over furring
strips.

Next, the adhesive should be placed on
the individual tile units, in daubs about the
size of a walnut. These daubs should be
placed approximately 3 inches in from each
corner. When the tile is pressed into position, the daubs of adhesive should be about
$\frac{1}{8}$ inch thick and spread to an approximate
diameter of $2\frac{3}{4}$ inches. Moderate pressure
should be applied directly over the adhesive; then the tile should be moved back
and forth to provide a good bond. When
the tile is positioned, all four corners
should be level, or the tile will give a droop-
ing appearance.

If it is necessary to reposition the tile,
the adhesive must be cleaned from both tile
and ceiling; then the tile should be rebut-
tered and repositioned on the ceiling.

## SUSPENDED ACOUSTICAL CEILINGS

A suspended acoustical ceiling consists of
four different components: wall angle
molding, main runners, cross-tees, and
2 × 4-foot panels.

Wall angle molding is an L-shaped strip
that is nailed around the perimeter of the
room. (See Figure 14–3A.) This molding is
available in 10-foot lengths, and is placed
at ceiling height.

Main runners are available in 12-foot
lengths, and provide the strength and level-
ness of a suspended ceiling. (See Figure
14–3B.) They are placed perpendicular to
the ceiling joists and are supported by 12-
gauge wire nailed to the joists at 4-foot intervals.

Cross-tees are available in 2- or 4-foot
lengths and are locked into the main runners to form the total grid. (See Figure
14–3C.)

The first step in installing a suspended
ceiling is to nail or screw the wall angle
molding to the wall. The molding should be
placed a minimum of 3 inches below the
joists, unless luminous panels and fluores-
cent lamps are to be mounted above the
grid. In this case the molding should be
placed at least 5 inches from the joists.
This dimension should be measured down
from the joists in each corner of the room.
Next, a line is chalked on the wall from
corner to corner. The molding can then be

FIGURE 14–3. The components that support the panels of a suspended acoustical ceiling: A, wall-angle
molding; B, main runners; C, cross tees

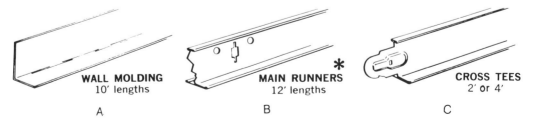

WALL MOLDING
10' lengths

A

MAIN RUNNERS
12' lengths

B

CROSS TEES
2' or 4'

C

FIGURE 14–4. Bending hanger wire

fastened to the wall, set on 2-foot centers. The inside corners of the wall molding should be overlapped, and the outside corners should be mitered.

After the wall angle molding has been installed, the main runners are suspended from the joists by 12-gauge wire. The main runner should be placed on 4-foot centers except at the border rows. The two border rows should be equal. For proper location of the main runners, chalklines should be marked across the joists. Next, hanger wires are nailed or screwed directly above this chalkline. These wires should be long enough to extend past the wall angle molding by 3 or 4 inches; this space will be used for tying back after the main runner has been set.

To make sure that all main runners are placed at the same height, a string can be stretcched above the bottom edge of the wall angle molding. The hanger wires can then be bent to the proper elevation, with the stretched string as a guide. (See Figure 14–4.) The main runners can now be positioned. One end is placed on the wall angle molding, while the other end is supported by a hanger wire. To assure the proper placement of the hanger wire in the main runner, wire holes should be located 3 inches on center The ends of each main runner have tabs for splicing when necessary. (See Figure 14–5.)

After the main runners have been positioned and secured, 4-foot cross-tees are placed between the main runners. These should be placed on 2-foot centers laterally from wall to wall, rather than in rows from front to rear of the room. Lateral installation locks the grid into position and prevents movement of main runners and cross-tees. If it is necessary to cut main runners or cross-tees, a tin snip or hacksaw can be used.

When the grid has been completed, the panels can be set into position; they should be slightly tilted and then carefully dropped into place. If a fire-rated ceiling is required or specified in the plans, hold-down clips should be positioned over the

FIGURE 14–5. Splicing main runners

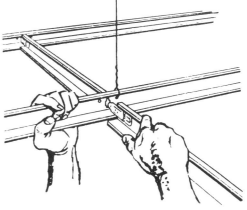

main runners and cross-tees. These hold the panels in position and keep them from fluttering. (See Figure 14–6.)

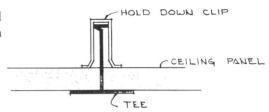

FIGURE 14–6 (right)   A hold-down clip, installed

## THE INTEGRAL FURRING CHANNEL INSTALLATION METHOD

For years the only method of installing ceiling tiles was to staple them to furring strips placed on 12-inch centers. But because of the tendency of furring strips to absorb moisture and thus change dimensional shape and size, they are often unsatisfactory. They are also tedious to place, and often quite expensive. To offset these problems a new system, it is referred to as the "Integral Furring Channel Method," was developed. Its components are molding, metal furring channels, cross-tees, and individual tile units.

For this method lines are first chalked around the perimeter of the room. The lines should be level and placed 2 inches below the existing ceiling. With these lines as a guide, molding is then nailed to the walls.

Chalklines are then placed on the ceiling to act as guides in placing the metal furring channels. The first line is placed 26 inches from the sidewall, and additional chalklines are placed 4 feet on center.

Next, the metal furring channels are nailed over the chalklines. (See Figure 14–7.) The nails should be either 6d or 8d common or 6d or 8d form nails, spaced on 48-inch centers. Because the furring channels are self-leveling, it is not necessary to shim them.

After all furring channels have been placed, cross-tees should be slightly bent

over the channels and clipped onto the furring strip. (See Figure 14–8.) When the cross-tee has been installed, it should be able to slide along the furring channel. Next, the ceiling tile is placed over the wall

FIGURE 14–7.   Placement of furring channels

FIGURE 14-8. Attachment of cross tees

FIGURE 14-9. Positioning ceiling tile

molding, and the cross-tee is slid into the proper position. (See Figure 14–9.) When it is properly positioned, the cross-tee will fit in a concealed slot on the leading edge of the tile. Additional rows of tile are then placed in the same manner.

This system can be used for both 12 × 48-inch and 12 × 12-inch ceiling tiles.

# CHAPTER 15

# DOORS AND WINDOWS

DOOR AND WINDOW UNITS ARE USUALLY millwork items that are assembled at a factory. Because they are manufactured under controlled working conditions, they are usually superior to those assembled at the job site. Moreover, since these units are preassembled, the need for highly skilled craftsmen at the job site is eliminated.

## DOORS

Several different types of doors are used in light construction, but the most popular types are wood flush, panel, folding, and sliding doors. Wood flush doors are popular as both interior and exterior doors, while panel doors are primarily used as exterior doors. Folding and sliding doors have gained wide acceptance in light construction and are often used as closet doors or room dividers.

### Wood Flush Doors

Wood flush doors are available in a wide variety of both solid and hollow-core construction; and in wood, hardboard, and plastic veneered faces. The core construction and veneered materials vary, but some of the more popular doors have glued-block cores, mat-formed composition cores, hollow ladder cores, and hollow mesh or cellular cores.

If a *glued-block core* is used, the blocks should all be of the same wood species and no wider than 2½ inches. (See Figure 15–1A.) The blocks are bonded together so that the end joints are staggered in adjacent rows. The core edges are trimmed with minimum ½-inch edge bands on the top, bottom, and side edges. The face veneer should be at least ½ inch thick, and is applied to both sides of the door.

*Mat-formed composition core* doors are constructed from particleboard or other dry-processed lignocellulose material. (See Figure 15–1B.) Surrounding the core are rails and stiles that are a minimum of 1⅛ inches thick. (If they are glued to the core, they only need to be ⅞ inch thick.)

FIGURE 15–1.  Types of wood flush doors: A, glue-block core; B, mat-formed composition-core door; C, hollow-ladder core door; D, hollow-mesh or cellular core door

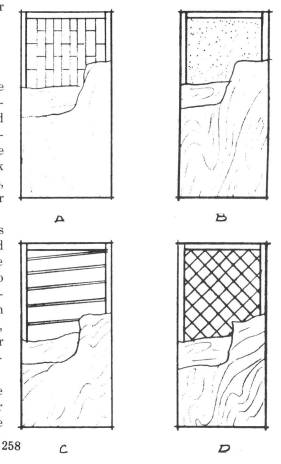

A          B

C          D

A *hollow-ladder core* is considered a hollow-core flush door, and is constructed by spacing wood or wood-derivative strips between two face panels. (See Figure 15–1C.) To increase the rigidity of the door, 2½-inch top and bottom rails, 1⅛-inch stiles, and 20-inch lock blocks are placed between the panels.

*Hollow mesh* or *cellular core* doors have uniform-spaced components made of wood or wood-derivative strips. (See Figure 15–1D.) The door has top and bottom rails 2½ inches wide, and stiles 1⅛ inches wide. Lock blocks are also required at the midlength point of the stile. The minimum length of each lock block is 20 inches, and the minimum combined width of the lock block and its adjacent stile is 4 inches.

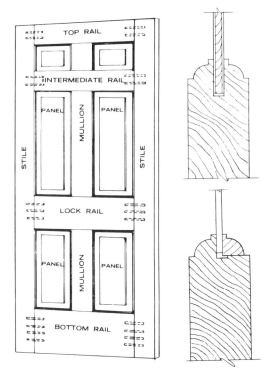

FIGURES 15–2 (above left) Six-panel door, 15–3 (top right) Ovolo sticking, and 15–4 (above right) Glass bead

## Panel Doors

A panel door is constructed of rails and stiles with panels of plywood, hardboard, or solid stock inserted between them. (See Figure 15–2.) The rails and stiles are usually assembled with dowels, ½ inch in diameter and approximately 5 inches long. The dowels should have glue grooves and/or indentations, and be sized for a drive fit.

Most rails and stiles of panel doors have ovolo sticking (convex molding). (See Figure 15-3.) If glass inserts are used, the glass bead should be ovolo to match the sticking. (See Figure 15-4.) Glazing is accomplished by running a bead of glazing or adhesive compound ⅛ inch in diameter around the perimeter of the glass rabbet. Wood glass beads are then placed to hold the glass in the proper position.

## Folding Doors

Folding doors operate on overhead tracks and rollers, and are popular units for closets and room dividers. (See Figure 15–5.) They are available in several different patterns, but the four most popular patterns are full-louvered, louvered with raised bottom panel, flush, and panel. The door units are also available in different widths and varying panel numbers. (See Table 15–1.)

Most folding doors are installed by initial placement of a track or tracks in a framed opening. (See Figure 15–6.) The top track is usually placed 1 inch back from the front opening, and is secured to the head jamb with 1½-inch screws. If a bottom track is not used, a bottom jamb bracket is fastened to each side jamb. (See Figure 15–7.) Once the tracks and jamb brackets

**Full Louvered**          **Louvered with raised bottom panel**

FIGURE 15–5.  Two types of folding door

**TABLE 15-1    WIDTHS AND PANEL NUMBERS OF FOLDING DOORS**

|  | Nominal Door Size | Actual Door Size | Net Clear Opening[1] |
|---|---|---|---|
| **Height:** | 6'8" | 6'6½" | 6'8¼" |
| **Width:** | | | |
| 2-panel[2] | 1'6" | 1'5½" | 1'6" |
|  | 2'0" | 1'11½" | 2'0" |
|  | 2'6" | 2'5½" | 2'6" |
|  | 3'0" | 2'11½" | 3'0" |
| 4-panel | 3'0" | 4'11" | 3'0" |
|  | 4'0" | 3'11" | 4'0" |
|  | 5'0" | 4'11" | 5'0" |
|  | 6'0" | 5'11" | 6'0" |
| 6-panel | 7'6" | 7'4½" | 7'5½" |
| 8-panel | 8'0" | 7'10" | 7'11½" |
|  | 10'0" | 9'10" | 9'11½" |
|  | 12'0" | 11'10" | 11'11½" |

[1] Specify right- or left-hand.
[2] Finished opening shall be net clear opening plus thickness of finished door and/or floor covering.
Reprinted with permission of H.C. Products Company.

Vertical adjustments are made by turning the hexhead bolt on the bottom pivot. The door knobs are located 34 inches from the bottom of the panel, and 1½ inches from the edge of the last panel.

## Sliding (Bypass) Doors

Gliding or bypassing doors are used extensively in closets; like folding doors, they operate on overhead tracks and rollers. A track is usually attached to the head jamb to install a sliding door. The track can be designed to cover the rollers (see Figure 15–8), or a piece of trim may be used to cover the track assembly. The standard by-pass track can be used with either ¾- or 1⅜-inch doors.

## Sliding Glass Doors

Sliding glass doors have found widespread

have been installed, the top and bottom assembly must be compressed, and then placed in the top and bottom pivot socket and assembly. The doors are adjusted horizontally by repositioning of the top pivot brackets and the bottom pivot sockets.

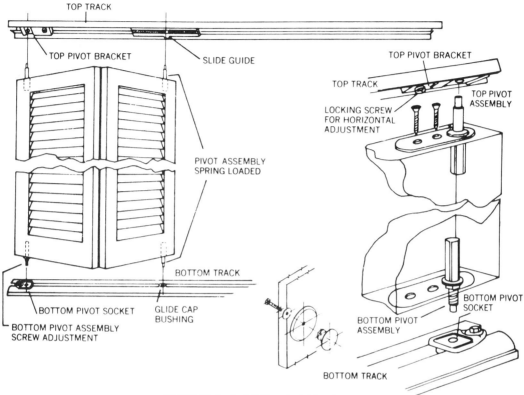

FIGURE 15-6. Installation of a folding door

acceptance for patios. The units are usually 6 feet 10 inches high and 6 feet wide, but these dimensions vary. Sliding glass unit frames are manufactured of aluminum, steel, or wood and are fitted with tempered glass.

## Garage Doors

Garage doors are usually constructed of wood, metal, of fiberglass and operate on an overhead track. (See Figure 15-9.) To allow overhead doors to be operated with a minimum amount of effort, they are usually counterbalanced to their weight. Most are equipped with full-floating, hardened steel, ball-bearing rollers that glide in vertical channels fastened to the jambs. Above each horizontal track, a strong helical-wound spring is mounted and attached to the door pulleys and lift cables.

FIGURE 15-7. Bottom jamb bracket

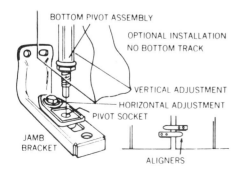

FIGURE 15–8. Cutaway views of a sliding-door installation: sliding door (above); sliding-door track, designed to cover the track assembly (right)

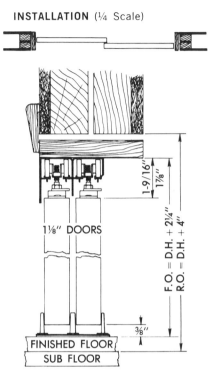

INSTALLATION (¼ Scale)

1-9/16″
1⅞″
F.O. = D.H. + 2¼″
R.O. = D.H. + 4″
1⅛″ DOORS
⅜″
FINISHED FLOOR
SUB FLOOR

## Door Sizes

The standard sizes of wood flush doors are shown in Table 15–2. The same sizes are considered standard for panel doors. The recommended *thickness* for exterior doors is 1¾ inches, while most interior doors are 1⅜ inch thick. The *width* of most exterior doors is 2 feet 8 inches or 3 feet, while the standard width of interior doors varies, depending on their usage. According to *Minimum Property Standards* the main entrance should have a doorway width of 3 feet; in other exterior entrances, a minimum doorway should be 2 feet 8 inches wide. In habitable rooms a minimum doorway width of 2 feet 6 inches is recommended; and all bathrooms should have a minimum doorway width of 2 feet. (*Minimum Property Standards* is published by the Department of Housing and Urban Development and is intended to provide a sound technical basis for mortgage insurance by giving minimum standards to assure well-planned, safe, and soundly constructed homes.) Most doors have standard *heights* of 6 feet 8 inches.

## Jamb Units

Door jambs are used to close the interior of rough openings, and also to receive the

**TABLE 15–2   STOCK SIZES OF WOOD FLUSH DOORS[1]**

| Door Sizes | |
|---|---|
| 1⅜″ Interior Flush Doors | 1¾″ Exterior Flush Doors |
| 1′-6″ × 6′-8″ | 2′-4″ × 6′-8″ |
| 2′-0″ × 6′-8″ | 2′-6″ × 6′-8″ |
| 2′-4″ × 6′-8″ | 2′-8″ × 6′-8″ |
| 2′-6″ × 6′-8″ | 3′-0″ × 6′-8″ |
| 2′-8″ × 6′-8″ | |
| 3′-0″ × 6′-8″ | |

[1] These same sizes are standard for panel doors.

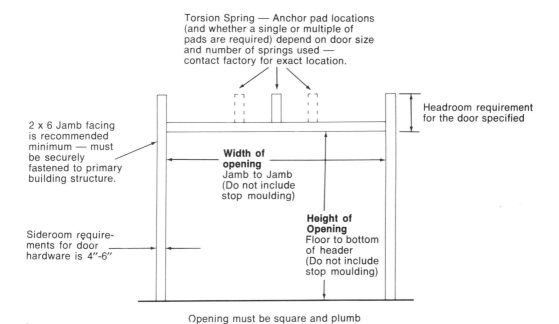

Torsion Spring — Anchor pad locations (and whether a single or multiple of pads are required) depend on door size and number of springs used — contact factory for exact location.

Headroom requirement for the door specified

2 x 6 Jamb facing is recommended minimum — must be securely fastened to primary building structure.

**Width of opening** Jamb to Jamb (Do not include stop moulding)

**Height of Opening** Floor to bottom of header (Do not include stop moulding)

Sideroom require- ments for door hardware is 4"-6"

Opening must be square and plumb

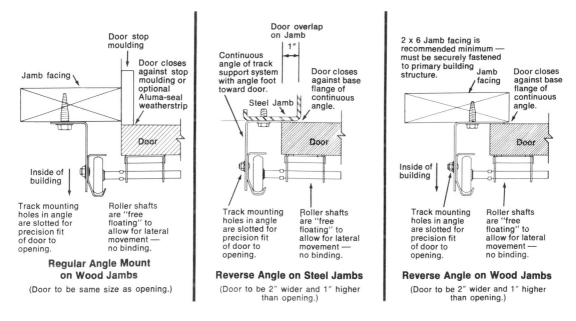

Door stop moulding

Door closes against stop moulding or optional Aluma-seal weatherstrip

Jamb facing

Inside of building

Door

Track mounting holes in angle are slotted for precision fit of door to opening.

Roller shafts are "free floating" to allow for lateral movement — no binding.

**Regular Angle Mount on Wood Jambs**

(Door to be same size as opening.)

Door overlap on Jamb

1"

Continuous angle of track support system with angle foot toward door.

Door closes against base flange of continuous angle.

Steel Jamb

Door

Track mounting holes in angle are slotted for precision fit of door to opening.

Roller shafts are "free floating" to allow for lateral movement — no binding.

**Reverse Angle on Steel Jambs**

(Door to be 2" wider and 1" higher than opening.)

2 x 6 Jamb facing is recommended minimum — must be securely fastened to primary building structure.

Jamb facing

Door closes against base flange of continuous angle.

Inside of building

Door

Track mounting holes in angle are slotted for precision fit of door to opening.

Roller shafts are "free floating" to allow for lateral movement — no binding.

**Reverse Angle on Wood Jambs**

(Door to be 2" wider and 1" higher than opening.)

FIGURE 15–9. Garage-door installations: jamb and framing details

hinges and strike plate for the door. Door jambs vary in design and do not lend themselves to rigid standardization, but two common types are the adjustable jamb unit (split jamb) and the nonadjustable jamb unit.

In an *adjustable jamb* unit, two jamb members are fitted together with an accurately machined rabbet along the two adjoining edges. (See Figure 15–10A.) At the intersection of the two jambs, a doorstop is fastened to the jamb section to hold the door. In some cases, the jamb members are fitted together with metal pins which extend into adjoining edges of the two jamb members. (See Figure 15–10B.) The pins are usually spaced on 12-inch centers and

should not be more than 8 inches from each end. The jamb sections are accurately bored for proper alignment of the faces of the two jambs. The pins should be approximately $\frac{3}{16}$ inch in diameter, with a drive fit into one jamb section and a movable snug fit into the adjacent jamb.

A *nonadjustable jamb* unit has a single jamb member the full width of the rough opening. (See Figure 15–11.) The casing may be fastened to one side of the jamb only, and furnished with the other side loose, or it can be furnished loose for both sides.

The side jamb and the head jamb are connected by a dado joint at least $\frac{3}{16}$ inch deep. The width of the dado should not be

FIGURE 15–10A.   The adjustable-jamp unit

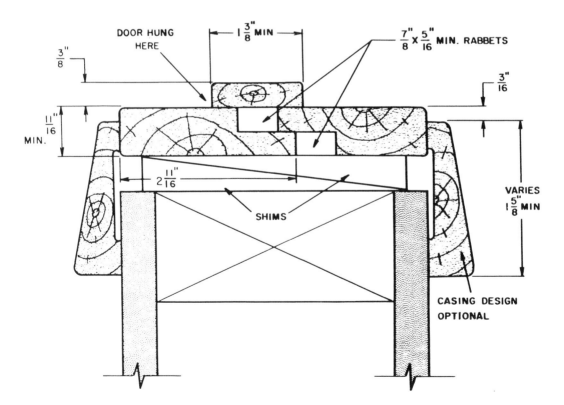

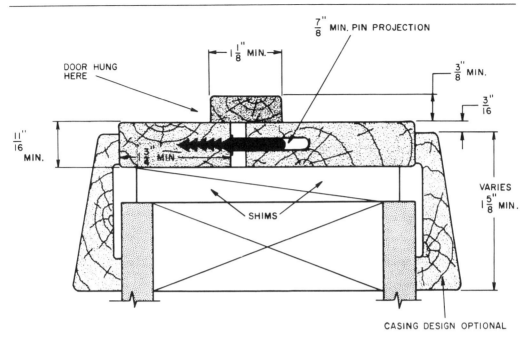

FIGURE 15–10B.  Split jamb with metal pins, another adjustable-jamb unit

FIGURE 15–11.  The nonadjustable-jamb unit

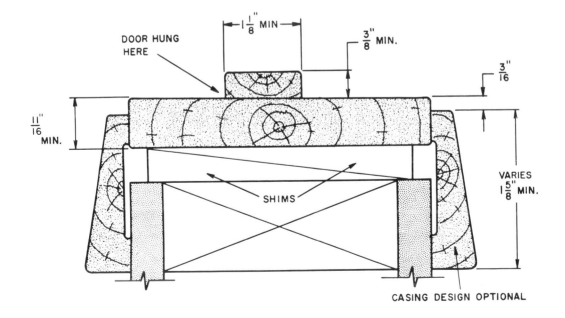

more than $\frac{1}{32}$ inch wider than the receiving member. The head and side jambs are fastened together with four 5d box nails placed in each end of the head jamb.

## Casing

To cover the joint between the jamb and the rough opening, casing must be placed around the head and side jambs. It should be a minimum of $\frac{1}{4}$ inch thick where it is fastened to the jamb and $\frac{1}{16}$ inch thick at its thickest part.

To connect the side casing to the head casing the two pieces of trim should be mitered. The miter joint must be cut accurately so that the finished joint will be true and tight, with faces in alignment. The casing should be set in $\frac{3}{16}$ inch from the face of the jamb, and secured with 6d finishing nails spaced 12 inches on center. (See Figure 15–12.) The nails should be driven and then set approximately $\frac{1}{16}$ inch below the casing surface.

## Installation of Hinges

To receive the hinges, both the door and the jamb must be routed to a depth equal to the thickness of the hinge. The hinge-pin center should be located at least $\frac{1}{2}$ inch from the edge of the jamb, so that the hinge can swing 180°.

For doors up to $2\frac{1}{4}$ inches thick, there

FIGURE 15–12.   Casing placement

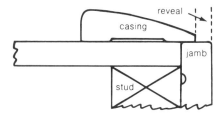

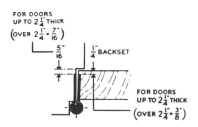

FIGURE 15–13.   Hinge backset

should be at least a $\frac{1}{4}$-inch backset for the hinges. But if the door is over $2\frac{1}{2}$ inches thick, the backset can be $\frac{3}{8}$ inch. (See Figure 15–13.)

The hinges should also be installed so that there is a $\frac{1}{16}$-inch space between jamb and door edge, a $\frac{1}{16}$-inch clearance between the face of the door and the edge of door stops, and a $\frac{1}{8}$-inch space between the top of the door and the head jamb.

The screws for the hinges can be either slotted or Phillips head, at least $\frac{3}{4}$ inches long, with full screw threads.

The bottom of the bottom hinge should be 10 inches from the finished floor; the top hinge should be 5 inches from jamb rabbet to the top edge of the barrel; and the third hinge should be centered between the top and bottom hinges. (See Figure 15–14.)

The sizes of the individual hinges vary according to the width and thickness of the door. Table 15–3 can be used to determine the correct height of the hinges.

## Door Locks

The two most popular types of passage door locks used in light construction are cylindrical and tubular locks. They are installed in open cutouts in the door and do not need to be disassembled when installed. (See Figure 15–15A,B.) Some doors are available with predrilled holes for the lock

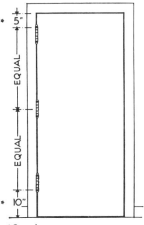

*Certain western states use as
standard 7" from top and 11"
from bottom

FIGURE 15–14.   Hinge placement

## TABLE 15–3   HEIGHT OF HINGES

*Note:* Height of hinge is always first dimension, not
including tips. Always use *ball-bearing hinges* on
doors equipped with *closers.*

| Thickness (inches) | Width of Doors (inches) | Height of Hinges[1] (inches) |
|---|---|---|
| **DOORS** | | |
| $\frac{3}{4}$ to $1\frac{1}{8}$ cabinet | to 24 | $2\frac{1}{2}$ |
| $\frac{7}{8}$ and $1\frac{1}{8}$ screen | | |
| or combination | to 36 | 3 |
| $1\frac{3}{8}$ | to 32 | $3\frac{1}{2}$ |
| | over 32 | 4 |
| | to 36 | $4\frac{1}{2}$ [2] |
| $1\frac{3}{4}$ | over 36 to 48 | 5[2] |
| | over 48 | 6[2] |
| 2, $2\frac{1}{4}$ and $2\frac{1}{2}$ | to 42 | 5 Extra |
| | over 42 | 6 Heavy |
| **TRANSOMS** | | |
| $1\frac{1}{4}$ and $1\frac{3}{8}$ | | 3 |
| $1\frac{3}{4}$ | | $3\frac{1}{2}$ |
| 2, $2\frac{1}{4}$ and $2\frac{1}{2}$ | | 4 |

[1] Width of hinges as necessary to clear trim (see lower left column).

[2] Extra heavy hinges should be specified for heavy doors and for
doors where high-frequency service is expected. The extra-heavy
hinges should be of $4\frac{1}{2}$", 5", and 6" sizes, as shown in the table.

Reprinted with permission of Stanley Hardware.

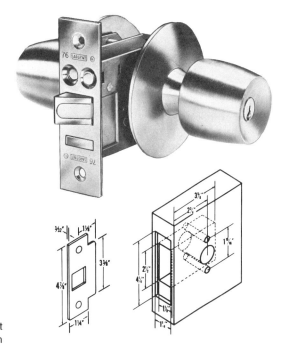

FIGURE 15–15.   Installation of cylindrical and
tubular locks

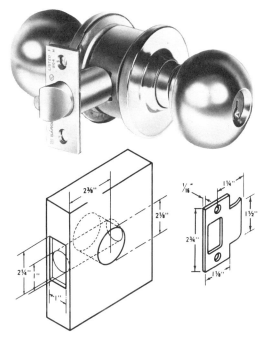

set; if they are not, the holes must be properly marked and bored. A template is usually furnished with the lock set and should be positioned correctly. The usual height of the lock set above the finish floor line is 38 inches. One hole should be bored through the face of the door, and another into the edge. The shallow mortise on the edge of the door is then laid out and cut with a wood chisel. Next, the lock can be installed in its proper position.

## Thresholds

To eliminate the possibility of air infiltration, all exterior doors require a threshold. (See Figure 15–16.) In most cases these are wooden or aluminum, and have a rubber or vinyl sealing strip. When the door is closed, it forces the sealing strip into a bulging position and eliminates any possibility of air infiltration. Because there is usually no heat transfer from room to room, interior doors do not require a threshold.

## Installation of a Prehung Door

In most cases prehung doors have split jambs and predrilled holes for the lock set,

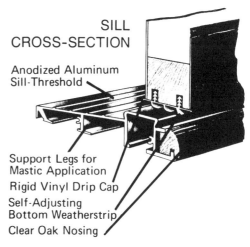

**SILL CROSS-SECTION**

Anodized Aluminum Sill-Threshold

Support Legs for Mastic Application

Rigid Vinyl Drip Cap

Self-Adjusting Bottom Weatherstrip

Clear Oak Nosing

FIGURE 15–16.   Cutaway view of a threshold

and need only to be placed in the rough opening, plumbed, and nailed in place.

The hinge side of the jamb should be nailed first. To plumb the jamb, shingle wedges are driven behind each hinge. Then shingle wedges should be driven behind the strike plate, to plumb the opposite side. The remaining portion of the split jamb and casing is placed in the rough opening adjacent to the previously placed set, and is secured with 6d finishing nails. To complete the installation, the lock sets should be installed in the predrilled holes.

## WINDOWS

Of the many window styles, there are three basic classifications: swinging; sliding; and fixed. *Swinging* windows can be hinged vertically or horizontally. If they are hinged on a vertical side, they are called casement windows; if they are hinged on the top or bottom they are called awning or hopper windows. *Sliding* windows can be double-hung or single-hung. A *fixed window* is

sometimes called a picture window. It is usually a large stationary window, that provides no ventilation.

## Casement Windows

Casement windows have a sash hinged on one side that allows the windows to swing outward by means of a crank. (See Figure

FIGURE 15–17. Casement windows

15–17.) When the window is fully open, it allows for 100 percent top-to-bottom ventilation.

Because of the unrestricted surface area of casement windows, they are very attractive in contemporary construction.

FIGURE 15–18. Double-hung windows

## Double-Hung Windows

*Double-hung windows* consist of two sashes that operate up and down in channels in the frame. (See Figure 15–18.) The sashes are held in place either by a friction fit, or by various balancing devices placed in the sash or casement. Both sashes are movable and allow for up to 50 percent ventilation.

Double-hung windows are the most popular because they lend themselves well to almost any type of architecture.

*Single-hung windows* are similar, but they have only one movable sash. The one that operates is usually the lower sash; it allows for possible 50 percent ventilation when the window is fully open.

## Sliding Windows

Sliding windows consist of two sashes that operate horizontally in a common frame, allowing for 50 percent ventilation. (See Figure 15–19.) They are available in a vari-

FIGURE 15–19. Sliding windows

ety of sizes, and can be combined with other sliding windows to provide a large panoramic view.

## Awning Windows

Awning windows can be placed so they swing up and out, in and down, or they can be placed on one side and used as casewindows. (See Figure 15-20A.) An awning window offers great flexibility for ventilating, opening fully or at any angle for air control. The window is often placed at the bottom of a fixed window to allow for

ventilation without obstructing the view. (See Figure 15-20B.)

## Fixed Windows

Fixed windows are large sheets of glass that are placed in a stationary sash. They can be used alone or in combination with a movable unit.

## Insulating Window Glass

Glass is a major source of heat transfer, but this can be greatly reduced by insulating glass, consisting of two panes separated by a blanket of dry air. Insulating glass reduces heating and air conditioning costs, reduces condensation, and eliminates the need for a storm sash. (See Figure 15-21.)

## Window Parts

A window is made up of several different parts. (See Figure 15-22.) The sash, one of the main features, is used to hold the glass. Other features include jamb liners, anchoring flanges, glass, and weatherstripping.

Window details are usually shown on architectural drawings to indicate how the

FIGURE 15-20.   Awning window (A), placed at the bottom of a fixed unit (B) (*a*) (*b*)

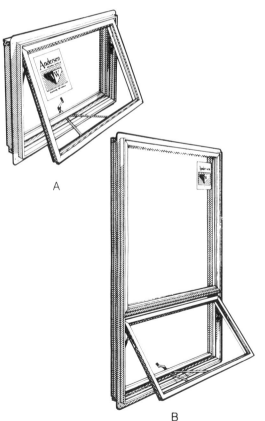

A

B

FIGURE 15-21.   Insulating glass

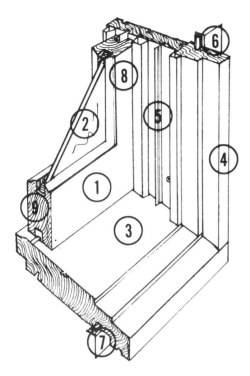

FIGURE 15–22. Typical window parts: (1) Sash protected with a patented 4-step polyurea factory finishing process. (2) Select quality welded insulating glass. (3) Wood sill covered with white rigid vinyl (PVC). (4) Wood outer frame member covered with rigid viny (PVC). (5) Rigid vinyl (PVC) jamb liner. (6) Vinyl anchoring flange and windbreak fits into groove of outer frame. (7) Soft vinyl sill windbreak. (8) Rigid vinyl rib on jamb liner fits into polypropylene covered urethane foam weatherstripping on sash. (9) Foam type weatherstripping on bottom and top rails.

elevation size are the unit dimension, rough opening, sash opening, and the unobstructed glass sizes.

## Installation of Windows

Before windows are installed, the rough opening should be checked for plumb, level, and correct size. In most cases the rough opening should be 1 inch wider and ½ inch higher than the window. The added space in the rough opening allows the window to be leveled and plumbed.

FIGURE 15–23. Mullion details

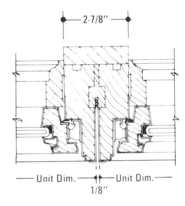

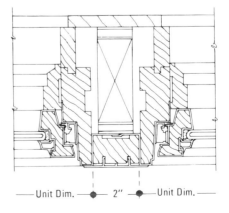

individual parts fit together. The head and sill detail is a full section taken longitudinally through the window. In addition, a plan section is often used to show the relationship of sash to the side jambs. Mullion details also indicate how multiple units are joined. (See Figure 15–23.)

Most detail catalogs also show the basic sizes of the individual windows in elevation. (See Figure 15–24.) Included in the

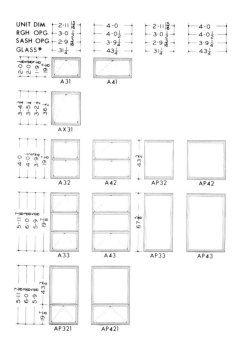

FIGURE 15–24.  Individual windows, in elevation

is usually fabricated from a 2 × 8, and the sides and top from 2 × 6s. To allow for the adjustment of the glass, the frame should be ½ inch larger in both height and width than the glass. Once the frame has been positioned, a backstop is nailed in place. (See Figure 15–25A.) A generous bed of

FIGURE 15–25.  Installation of a fixed window unit: A, placement of back stop; B, positioning the glass; C, placement of face stop

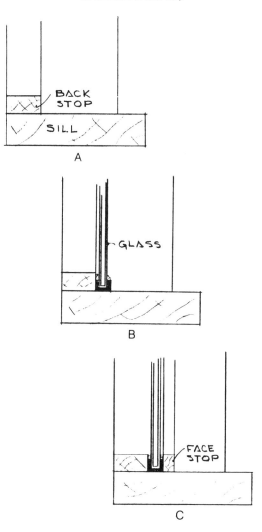

After the rough opening has been checked, the window is placed in the opening and the sill is checked with the level. If the sill is level, the anchoring flange is temporarily nailed to the wall frame. The side jambs are then plumbed and the anchoring flange is secured to the wall frame with aluminum or galvanized nails spaced on 16-inch centers.

If the sill is not level, wedge blocks are used to level it. Because of the weight of the window unit, several wedges should be placed under the sill.

## Installation of a Fixed Unit

Before glass can be installed for a fixed window unit, a frame should be constructed and installed in the rough opening. The sill glazing compound is then placed next to the backstop and on the top and sides of the frame to create a weather-tight seal.

To assure uniform setting in the frame, setting blocks are sometimes snapped onto the glass unit. The glass is then positioned onto the sill and firmly pressed against the backstop. (See Figure 15–25B.) To complete the installation of a fixed unit, a face stop is nailed in place. (See Figure 15–25C.)

## SCREENS

Windows that have operating sashes are often covered with screen panels, to allow for ventilation and retard the entrance of insects. To be an effective barrier, the screens should have a minimum of 252 openings per square inch and be constructed of a noncorrosive material.

# CHAPTER 16

# CABINET CONSTRUCTION

CABINETS ARE CONSIDERED ONE OF THE MOST important built-in features of a building. For maximum flexibility and maximum usage, they should be carefully designed and properly constructed.

There are several different types of cabinet construction. They may be built on the job, or ordered from a supplier, or built to architect's specifications. Regardless of the type of cabinet construction, it is the responsibility of the carpenter to see that the cabinets are properly installed.

## COMMON WOOD JOINTS

The strength and appearance of finished cabinets is often decided by the joints used in construction. A joint is defined as the place where two separate pieces of wood are joined. Although there are many different types of joints, only ten or twelve basic joints are ordinarily used in cabinet construction.

### Butt Joints

The butt joint is the simplest of the joints, and the easiest to construct. This is done by placing the square end of one member against the end, edge, or surface of another member. (See Figure 16–1A.) Glue, nails, screws, and dowels are often used to fasten the two members together. The butt joint is considered to be one of the weakest types, and should not be used where strength is a primary consideration.

### Dowel Joints

A dowel joint (see Figure 16–1B.) increases the strength of a butted joint and is an alternative to the conventional mortise-and-tenon joint. The dowels are usually constructed from birch or maple, and have a diameter ranging from ¼ to 1½ inches.

To permit a better bond between the dowel and the intersecting member, many dowels have spiral rings cut into them that allow the glue to spread evenly and that let trapped air escape. In cabinet construction, dowels are often used to connect rails and stiles.

### Mortise-and-Tenon Joints

The conventional mortise-and-tenon joint has a rectangular opening, called the mortise, cut into one member, and a tenon (projection) cut on the intersecting member. (See Figure 16–1C.) The tenon should be about one-half the thickness of the stock and 1 to 2 inches long. Variations of the mortise-and-tenon joint include the barefaced mortise-and-tenon, the open mortise-and-tenon, and the haunched mortise-and-tenon.

### Miter Joints

A miter joint is the intersection of two members that are of equal width and are cut at an angle less than 90°. (See Figure 16–1D.) Unless this joint is strengthened, it is relatively weak. The joint is often reinforced with dowels, splines, glued blocks, or corrugated fasteners. A *spline* is a thin strip of wood, metal, or plastic that is inserted in a groove to strengthen a joint. (See Figure 16–1E.) A *glue block* is a small triangular-shaped piece of wood that also supports and strengthens a joint. (See Figure 16–1F.) *Corrugated fasteners* are small pieces of metal that are driven into the in-

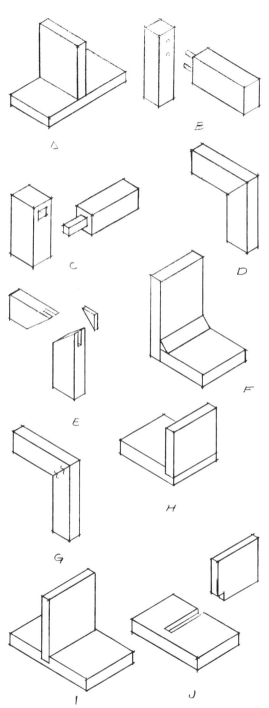

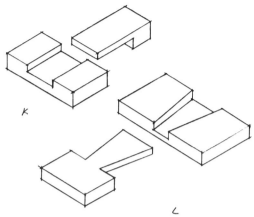

FIGURE 16-1. Common wood joints: A, butt joint; B, dowel joint; C, mortise-and-tenon joint; D, miter joint; E, spline; F, glue block; G, corrugated fasteners; H, rabbet joint; I, through dado; J, stop dado; K, end lap; L, dovetail joint

tersecting boards to reinforce a weak joint. (See Figure 16–1G.)

## Rabbet Joints

A rabbet joint is constructed by cutting an L-shaped groove across the edge or end of one member and inserting the square end of another member into the joint. (See Figure 16–1H.) The width of the rabbet is usually cut the same width as the stock, and the depth is cut to one-half the stock width. The two intersecting members are usually secured with glue and nails.

## Dado Joints

A dado joint is a groove cut across grain to receive the butt end or edge of an intersecting member. (See Figure 16–1I.) The width of the dado should equal the width of the intersecting member, and should be cut to a depth equal to one-half the width of the stock. A variation is the stop or blind dado, which is cut only partway across the stock.

(See Figure 16–1J.) The blind dado is used when the case body edge is exposed.

## Lap Joints

A lap joint is constructed by cutting away half the thickness of two intersecting members so that when they are joined, their thickness equals that of one member. (See Figure 16–1K.) The two most popular vari-ations of the lap joint are the end-lap and the cross-lap joint.

## Dovetail Joints

The dovetail joint is often found in drawer construction and is considered to be one of the most difficult joints to construct. (See Figure 16–1L.) It gets its name from its resemblance to the spread of a dove's tail.

## BASE UNIT CONSTRUCTION

The base cabinet unit is usually built to a height of 36 inches and a width of 24 inches. Although these dimensions may vary, they are the ones that have been generally adopted by most cabinet builders.

Usually, a base is started with a frame of 2 × 4s strengthened by 2 × 4 lookouts placed on 24-inch centers. (See Figure 16–2.) The total width of the base should be no more than 21 inches. The shortened base compared to the overall width of the cabinet allows for a 3-inch toe space and a 1-inch overhang of the cabinet top.

Once the base has been completed, the sides are cut from ¾-inch stock: in most cases plywood is preferred. The sides are usually cut to a height of 35¼ inches and a width of 23¼ inches. A notch is cut from each side to allow for the toe space, and dados are cut into the sides to support the shelves. The sides are then fastened to the base, and the base is secured to the floor. If the floor is not level, it may be necessary to place shims under the cabinet base.

At this point, shelves are placed in their proper positions and fastened to the sides. Then the rails and stiles are added to the cabinet. (See Figure 16–3.) These can either be preassembled, or assembled on the cabinet. The horizontal members are called rails and the vertical members are called stiles. A variety of joints can be used to connect the rails and stiles, but mortise-and-tenon joints or dowel joints are usually preferred.

For base units, a cross-rail is usually placed 4 inches below the top rail to divide the area between the drawer space and the door. A mullion divides the area between two doors; it makes the installation of doors much easier.

FIGURE 16–2.   Cabinet base

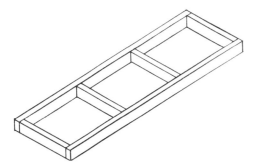

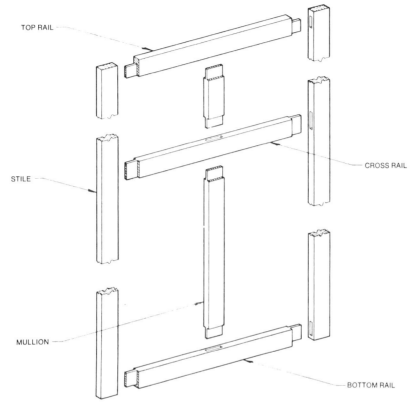

FIGURE 16–3.   Rail and stile assembly: typical frame parts

## WALL UNITS

Wall units are usually 12 inches wide and 30 inches high, and are built much like base units. (See Figure 16–4.) The primary difference between a base unit and a wall unit is the hanging strip placed on the back side of the wall unit. This strip, usually a 1 × 4, is used to help fasten the wall unit to the wall. For maximum strength, screws should be inserted directly over a framing member.

A minimum 18-inch space should exist

FIGURE 16–4 (right)   Wall-hung unit

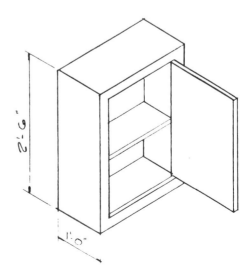

between the top of the base unit and the bottom of the wall-hung unit. The space between the ceiling and the top of the wall-hung unit varies depending on the height of the wall, but in most cases the distance is 12 inches. This area above the wall unit can be left open, but usually the ceiling is furred down to close it.

## DOORS

Doors used in cabinet construction are either flush or lipped. (See Figure 16–5.) *Flush* doors are the most difficult to place, for they must be accurately fitted in an opening with about $\frac{1}{16}$-inch clearance on each side. *Lip* doors have a $\frac{3}{8}$-inch rabbet cut around the edges to allow for a slight tolerance in the fit. Because of the ease of fitting the lip door, it is usually the most popular. Cabinet doors are also designed in various styles. (See Figure 16–6.) The styles can vary from the simplest routered edge to an elaborate overlay.

Various types of hinges are used to secure the door to the framework. (See Figure 16–7A,B.) If a door is lipped, the hinge is offset to fit the rabbet and fastened to the door with wood screws. When flush doors are used, the hinge can be offset to wrap around the back side of the door, or a butterfly hinge can be fastened directly to the stile and the door.

Cabinet catches are usually placed behind the door to keep it in a closed position. The two most popular types of

FIGURE 16–7. Cabinet door hinges and catches: A, lipped, and B, flush cabinet door hinges; C, double-roller friction catch for lipped, flush, and overlay cabinet doors; D, double magnetic catch for lipped and flush cabinet doors

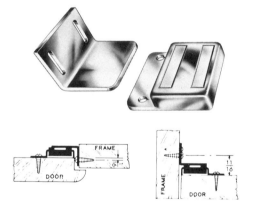

FIGURE 16–5. Flush and lipped doors, showing magnetic-catch installation

FIGURE 16–6. Cabinet doors

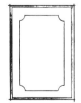
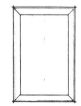

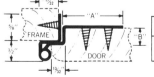

| | door inset "B" | "A" |
|---|---|---|
| | $\frac{3}{8}$" | $1\frac{7}{16}$" |
| | $\frac{3}{4}$" | $1\frac{1}{16}$" |

A

FIGURE 16–7. (cont.)

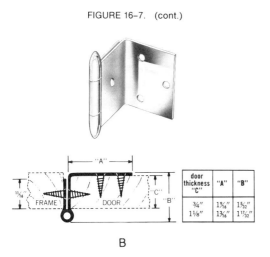

B

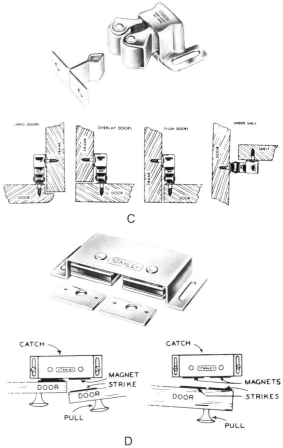

C

D

catches are friction and magnetic. A friction catch relies on the friction of two separate pieces. (See Figure 16–7C.) A magnetic catch has a floating magnet that attracts the strike plate. (See Figure 16–7D.) Of course the doors do not need catches if the hinges have automatic closing devices.

Most doors are also provided with a pull or knob, the type varying with the architectural style of the building and quality of the cabinets.

## DRAWERS

Drawers are also classified as either flush or lip. The lipstyle is commonly preferred in cabinet construction because it is easier to build and fit. The drawer front is first cut to proper size. The width and height of the front are usually both ⅝ inch larger than the opening, thus allowing for a ⅜-inch rabbet around the drawer front. When the back is cut, a groove is added to receive the bottom. The sides also have a groove cut in them, and in most cases a dado is cut into the sides to receive the back. (See Figure

16–8.) Once the four sides have been properly cut, a bottom is cut to the correct size. The individual parts can then be assembled, squared, and fastened together.

To keep drawers in proper alignment and operating efficiently, commercial drawer guides are ordinarily used. But three types of drawer guides can be fabricated at the job site: side, center, and corner guides. To construct a side guide, a runner is nailed on the side of the cabinet or the drawer. The runner is then placed in

FIGURE 16–8  (right)    Drawer side

FIGURE 16–9  (center)    The three types of drawer guide: A, side guide; B, center guide; C, corner guide

FIGURE 16–10  (bottom)    Commercial drawer guides

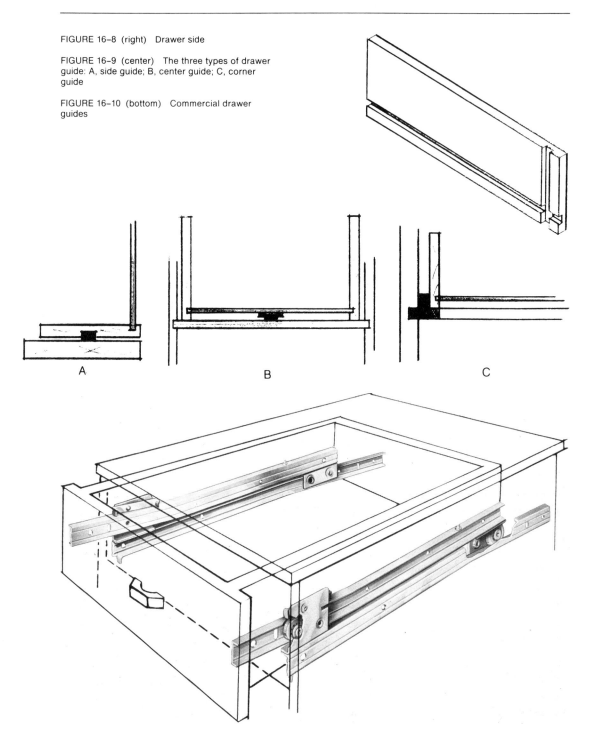

A

B

C

an oversized dado, allowing the drawer to move freely. (See Figure 16–9A.) A center drawer guide is built on the bottom of a drawer and freely operates on a runner that has been placed directly below the center guide. (See Figure 16–9B.) Corner guides can be built on either side of the drawer with a $\frac{1}{16}$-inch clearance allowed between the drawer and the guide. (See Figure 16–9C.) To prevent the drawer from tilting once it has been pulled out, a kicker is placed above the drawer.

Commercial drawer guides are also available that operate either on a single or a double track. (See Figure 16–10.) A single-track drawer slide is placed under the drawer, fitting any length or width. To keep the track in proper alignment a nylon roller must be mounted to the back side of the drawer and inserted in the single track. If a double-track guide is used, tracks are attached on each side of the drawer, and a matching guide is installed in the cabinet adjacent to each track.

## SURFACING FOR CABINET TOPS

In most cases, a high-pressure decorative plastic laminate is used as a surfacing material for cabinet tops. The material, usually $\frac{1}{16}$ inch thick, is highly resistant to stains, wear, and ordinary household chemicals.

Three basic materials are involved in installation of plastic laminates: the plastic laminate, the adhesive, and the substrate or supporting material.

To meet various specific needs there are six different types of *plastic laminates:* general-purpose, vertical-surface, postforming, hardboard-core, fire-rated, and backing sheets. Vertical-surface laminate is a thin sheet of material that can be used to cover cabinet doors and vertical surfaces; it is much more economical than a general-purpose laminate. Postforming laminates are relatively flexible, allowing for simple bends that cover unsightly edges and eliminate the need for a seam at the intersection of horizontal and vertical planes. The hardboard-core laminate is the only one that has structural strength and is used to fabricate furniture over a framework. The plastic laminate used for horizontal application should be of the general-purpose or hardboard-core type. These two types of laminates combine good mechanical strength and the ability to hide small irregularities.

The *adhesives* that can be used to bond the laminate to the substrate are contact, casein, urea, resorcinal, polyvinyl acetate (white glue), and epoxy. None of these adhesives needs clamping or pressing, so they provide a versatile means for installing plastic laminates. Regardless of the type of adhesive used, the manufacturer's instructions for application should be followed.

The *substrate* can be a good-quality plywood, particleboard, or hardboard. After the substrate has been securely fastened to the cabinet, blind edging is mounted beneath or along the edge of the substrate. (See Figure 16–11.) Because this edging is a major problem area in on-the-job installations, it should be well secured to the substrate with both nails and glue.

Before the laminate is placed, the substrate must be sanded smooth and free of dirt, grease, wax, or any foreign matter that might protrude through the decorative surface.

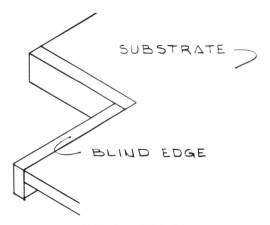

FIGURE 16–11.  Blind edging

The laminate should be allowed to condition for 48 hours at a temperature of 75° F and a relative humidity of 45 to 55 percent.

After the laminate has been conditioned properly, it can be cut to rough size with a handsaw, table saw, portable saw, portable router, or tin snips. The laminate should be supported near the cut, and overcut ⅛ inch to ¼ inch. If any cutouts, such as range openings are necessary, they should have a minimum corner radius of ⅛ inch.

Next, the plastic laminate should be tested for correct fit and then, if contact bond cement is used, both the substrate and backside of the laminate should be completely covered with adhesive. Spot bonding should never be used. The adhesive can be spread with a brush, roller, or metal spreader notched to apply thin even ribbons of mastic.

After the contact cement has been applied, a slip sheet is placed on top of the substrate. This sheet is used so that the laminate can be positioned correctly before the bond is made; once the two surfaces come in contact, the laminate cannot be shifted. A slip sheet can be constructed from brown paper, scraps of laminate, or thin strips of wood. When it has been properly positioned, the slip sheet should be slowly withdrawn so that one edge can be bonded. Then the remaining portion of the slip sheet is removed and pressure is applied to the cabinet top, usually with a hand roller or a rubber mallet.

To complete the application of plastic laminate, the edge or edges should be trimmed with a router and smoothed with a mill file so that there is a slight bevel.

Problems that might arise after the plastic laminate has been fabricated include delamination of the plastic from the substrate, cracking of the laminate at corners and around cutouts, blistering or bubbling of the laminate, opening of joints, or cracking of the laminate in the center of the sheet.

*Delamination of the plastic from the substrate* is often caused by poor bonding procedures, such as failure to make a uniform glue line, apply uniform pressure, or thoroughly clean the two mating surfaces. When only the edges have delaminated, the self-edge may be too high or the adhesive may have been insufficient, or too heavy.

If the laminate is cracking at corners and around cutouts, it has not been properly conditioned and the edges may not be rounded.

The blistering of plastic laminate is usually caused by a starved glue line, poor conditioning, and improperly applied pressure.

Opening of joints is also caused by poor conditioning and bonding.

When cracks occur in the center of the plastic laminate, they can often be attributed either to a wide span of laminate or spot gluing.

**CHAPTER 17**

# STAIRS

THE MAIN STAIRWAY DESIGN CAN VARY considerably, of course, but in most cases it is either a straight-run type, a winder-and-straight-run type, or a flat-landing–split-run type. (See Figure 17–1.) A *straight-run* stair is continuous and stretches from one level to another without landings or turns. A *winder-and-straight-run* stair is constructed with a maximum of two risers across the turn and a straight run of ten treads. A *flat-landing–split-run,* commonly referred to as a double L- or a U-type stair, is constructed with a landing where the direction of the stair run is changed.

FIGURE 17–1 (right, below, and below right) Stairway designs

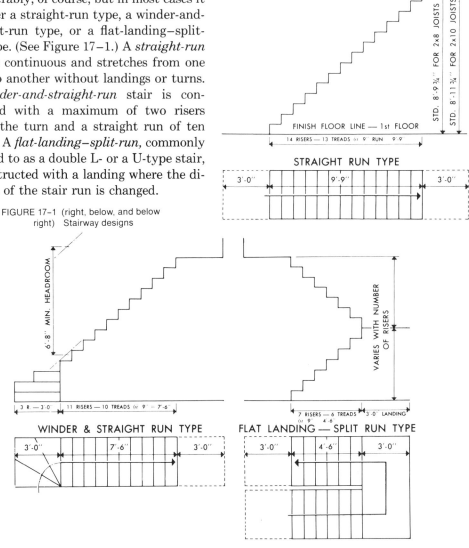

## STAIR TERMS AND MINIMUM DIMENSIONS

To prevent and eliminate possible danger, stairways should be planned carefully. The main stair should be provided with a mini-

mum headroom of 6 feet 8 inches, and for basement stairs, 6 feet 4 inches. The headroom is the vertical distance from a

FIGURE 17–2. Nosing styles

line along the front edges of the tread to a parallel line that intersects the bottom of the header.

The main stair should have a minimum width of 2 feet 8 inches clear of the handrail. The minimum clearance for basement stairs is 2 feet 6 inches. If a door is located at the top and swings toward the stair, a 2-foot 6-inch landing should be provided.

Stairs are usually constructed of three main parts—the treads, the risers, and the stringers. The *treads* are the parts of the stair that are stepped on and should be a minimum of 9 inches deep. The total depth

of the tread, including the nosing, is called the *unit run.* The sum of all the treads is called the *total run.* The nosing is the front part of the tread which extends past the riser. It usually projects 1⅛ to 1½ inches past the riser, but seldom should project more than 1¾ inches. To complement or enhance a particular architectural style, the nosing can be finished in several different ways. (See Figure 17–2.)

The *riser* is the vertical board that is placed directly behind the tread; it should not be taller than 8¼ inches. The height of the riser is called the *unit rise.* The sum of all the risers is called the *total rise.*

A *stringer* is the inclined piece of a stairway to which the risers and treads are fitted. The stringers are usually cut from 2 × 12 stock and stretch from one level to another. The opening in which the stringers are placed is called the stairwell, and should consist of well-braced framing members.

To protect the occupants from falls, the stairs should have a continuous handrail on at least one side, as well as railings at open portions of sides where the change of level exceeds 2 feet.

## TREAD AND RISER DESIGN

One of the most important features of stairs is the relationship between the treads and the risers. Without the correct proportions between the two features, the stairs could be dangerous. Three formulas can be used to calculate the riser-tread ratio and make sure the treads and risers are correctly designed:

■ *Formula 1:* The sum of the dimensions of a tread run and a unit rise should

equal 17 to 18 inches. For example, a unit rise of 7 inches would require a tread run of 10 or 11 inches.

■ *Formula 2:* The sum of one tread run and two unit rises should equal 25 inches. For example, risers 7½ inches tall would require a tread run of 10 inches, because 7½ inches plus 7½ inches plus 10 inches equals 25 inches.

■ *Formula 3.:* When the unit run is multiplied by the unit rise, the product should equal approximately 75 inches. For example, 10 inches multiplied by 7½ inches equals 75 inches.

To calculate the number and size of treads and risers, the total rise is first divided by 7. The quotient obtained is the number of risers required in the stair. If the quotient is a fraction, it should be rounded off, because there must of course be a whole number of risers. To find the height of each riser, the total rise is divided by the number of risers. For example, if the total rise were 8 feet 10 inches, the number of risers and height of the risers would be:

$$8 \text{ feet } 10 \text{ inches (106 inches)} \div 7 = 15.14$$
$$(15 \text{ risers})$$
$$106 \text{ (total rise)} \div 15 \text{ (no. of risers)} = 7.06$$
$$\text{inches (riser height)}$$

There will always be one less tread than the total number of risers. Therefore, if there are fifteen risers in a stair, there will be fourteen treads. If the riser height is 7 inches and Formula 1 is followed, the tread run should equal 10 inches.

To find the stairwell length, the tread unit run should be multiplied by the number of risers.

## TYPES OF STRINGERS

The stringer is the supporting member of the stair and can be classified as *housed, semihoused, mitered, open,* or *plain.*

A *housed* stringer has grooves cut into it so that it may receive the treads, risers, and wedges. It can be purchased already cut, or can be cut at the job site. The groove is cut wider at the back so that after each riser and tread is positioned, a wedge can be inserted and glued in place. (See Figure 17–3A.) Once the treads and risers have been properly placed, glue blocks are usually fastened on the underside of the stairway in each of the 90° turns created by the treads and risers, to eliminate squeaks. For added solidity, the treads and risers can be rabbeted and glued into each other. (See Figure 17–3B.).

A *semihoused* stringer consists of a rough-cut stringer attached to a solid-finish stringer. (See Figure 17–3C.) The treads and risers are nailed to the rough riser and

tread cuts, and are butted to the finish stringer. The finish stringer is usually built from 2-inch stock that is wide enough to cover the intersection of the tread and riser, and wide enough to project at least 2 inches past the tread nosing.

A *mitered* stringer is seldom used; but when it is, each riser is mitered into the stringer so that no end grain shows. (See Figure 17–3D.) The treads are nailed to the level cut, and are allowed to project past the stringer a distance equal to the nosing.

An *open* stringer is used on open stairways, and is constructed by allowing the risers and treads to project beyond the stringer. (See Figure 17–3E.) This particular type of stair construction is usually not recommended unless the stairway is in an inconspicuous place.

A *plain* stringer is cut only at the top and bottom ends. The treads are fastened to the stringer with cleats nailed to the

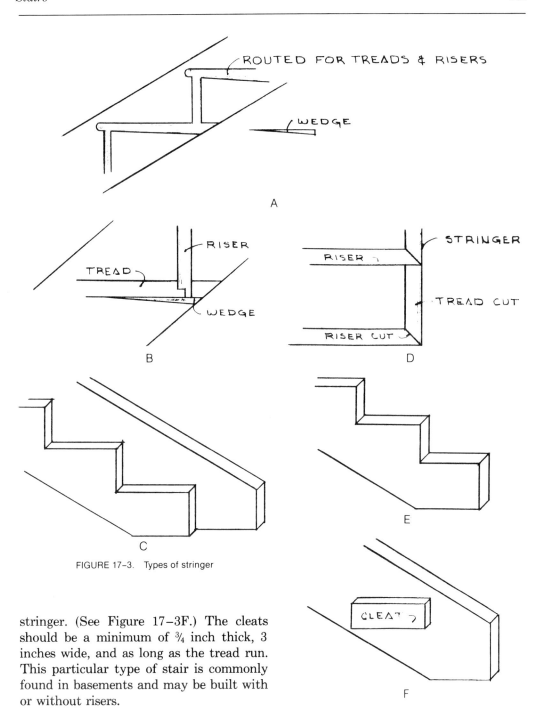

FIGURE 17-3. Types of stringer

stringer. (See Figure 17–3F.) The cleats should be a minimum of ¾ inch thick, 3 inches wide, and as long as the tread run. This particular type of stair is commonly found in basements and may be built with or without risers.

## STRINGER LAYOUT

Before the stringer can be laid out, the riser height must first be determined. The total rise can be determined by placing a story pole in the stairwell opening and marking the height of the finished floor. (A story pole is a rod or pole that is used to measure the height of the finished floor. It can be a 1 × 4, 1 × 6, or 2 × 4.) The total rise can then be measured along the story pole. From the total rise, the riser height, riser number, tread unit run, and tread number can be calculated.

After the tread run and riser height have been determined, a framing square is placed on the stringer stock. The square should be held with blade in the left hand, the tongue in the right hand, and the outside point of the corner pointing toward the layout person. (See Figure 17–4.)

When the framing square is correctly placed on the stringer stock, the unit run of the tread is indicated by the previously determined dimension on the blade, and the riser height is marked from the dimension on the tongue. When the tread run and riser are marked they should be laid out accurately, because a small error on one step could become a large error when projected over several steps.

The mark against the tongue gives the plumb cut of the first riser. The mark against the blade indicates the level cut at the bottom of the stringer, and should be

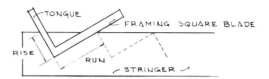

FIGURE 17–4. Stringer layout

carried across the full width of the stringer. After the level cut and the plumb cut of the first riser have been marked, the framing square is repositioned along the stringer. The unit run of the first tread and second riser is laid out by positioning the framing square to form a right angle with the plumb cut of the first riser. The dimensions on the blade and tongue of the framing square should again correspond to the unit run of the tread and the height of the riser.

The remainder of the treads and risers are laid out in the same way. To keep up with the number of risers laid out, each riser is usually numbered. There will be one less riser in the stringer than the total number of risers calculated, because the top riser is always formed by the joist on the upper fooor.

An adjustment must be made to the bottom of the stringer before it is cut. Since the height of the first riser will increase when the tread is placed, a proportional amount should be cut from the bottom of the stringer.

## NEWELS AND HANDRAILS

To provide for maximum protection, all stairways should have handrails. If the stairway is closed, the rail is attached to the wall with suitable metal brackets. If it is open, the handrail is supported by balusters and a newel post. The rail should be set a minimum of 2 feet 8 inches above the tread at the riser line. The balusters should be doweled or dovetailed into the treads, and in some cases are covered by pieces of return nosing. (See Figure 17–5.)

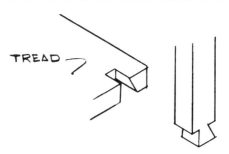

FIGURE 17–5 (right)   Dovetail baluster

# INTERIOR TRIM AND WOOD MOLDINGS

WOOD MOLDING IS A THIN STRIP OF WOOD milled with a plain or curved surface. Several areas in a building require wood molding, the most common areas being the floor lines, windows, and ceilings.

Base molding is used at the *floor line* to protect the bottom portion of the wall and sometimes to cover the raw edges of flooring materials. (See Figure 18–1A.) A base shoe is also often placed at the intersection of the base and the floor, to protect the base molding and conceal any uneven lines.

Windows are usually trimmed with casing, stool, and apron moldings (see Figure 18–1B); or a picture-framing technique is used to cover the termination of the interior wall finish material.

Ceiling moldings cover the intersection where the wall and ceiling meet. (See Figure 18–1C.)

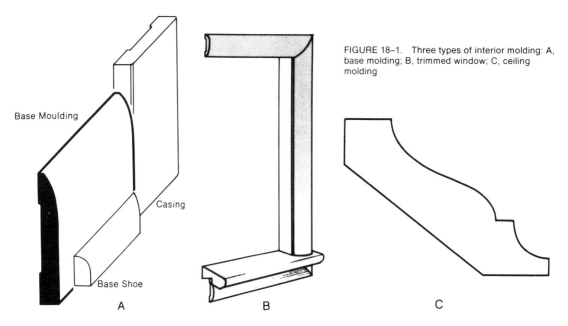

Base Moulding

Casing

Base Shoe

A

B

C

FIGURE 18–1. Three types of interior molding: A, base molding; B, trimmed window; C, ceiling molding

## BASE MOLDING

Base molding is available in various styles, which can also be combined to form various shapes. (See Figure 18–2.) It can be purchased in lengths from 3 to 20 feet. When molding is purchased, the length should include an added dimension for each miter. To compensate for a miter, the width of the molding should be added to its required length. For example, if the molding is 3 inches wide and requires three miters, add 9 inches to the total length, and then round the length off to the next foot. The thickness and the width must also be specified.

When molding is purchased, the thickness is specified first, the width second, and

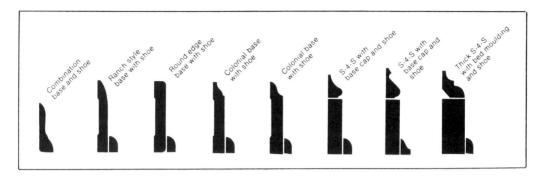

FIGURE 18–2. Base-molding variations

the length last. The thickness is determined by measuring from the top to the bottom extremity. The width is also measured at the widest points.

Base molding is placed continuously around a room, with the external corners mitered and the internal joint coped. To construct a miter joint intersecting pieces must be cut to appropriate opposite angles —usually 45° angles.

A coped joint (see Figure 18–3) is constructed by first trimming the molding at a 45° angle. When it is cut, it presents an ex-

posed profile of the installed molding. With this profile as a guide, the molding is undercut with a coping saw. It should be overcut $\frac{1}{16}$ to $\frac{1}{8}$ inch, to assure a tight joint.

When the molding is installed, the first piece should be cut so that it forms a butt joint with the intersecting wall. Then the adjoining piece of molding is coped, as discussed earlier, so that the two pieces of molding form a tight joint. The molding is secured with 6d or 8d finishing nails that penetrate either the studs or the base plate. Base molding is usually butted to the door casing, but the base shoe is cut back at a 45° angle. (See Figure 18–4.)

In many cases a large span must be covered; if the molding is not long enough it must be spliced. The most popular means of splicing molding is to miter the joining ends at 45° angles from front to back. One piece of molding will then overlap the other, creating a vertical seam.

FIGURE 18–3. Fitting a coped joint

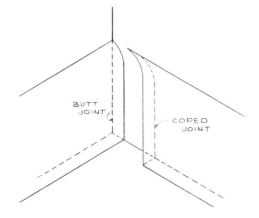

FIGURE 18–4. Application of the baseboard and base shoe

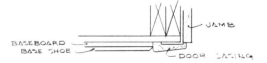

## WINDOW TRIM

Windows are usually trimmed out by means of one of two techniques: either with casing, stool, and apron (see Figure 18–5A), or with a picture-framing technique (see Figure 18–5B).

For the first method, the stool is first

FIGURE 18–5.  Window trim: A, conventional framing; B, picture framing

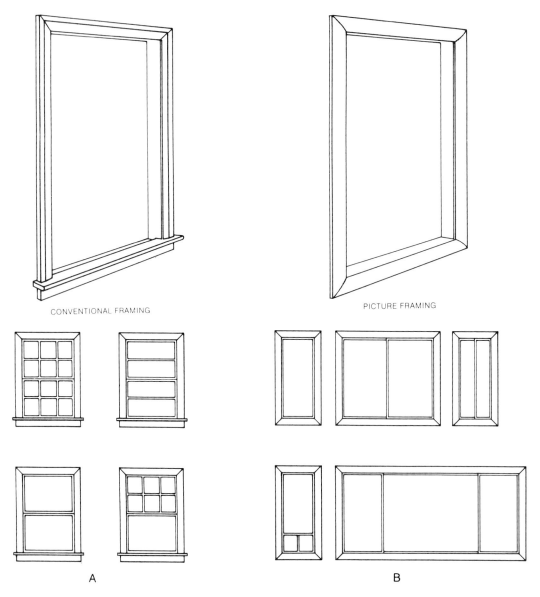

CONVENTIONAL FRAMING

PICTURE FRAMING

A

B

fitted to the unfinished window sill. The stool usually projects ¾ inch past the side casing, and in some cases is placed in a bed of caulking compound or white lead. Next, three pieces of casing are placed around the window. The bottoms of the two side casings butt the stool, while the intersection of the side and head casing is connected by a miter joint. To complete the job, an apron is placed beneath the stool. The apron conceals the joint made by the intersection of the interior wall finish. The length of the apron should equal the distance between the outside edges of the side casing.

For the *picture-framing technique,* the window opening is framed by the application of four pieces of casing. Each piece is cut with two 45° miters so that the intersecting pieces can be tightly fitted. When the casing is applied, it is not placed flush with the jamb, but back about ¼ inch.

Regardless of the technique used, the nails should be set and the trim lightly sanded.

## TRIMMING CEILINGS AND CORNERS

Ceilings are trimmed to close the joint between the wall and the ceiling, and to provide an attractive architectural feature. Several types of molding can be used as ceiling trim, but the most popular are cove, bed, and crown molding. (See Figure 18–6.) As with base molding, several different pieces of molding can be used together.

Ceiling molding is fitted by first butting a squared end into a corner and securing it to the wall with 6d finishing nails. The intersecting pieces of molding are then mitered at one end and coped to fit the existing profile. If it is necessary to trim around an exterior corner, the intersecting pieces of molding should be mitered at a 45° angle and fitted together. When the molding is secured, the finish nails should be driven into both the top plate and the vertical and horizontal framing members.

FIGURE 18–6.    Ceiling moldings

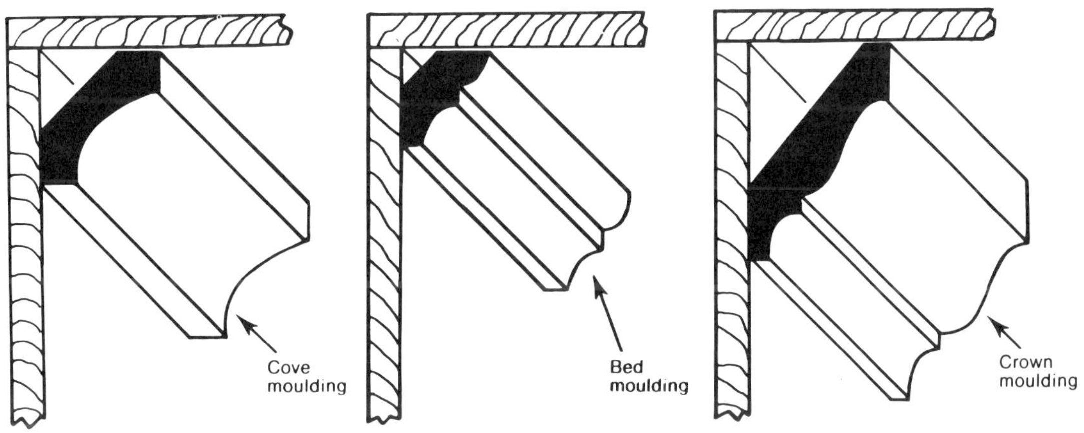

Cove moulding

Bed moulding

Crown moulding

## CHAIR RAILS

Chair rails are thin strips of molding used to protect the wall. (See Figure 18–7.) They are usually placed either one-third the distance from the floor to the ceiling, or at a level equal to the height of a chair.

Before the chair rail can be installed, a chalkline must be placed at the appropriate level. This line is used as a guide in the placement of the molding. The molding should be secured to the wall with 6d finishing nails. The interior corners of the molding should be coped, and the outside corners should be mitered.

FIGURE 18–7.   Chair rails

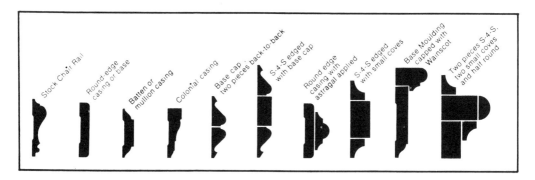

# CHAPTER 19

# THERMAL INSULATION

BECAUSE OF THE ENERGY CRISIS AND THE need for conservation, thermal insulation is becoming increasingly important. It provides an effective barrier against heat transfer and, when properly designed and installed, reduces the consumption of fuel.

The primary reason for using insulation is to reduce the transfer or movement of heat, either from the inside of a building to the outside or from the outside to the inside. Temperature inside tries to adjust to existing outside temperatures, and thus heat will move through the walls, floors, and ceiling.

The transfer of heat is accomplished by one or more of three methods: convection, conduction, and radiation. (See Figure 19–1.) *Convection* refers to heat carried by the movement of air, such as would occur as air passes over a hot radiator.

*Conduction* refers to heat being passed through a material. The transfer of heat in a steel rod that has been placed in a fire is an example of conduction.

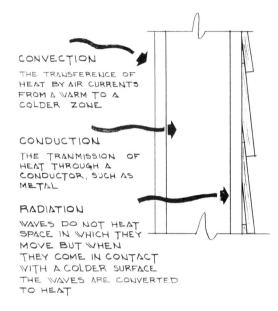

CONVECTION

THE TRANSFERENCE OF HEAT BY AIR CURRENTS FROM A WARM TO A COLDER ZONE

CONDUCTION

THE TRANMISSION OF HEAT THROUGH A CONDUCTOR, SUCH AS METAL

RADIATION

WAVES DO NOT HEAT SPACE IN WHICH THEY MOVE BUT WHEN THEY COME IN CONTACT WITH A COLDER SURFACE THE WAVES ARE CONVERTED TO HEAT

FIGURE 19–1.   Methods of heat transfer

*Radiation* is the transfer of heat through inversion space. Light rays from the sun are examples of heat transfer by radiation.

## THERMAL PROPERTIES OF MATERIALS

Heat is a form of energy that is measured in British thermal units (Btu's). A Btu is the amount of heat necessary to raise 1 pound of water 1° F.

When it is necessary to calculate the air-to-air heat transmission factor, the *overall coefficient of heat transmission* is represented by the capital letter $U$. The overall coefficient of heat transmission is the amount of heat in Btu's which will pass through a 12-inch square of wall, ceiling, or floor in 1 hour per degree of temperature difference between the air on the warm side and the air on the cool side. (See Figure 19–2.) The 12-inch section that is used to calculate the heat transfer includes all the building materials in that particular section.

FIGURE 19–2.   Overall heat-transmission coefficient

HEAT FLOW

PER HOUR PER °F (AIR TO AIR)

U

OVERALL HEAT TRANSMISSION COEFFICIENT

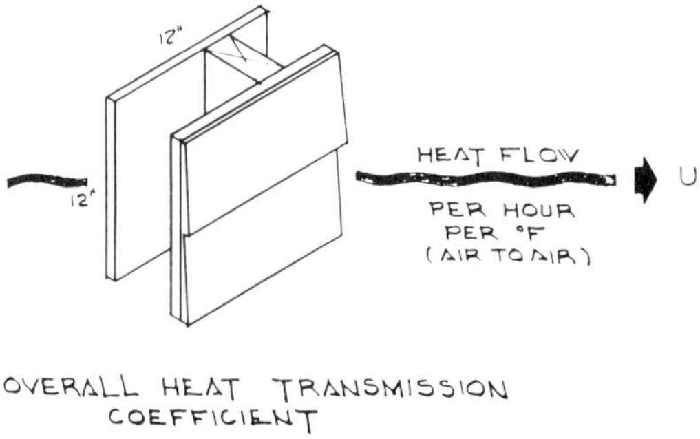

**TABLE 19–1    THERMAL PROPERTIES OF VARIOUS BUILDING MATERIALS, PER INCH OF THICKNESS**

| Material | Thermal Conductivity K | Thermal Resistance R | Efficiency as an Insulator Percent |
|---|---|---|---|
| Wood | 0.80 | 1.25 | 100.0 |
| Air space[1] | 1.03 | 0.97 | 77.6 |
| Cinder block | 3.6 | 0.28 | 22.4 |
| Common brick | 5.0 | 0.20 | 16.0 |
| Face brick | 9.0 | 0.11 | 8.9 |
| Concrete (sand and gravel) | 12.0 | 0.08 | 6.4 |
| Stone (lime or sand) | 12.5 | 0.08 | 6.4 |
| Steel | 312.0 | 0.0032 | 0.25 |
| Aluminum | 1416.0 | 0.00070 | 0.06 |

[1] Thermal properties are for air in a space and apply for air spaces ranging from ¾ to 4 inches in thickness.

Reprinted with permission of National Forest Products Association.

The $R$ value that is sometimes also used in the calculation of heat transfer is the *resistance to heat flow,* the reciprocal of conductivity. Thus, good insulation will have a high $R$ value.

The $K$ value is the Btu per hour, per square foot, per degree of difference Fahrenheit, per inch of thickness for any one material. The primary difference between the $K$ factor and the $U$ factor is that the $K$ factor is the coefficient of heat transmission for only *one* material, while the $U$ factor is composed of a combination of $K$ factors.

Table 19–1 indicates the thermal properties of various building materials per inch of thickness. A comparison of the relative insulation values of the different materials will show that the insulating properties of wood are far better than the other listed materials.

Table 19–2 lists the $U$ values and insulation requirements for different standards of comfort. The insulation is listed by $R$ value instead of by specification of the thickness of the insulation, because insulating materials vary in density and heat conductivity.

Table 19–3 shows the calculated $U$ values for wood-frame walls with various combinations of siding, sheathing, and interior

**TABLE 19–2    U VALUES AND INSULATION REQUIREMENTS**

| | U Value | Insulation R Number |
|---|---|---|
| **ALL-WEATHER COMFORT STANDARD** | | |
| Ceilings | 0.05 | R19 |
| Walls | 0.07 | R11 |
| Floors over unheated spaces | 0.07 | R13 |
| **MODERATE COMFORT AND ECONOMY STANDARD** | | |
| Ceilings | 0.07 | R13 |
| Walls | 0.09 | R8 |
| Floors over unheated spaces | 0.09 | R9 |
| **MINIMUM COMFORT STANDARD** | | |
| Ceilings | 0.10 | R9 |
| Walls | 0.11 | R7 |
| Floors over unheated spaces | 0.11 | R7 |

Reprinted with permission of National Forest Products Association.

## TABLE 19–3    U VALUES FOR WOOD-FRAME WALLS

For 2 × 4-inch studs spaced 12 to 24 inches on centers with flexible insulation stapled to faces of studs or recessed between studs to form an air space on each side of insulation. Inside surface of insulation may be reflective or nonreflective and of thicknesses indicated.

*Note:* Plaster finishes over gypsum lath do not significantly alter the U values from those shown.

| Interior Finish | with No Insulation, Sheathing Thickness to be: | | | with 1-inch Nonreflective Insulation (to stud face), Sheathing Thickness to be: | | | with 1-inch Nonreflective Insulation (recessed), Sheathing Thickness to be: | | |
|---|---|---|---|---|---|---|---|---|---|
| | $25/32''$ Wood & Bldg. Paper | $3/8''$ Ply-wood | $1/2''$ Insula-tion Board | $25/32''$ Wood & Bldg. Paper | $3/8''$ Ply-wood | $1/2''$ Insula-tion Board | $25/32''$ Wood & Bldg. Paper | $3/8''$ Ply-wood | $1/2''$ Insula-tion Board |
| **EXTERIOR FINISH $1/2'' \times 8''$ LAP SIDING** | | | | | | | | | |
| $1/2''$ Gypsum-board | 0.24 | 0.28 | 0.23 | 0.13 | 0.14 | 0.12 | 0.11 | 0.12 | 0.11 |
| $1/2''$ Gypsum-board, foil-backed | 0.17 | 0.19 | 0.17 | 0.13 | 0.14 | 0.12 | 0.09 | 0.10 | 0.09 |
| $3/4''$ Wood paneling | 0.22 | 0.25 | 0.20 | 0.12 | 0.13 | 0.11 | 0.11 | 0.12 | 0.10 |
| $1/2''$ Rigid insulation board | 0.19 | 0.22 | 0.18 | 0.11 | 0.12 | 0.11 | 0.10 | 0.11 | 0.10 |
| $1/4''$ Plywood paneling | 0.25 | 0.29 | 0.23 | 0.13 | 0.14 | 0.12 | 0.12 | 0.13 | 0.11 |
| **EXTERIOR FINISH $3/4'' \times 10''$ LAP SIDING** | | | | | | | | | |
| $1/2''$ Gypsum-board | 0.23 | 0.26 | 0.22 | 0.12 | 0.14 | 0.12 | 0.11 | 0.12 | 0.11 |
| $1/2''$ Gypsum-board, foil-backed | 0.17 | 0.18 | 0.16 | 0.12 | 0.13 | 0.12 | 0.09 | 0.10 | 0.09 |
| $3/4''$ Wood paneling | 0.21 | 0.23 | 0.19 | 0.12 | 0.13 | 0.11 | 0.10 | 0.11 | 0.10 |
| $1/2''$ Rigid insulation board | 0.19 | 0.21 | 0.18 | 0.11 | 0.12 | 0.11 | 0.10 | 0.11 | 0.10 |
| $1/4''$ Plywood paneling | 0.24 | 0.27 | 0.22 | 0.13 | 0.14 | 0.12 | 0.11 | 0.12 | 0.11 |
| **EXTERIOR FINISH $1'' \times 8''$ DROP SIDING** | | | | | | | | | |
| $1/2''$ Gypsum-board | 0.24 | 0.28 | 0.23 | 0.13 | 0.14 | 0.12 | 0.11 | 0.12 | 0.11 |
| $1/2''$ Gypsum-board, foil-backed | 0.17 | 0.19 | 0.17 | 0.13 | 0.14 | 0.12 | 0.09 | 0.10 | 0.09 |
| $3/4''$ Wood paneling | 0.22 | 0.25 | 0.21 | 0.12 | 0.13 | 0.12 | 0.11 | 0.12 | 0.10 |
| $1/2''$ Rigid insulation board | 0.20 | 0.22 | 0.19 | 0.11 | 0.12 | 0.11 | 0.10 | 0.11 | 0.10 |
| $1/4''$ Plywood paneling | 0.25 | 0.29 | 0.24 | 0.13 | 0.14 | 0.13 | 0.12 | 0.13 | 0.11 |

**TABLE 19–3** *(cont.)*

| Interior Finish | with No Insulation, Sheathing Thickness to be: | | | with 1-inch Nonreflective Insulation (to stud face), Sheathing Thickness to be: | | | with 1-inch Nonreflective Insulation (recessed), Sheathing Thickness to be: | | |
|---|---|---|---|---|---|---|---|---|---|
| | $^{25}/_{32}''$ Wood & Bldg. Paper | $^{3}/_{8}''$ Ply-wood | $^{1}/_{2}''$ Insula-tion Board | $^{25}/_{32}''$ Wood & Bldg. Paper | $^{3}/_{8}''$ Ply-wood | $^{1}/_{2}''$ Insula-tion Board | $^{25}/_{32}''$ Wood & Bldg. Paper | $^{3}/_{8}''$ Ply-wood | $^{1}/_{2}''$ Insula-tion Board |
| **EXTERIOR FINISH, WOOD SHINGLES, SINGLE, ($7^{1}/_{2}''$ EXPOSURE)** | | | | | | | | | |
| $^{1}/_{2}''$ Gypsumboard | 0.24 | 0.28 | 0.22 | 0.13 | 0.14 | 0.12 | 0.11 | 0.12 | 0.11 |
| $^{1}/_{2}''$ Gypsumboard, foil-backed | 0.17 | 0.19 | 0.16 | 0.13 | 0.14 | 0.12 | 0.09 | 0.10 | 0.09 |
| $^{3}/_{4}''$ Wood paneling | 0.21 | 0.24 | 0.20 | 0.12 | 0.13 | 0.12 | 0.11 | 0.11 | 0.10 |
| $^{1}/_{2}''$ Rigid insulation board | 0.19 | 0.22 | 0.18 | 0.11 | 0.12 | 0.11 | 0.10 | 0.11 | 0.10 |
| $^{1}/_{4}''$ Plywood paneling | 0.25 | 0.29 | 0.23 | 0.13 | 0.14 | 0.12 | 0.11 | 0.13 | 0.11 |
| **EXTERIOR FINISH, WOOD SHINGLES, DOUBLE (12'' EXPOSURE)** | | | | | | | | | |
| $^{1}/_{2}''$ Gypsumboard | 0.22 | 0.25 | 0.21 | 0.12 | 0.13 | 0.12 | 0.11 | 0.12 | 0.11 |
| $^{1}/_{2}''$ Gypsumboard, foil-backed | 0.16 | 0.18 | 0.16 | 0.12 | 0.13 | 0.12 | 0.09 | 0.10 | 0.09 |
| $^{3}/_{4}''$ Wood paneling | 0.20 | 0.23 | 0.19 | 0.12 | 0.12 | 0.11 | 0.10 | 0.11 | 0.10 |
| $^{1}/_{2}''$ Rigid insulation board | 0.18 | 0.20 | 0.17 | 0.11 | 0.12 | 0.11 | 0.10 | 0.11 | 0.10 |
| $^{1}/_{4}''$ Plywood paneling | 0.23 | 0.26 | 0.22 | 0.12 | 0.13 | 0.12 | 0.11 | 0.12 | 0.11 |
| **EXTERIOR FINISH, $^{3}/_{4}''$ BOARD AND BATTEN** | | | | | | | | | |
| $^{1}/_{2}''$ Gypsumboard | 0.24 | 0.27 | 0.22 | 0.13 | 0.14 | 0.12 | 0.11 | 0.12 | 0.11 |
| $^{1}/_{2}''$ Gypsumboard, foil-backed | 0.17 | 0.19 | 0.16 | 0.13 | 0.14 | 0.12 | 0.09 | 0.10 | 0.09 |
| $^{3}/_{4}''$ Wood paneling | 0.21 | 0.24 | 0.20 | 0.12 | 0.13 | 0.11 | 0.11 | 0.11 | 0.10 |
| $^{1}/_{2}''$ Rigid insulation board | 0.19 | 0.21 | 0.18 | 0.11 | 0.12 | 0.11 | 0.10 | 0.11 | 0.10 |
| $^{1}/_{4}''$ Plywood paneling | 0.24 | 0.28 | 0.23 | 0.13 | 0.14 | 0.12 | 0.11 | 0.12 | 0.11 |
| **EXTERIOR FINISH, $^{1}/_{2}''$ PLYWOOD** | | | | | | | | | |
| $^{1}/_{2}''$ Gypsumboard | 0.25 | 0.30 | 0.24 | 0.13 | 0.14 | 0.13 | 0.12 | 0.13 | 0.11 |
| $^{1}/_{2}''$ Gypsumboard, foil-backed | 0.18 | 0.20 | 0.17 | 0.13 | 0.14 | 0.13 | 0.09 | 0.10 | 0.09 |
| $^{3}/_{4}''$ Wood paneling | 0.23 | 0.26 | 0.21 | 0.12 | 0.13 | 0.12 | 0.11 | 0.12 | 0.11 |

TABLE 19–3 *(cont.)*

| Interior Finish | with No Insulation, Sheathing Thickness to be: | | | with 1-inch Nonreflective Insulation (to stud face), Sheathing Thickness to be: | | | with 1-inch Nonreflective Insulation (recessed), Sheathing Thickness to be: | | |
|---|---|---|---|---|---|---|---|---|---|
| | 25/32" Wood & Bldg. Paper | 3/8" Ply-wood | 1/2" Insula-tion Board | 25/32" Wood & Bldg. Paper | 3/8" Ply-wood | 1/2" Insula-tion Board | 25/32" Wood & Bldg. Paper | 3/8" Ply-wood | 1/2" Insula-tion Board |
| **EXTERIOR FINISH, 1/2" PLYWOOD** | | | | | | | | | |
| 1/2" Rigid insulation board | 0.20 | 0.23 | 0.19 | 0.12 | 0.12 | 0.11 | 0.10 | 0.11 | 0.10 |
| 1/4" Plywood paneling | 0.16 | 0.31 | 0.25 | 0.13 | 0.14 | 0.13 | 0.12 | 0.13 | 0.11 |
| **EXTERIOR FINISH, BRICK VENEER** | | | | | | | | | |
| 1/2" Gypsumboard | 0.27 | 0.31 | 0.25 | 0.13 | 0.15 | 0.13 | 0.12 | 0.13 | 0.12 |
| 1/2" Gypsumboard, foil-backed | 0.18 | 0.20 | 0.18 | 0.13 | 0.15 | 0.13 | 0.10 | 0.10 | 0.09 |
| 3/4" Wood paneling | 0.23 | 0.27 | 0.22 | 0.13 | 0.14 | 0.12 | 0.11 | 0.12 | 0.11 |
| 1/2" Rigid insulation board | 0.21 | 0.24 | 0.20 | 0.12 | 0.13 | 0.11 | 0.11 | 0.11 | 0.10 |
| 1/4" Plywood paneling | 0.27 | 0.33 | 0.26 | 0.14 | 0.15 | 0.13 | 0.12 | 0.13 | 0.12 |

| with 1-inch Reflective Insulation (recessed), Sheathing Thickness to be: | | | with 2-inch Nonreflective Insulation (to stud face), Sheathing Thickness to be: | | | with 2-inch Nonreflective Insulation (recessed), Sheathing thickness to be: | | | with 2-inch Reflective Insulation (recessed), Sheathing Thickness to be: | | | with 3-5/8-inch Full Thick Insulation, Sheathing Thickness to be: | | |
|---|---|---|---|---|---|---|---|---|---|---|---|---|---|---|
| 25/32" Wood & Bldg. Pa-per | 3/8" Ply-wood | 1/2" Insu-lation Board | 25/32" Wood & Bldg. Pa-per | 3/8" Ply-wood | 1/2" Insu-lation Board | 25/32" Wood & Bldg. Pa-per | 3/8" Ply-wood | 1/2" Insu-lation Board | 25/32" Wood & Bldg. Pa-per | 3/8" Ply-wood | 1/2" Insu-lation Board | 25/32" Wood & Bldg. Pa-per | 3/8" Ply-wood | 1/2" Insu-lation Board |
| 0.09 | 0.10 | 0.09 | 0.09 | 0.09 | 0.09 | 0.08 | 0.09 | 0.08 | 0.07 | 0.07 | 0.07 | 0.06 | 0.06 | 0.06 |
| 0.09 | 0.09 | 0.09 | 0.09 | 0.09 | 0.09 | 0.07 | 0.07 | 0.07 | 0.07 | 0.07 | 0.07 | 0.06 | 0.06 | 0.06 |
| 0.09 | 0.09 | 0.08 | 0.08 | 0.09 | 0.08 | 0.08 | 0.08 | 0.08 | 0.06 | 0.07 | 0.06 | 0.06 | 0.06 | 0.06 |
| 0.08 | 0.09 | 0.08 | 0.08 | 0.08 | 0.08 | 0.07 | 0.08 | 0.07 | 0.06 | 0.07 | 0.06 | 0.06 | 0.06 | 0.06 |
| 0.09 | 0.10 | 0.09 | 0.09 | 0.09 | 0.08 | 0.08 | 0.09 | 0.08 | 0.07 | 0.07 | 0.07 | 0.06 | 0.06 | 0.06 |
| 0.09 | 0.09 | 0.09 | 0.09 | 0.09 | 0.08 | 0.08 | 0.08 | 0.08 | 0.07 | 0.07 | 0.07 | 0.06 | 0.06 | 0.06 |
| 0.09 | 0.09 | 0.08 | 0.09 | 0.09 | 0.08 | 0.07 | 0.07 | 0.07 | 0.07 | 0.07 | 0.06 | 0.06 | 0.06 | 0.06 |
| 0.09 | 0.09 | 0.08 | 0.08 | 0.09 | 0.08 | 0.08 | 0.08 | 0.07 | 0.06 | 0.07 | 0.06 | 0.06 | 0.06 | 0.06 |
| 0.08 | 0.09 | 0.08 | 0.08 | 0.08 | 0.08 | 0.07 | 0.08 | 0.07 | 0.06 | 0.07 | 0.06 | 0.06 | 0.06 | 0.06 |
| 0.09 | 0.09 | 0.09 | 0.09 | 0.09 | 0.08 | 0.08 | 0.08 | 0.08 | 0.07 | 0.07 | 0.07 | 0.06 | 0.06 | 0.06 |
| 0.09 | 0.10 | 0.09 | 0.09 | 0.09 | 0.09 | 0.08 | 0.09 | 0.08 | 0.07 | 0.07 | 0.07 | 0.06 | 0.06 | 0.06 |
| 0.09 | 0.09 | 0.09 | 0.09 | 0.09 | 0.09 | 0.07 | 0.07 | 0.07 | 0.07 | 0.07 | 0.07 | 0.06 | 0.06 | 0.06 |

TABLE 19–3 *(cont.)*

| with 1-inch Reflective Insulation (recessed), Sheathing Thickness to be: | | | with 2-inch Nonreflective Insulation (to stud face), Sheathing Thickness to be: | | | with 2-inch Nonreflective Insulation (recessed), Sheathing thickness to be: | | | with 2-inch Reflective Insulation (recessed), Sheathing Thickness to be: | | | with 3-⅝-inch Full Thick Insulation, Sheathing Thickness to be: | | |
|---|---|---|---|---|---|---|---|---|---|---|---|---|---|---|
| $^{25}/_{32}$" Wood & Bldg. Paper | ⅜" Plywood | ½" Insulation Board | $^{25}/_{32}$" Wood & Bldg. Paper | ⅜" Plywood | ½" Insulation Board | $^{25}/_{32}$" Wood & Bldg. Paper | ⅜" Plywood | ½" Insulation Board | $^{25}/_{32}$" Wood & Bldg. Paper | ⅜" Plywood | ½" Insulation Board | $^{25}/_{32}$" Wood & Bldg. Paper | ⅜" Plywood | ½" Insulation Board |
| 0.09 | 0.09 | 0.08 | 0.08 | 0.09 | 0.08 | 0.08 | 0.08 | 0.08 | 0.07 | 0.07 | 0.06 | 0.06 | 0.06 | 0.06 |
| 0.08 | 0.09 | 0.08 | 0.08 | 0.08 | 0.08 | 0.07 | 0.08 | 0.07 | 0.06 | 0.07 | 0.06 | 0.06 | 0.06 | 0.06 |
| 0.09 | 0.10 | 0.09 | 0.09 | 0.09 | 0.09 | 0.08 | 0.09 | 0.08 | 0.07 | 0.07 | 0.07 | 0.06 | 0.06 | 0.06 |
| 0.09 | 0.10 | 0.09 | 0.09 | 0.09 | 0.09 | 0.08 | 0.08 | 0.08 | 0.07 | 0.07 | 0.07 | 0.06 | 0.06 | 0.06 |
| 0.09 | 0.09 | 0.09 | 0.09 | 0.09 | 0.09 | 0.07 | 0.07 | 0.07 | 0.07 | 0.07 | 0.07 | 0.06 | 0.06 | 0.06 |
| 0.09 | 0.09 | 0.08 | 0.08 | 0.09 | 0.08 | 0.08 | 0.08 | 0.08 | 0.07 | 0.07 | 0.06 | 0.06 | 0.06 | 0.06 |
| 0.08 | 0.09 | 0.08 | 0.08 | 0.08 | 0.08 | 0.07 | 0.08 | 0.07 | 0.06 | 0.07 | 0.06 | 0.06 | 0.06 | 0.06 |
| 0.09 | 0.10 | 0.09 | 0.09 | 0.09 | 0.09 | 0.08 | 0.09 | 0.08 | 0.07 | 0.07 | 0.07 | 0.06 | 0.06 | 0.06 |
| 0.09 | 0.09 | 0.09 | 0.08 | 0.09 | 0.08 | 0.08 | 0.08 | 0.08 | 0.07 | 0.07 | 0.07 | 0.06 | 0.06 | 0.06 |
| 0.09 | 0.09 | 0.08 | 0.08 | 0.09 | 0.08 | 0.07 | 0.07 | 0.07 | 0.07 | 0.07 | 0.06 | 0.06 | 0.06 | 0.06 |
| 0.08 | 0.09 | 0.08 | 0.08 | 0.09 | 0.08 | 0.08 | 0.08 | 0.07 | 0.06 | 0.07 | 0.06 | 0.06 | 0.06 | 0.06 |
| 0.08 | 0.09 | 0.08 | 0.08 | 0.08 | 0.08 | 0.07 | 0.08 | 0.07 | 0.06 | 0.06 | 0.06 | 0.06 | 0.06 | 0.06 |
| 0.09 | 0.09 | 0.09 | 0.09 | 0.09 | 0.08 | 0.08 | 0.08 | 0.08 | 0.07 | 0.07 | 0.07 | 0.06 | 0.06 | 0.06 |
| 0.09 | 0.09 | 0.09 | 0.09 | 0.09 | 0.08 | 0.08 | 0.08 | 0.08 | 0.07 | 0.07 | 0.07 | 0.06 | 0.06 | 0.06 |
| 0.09 | 0.09 | 0.09 | 0.09 | 0.09 | 0.08 | 0.07 | 0.07 | 0.07 | 0.07 | 0.07 | 0.07 | 0.06 | 0.06 | 0.06 |
| 0.09 | 0.09 | 0.08 | 0.08 | 0.09 | 0.08 | 0.08 | 0.08 | 0.08 | 0.07 | 0.07 | 0.06 | 0.06 | 0.06 | 0.06 |
| 0.08 | 0.09 | 0.08 | 0.08 | 0.08 | 0.08 | 0.07 | 0.08 | 0.07 | 0.06 | 0.07 | 0.06 | 0.06 | 0.06 | 0.06 |
| 0.09 | 0.10 | 0.09 | 0.09 | 0.09 | 0.08 | 0.08 | 0.08 | 0.08 | 0.07 | 0.07 | 0.07 | 0.06 | 0.06 | 0.06 |
| 0.09 | 0.10 | 0.09 | 0.09 | 0.09 | 0.09 | 0.08 | 0.09 | 0.08 | 0.07 | 0.07 | 0.07 | 0.06 | 0.06 | 0.06 |
| 0.09 | 0.09 | 0.09 | 0.09 | 0.09 | 0.09 | 0.07 | 0.07 | 0.07 | 0.07 | 0.07 | 0.07 | 0.06 | 0.06 | 0.06 |
| 0.09 | 0.09 | 0.09 | 0.08 | 0.09 | 0.08 | 0.08 | 0.08 | 0.08 | 0.07 | 0.07 | 0.07 | 0.06 | 0.06 | 0.06 |
| 0.08 | 0.09 | 0.08 | 0.08 | 0.09 | 0.08 | 0.08 | 0.08 | 0.07 | 0.06 | 0.07 | 0.06 | 0.06 | 0.06 | 0.06 |
| 0.09 | 0.10 | 0.09 | 0.09 | 0.09 | 0.09 | 0.08 | 0.09 | 0.08 | 0.07 | 0.07 | 0.07 | 0.06 | 0.06 | 0.06 |
| 0.09 | 0.10 | 0.09 | 0.09 | 0.09 | 0.09 | 0.08 | 0.09 | 0.08 | 0.07 | 0.07 | 0.07 | 0.06 | 0.06 | 0.06 |
| 0.09 | 0.10 | 0.09 | 0.09 | 0.09 | 0.09 | 0.07 | 0.07 | 0.07 | 0.07 | 0.07 | 0.07 | 0.06 | 0.06 | 0.06 |
| 0.09 | 0.09 | 0.09 | 0.09 | 0.09 | 0.08 | 0.08 | 0.08 | 0.08 | 0.07 | 0.07 | 0.07 | 0.06 | 0.06 | 0.06 |
| 0.09 | 0.09 | 0.08 | 0.08 | 0.09 | 0.08 | 0.08 | 0.08 | 0.07 | 0.06 | 0.07 | 0.06 | 0.06 | 0.06 | 0.06 |
| 0.09 | 0.10 | 0.09 | 0.09 | 0.10 | 0.09 | 0.08 | 0.09 | 0.08 | 0.07 | 0.07 | 0.07 | 0.06 | 0.06 | 0.06 |

Reprinted with permission of National Forest Products Association.

finish. One conclusion to be drawn from interpreting this table is that heat transmission decreases as the insulation thickness increases, but not in direct relation to the increase.

Table 19–4 is used to list *U* values for masonry walls that have no insulation, for walls that have ¾-inch insulation placed between nominal 1-inch furring strips, and for walls that have 1⅝-inch insulation placed between nominal 2-inch furring strips.

Table 19–5 lists *U* values for wood-joisted floors over unheated crawl spaces. In

## TABLE 19–4  U VALUES FOR MASONRY WALLS

With interior finish applied over furring strips and insulation between the strips.

| Exterior Wall | Interior Finish | None | Insulation between furring strips | |
|---|---|---|---|---|
| | | | ¾-inch Flexible | 1⅝-inch Flexible |
| 8″ Brick (face and common) | ½″ Gypsumboard | 0.29 | 0.19 | 0.12 |
| | ½″ Gypsumboard, foil-backed | 0.19 | 0.19 | 0.12 |
| | ¾″ Wood paneling | 0.25 | 0.17 | 0.11 |
| | ½″ Rigid insulation board | 0.22 | 0.16 | 0.11 |
| | ¼″ Plywood | 0.30 | 0.19 | 0.12 |
| 8″ Cinder block | ½″ Gypsumboard | 0.25 | 0.17 | 0.11 |
| | ½″ Gypsumboard, foil-backed | 0.18 | 0.17 | 0.11 |
| | ¾″ Wood paneling | 0.22 | 0.16 | 0.11 |
| | ½″ Rigid insulation board | 0.20 | 0.15 | 0.10 |
| | ¼″ Plywood | 0.26 | 0.18 | 0.11 |
| 8″ Brick and block (face brick and cinderblock) | ½″ Gypsumboard | 0.26 | 0.18 | 0.11 |
| | ½″ Gypsumboard, foil-backed | 0.18 | 0.18 | 0.11 |
| | ¾″ Wood paneling | 0.23 | 0.16 | 0.11 |
| | ½″ Rigid insulation board | 0.20 | 0.15 | 0.10 |
| | ¼″ Plywood | 0.26 | 0.18 | 0.11 |
| 8″ Pumice aggregate block | ½″ Gypsumboard | 0.23 | 0.16 | 0.11 |
| | ½″ Gypsumboard, foil-backed | 0.17 | 0.16 | 0.11 |
| | ¾″ Wood paneling | 0.21 | 0.15 | 0.10 |
| | ½″ Rigid insulation board | 0.19 | 0.14 | 0.10 |
| | ¼″ Plywood | 0.24 | 0.17 | 0.11 |

Reprinted with permission of National Forest Products Association.

## TABLE 19–5  U VALUES FOR WOOD-JOISTED FLOORS OVER UNHEATED CRAWL SPACES

For 2-inch joists spaced 12 to 24 inches on centers with flexible insulation between joists, reflective side up.

| Finish Floor | Subfloor | None | Flexible Insulation Between Joists | | | | | |
|---|---|---|---|---|---|---|---|---|
| | | | 1-inch | | 2-inch | | 3⅝-inch | |
| | | | Nonrefl. | Reflect. | Nonrefl. | Reflect. | Nonrefl. | Reflect. |
| ²⁵⁄₃₂″ Hardwood | ²⁵⁄₃₂″ wood or ¾″ plywood | 0.28 | 0.12 | 0.09 | 0.08 | 0.07 | 0.06 | 0.05 |
| | ⅝″ wood or ⅝″ plywood | 0.29 | 0.12 | 0.09 | 0.08 | 0.07 | 0.06 | 0.05 |
| | ½″ plywood | 0.31 | 0.12 | 0.09 | 0.08 | 0.07 | 0.06 | 0.05 |

Reprinted with permission of National Forest Products Association.

**TABLE 19-6    U VALUES FOR WOOD-JOISTED CEILINGS UNDER UNHEATED ATTICS**

For 2-inch joists spaced 12 to 24 inches on centers with flexible insulation between joists and stapled to bottom faces, or with loose fill installed between joists.

| Floor on Joists | Ceiling Finish | Insulation Between Joists | | | | |
|---|---|---|---|---|---|---|
| | | None | 1-inch Flexible | 2-inch Flexible | 3⅝-inch Flexible[1] | 6-inch Loose Fill |
| None | ¾″ Metal lath and plaster | 0.74 | 0.20 | 0.11 | 0.07 | 0.05 |
| | ½″ Gypsumboard or ⅜″ Gypsum lath and ½″ plaster | 0.60 | 0.19 | 0.11 | 0.07 | 0.05 |
| | ½″ Insulation board | 0.38 | 0.16 | 0.10 | 0.06 | 0.04 |
| 25/32″ T and G Wood Boards | ¾″ Metal lath and plaster | 0.31 | 0.14 | 0.09 | 0.06 | 0.04 |
| | ½″ Gypsumboard or ⅜″ Gypsum lath and ½″ plaster | 0.29 | 0.14 | 0.09 | 0.06 | 0.04 |
| | ½″ Insulation board | 0.22 | 0.12 | 0.08 | 0.06 | 0.04 |

[1] U Values also apply to 4-inch loose fill

Reprinted with permission of National Forest Products Association.

most cases 1 inch of insulation placed between the floor joists is considered to be sufficient to meet minimum comfort standards, but for increased comfort standards 2 inches of insulation should be used.

Table 19-6 lists $U$ values for wood-

FIGURE 19-3.    Chart for determining the $U$ value for walls for various climatic conditions and for determining differences between inside air and surface temperatures

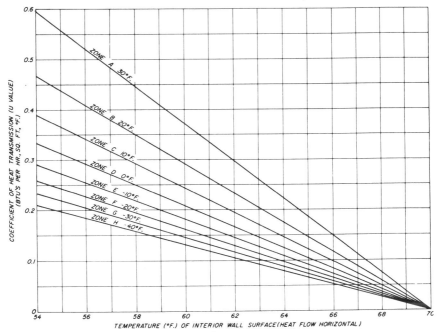

joisted ceilings of various finishes under unheated attics.

Figures 19–3, 19–4, and 19–5 illustrate the way to determine the adequate $U$ value for walls and ceilings in various climate conditions. The first step is to determine the zone where the building will be located. (See Figure 19–5.) Then an acceptable difference between the wall surface temperature and the air temperature in the room is subtracted from 70° F, which is assumed to be the most comfortable temperature for adults. This value represents the surface temperature as depicted on the bottom scale of Figure 19–3. The surface tempera-

ture is then followed up until it intersects the established zone line; the necessary $U$ value is then found by projecting across the chart at a right angle.

*For example:* Assume that a house will be built in zone B and the minimum wall surface temperature will not be more than 5° F. The surface temperature will then be 65° F. The point where the 65° line intersects the line for zone B in Figure 19–3, indicates the $U$ value, 0.150. The calculations for determining the $U$ value for ceilings is determined in much the same manner as were the walls, but Figure 19–4 is used rather than Figure 19–5.

FIGURE 19–4.   Chart for determining the $U$ value for ceilings (heat flow upwards) for various climatic conditions and for determining differences between inside air and surface temperatures

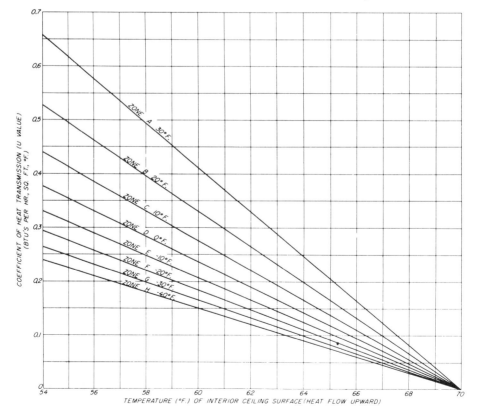

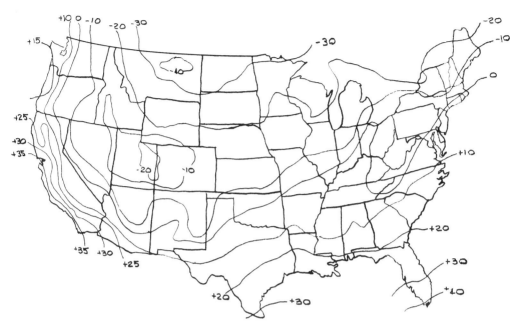

FIGURE 19–5. Approximate outside design temperature zones for houses or similar dwellings

## WHERE TO INSULATE

Insulation is used to retard the flow of heat through ceilings, walls, and floors where temperatures vary on opposite sides of the surface. One such area, often overlooked, is that of the foundation and floors. To insulate the perimeter of slabs, rigid insulation such as polystyrene foam boards can be applied both vertically and horizontally. When the insulation is placed vertically, it can be placed against the interior or exterior of the foundation during backfilling. The boards can either be laid dry against the foundation or attached to it with a mastic. If the insulation projects above grade, its exposed surface should be covered with sheet metal, cement asbestos board, or a similar finish bonded to the insulation. When rigid insulation is placed horizontally, it should be placed under a polyethylene vapor barrier and then covered with concrete.

The way to insulate a crawl space is to place rigid insulation against the inside of the foundation wall and cover the soil with 4-mil polyethylene. The joints of the polyethylene should be lapped and sealed with an asphaltic compound. The floors over crawl spaces should also be insulated. The insulation most often used is the batt type, which is placed between the joists. It can be held in place by wire mesh stapled to the edges of the joists or by pieces of heavygauge wire wedged between the joists. (See Figure 19–6.)

Because of the poor thermal resistance qualities of masonry, the insulation of

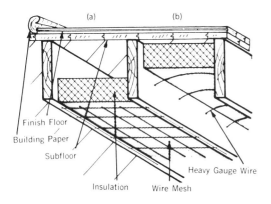

FIGURE 19-6. Floor insulation

heated basements is of extreme importance. To insulate the foundation wall, furring strips should be attached to the masonry wall and blanket insulation stapled to them. If the insulation has its own vapor barrier, the flanges are stapled to the furring strips; if it does not, a polyethylene film or foil-backed gypsumboard should be placed over the furring strips.

When insulation is placed above the ceiling and below the roof, an air space must be left between the insulation and the roof sheathing. This air space permits attic air circulation. If the insulation has a vapor barrier, the flanges are stapled to the bottom edges of the joists. When loose fill insulation is used, it is installed after the finished ceiling has been installed.

If a structure has a flat roof, a vapor barrier and then rigid insulation may be placed on top of the roof sheathing. One method of installing the insulation is to (1) nail the insulation to the rafters; (2) then nail 1-inch strips over the insulation; (3) nail sheathing over the strips; and (4) place the shingles over the sheathing.

All exterior walls should be properly insulated between heated rooms and unheated areas such as garages and porches. If batt insulation is used, the flange can be stapled to the inside edge of the stud, or the insulation can be recessed to provide a minimum air space of ¾ inch. (See Figure 19-7.)

FIGURE 19-7. Installation of batt insulation

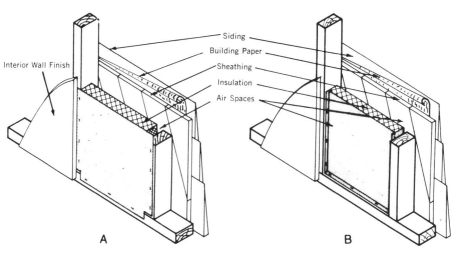

## TYPES OF INSULATION

There are three basic types of insulation: mineral (glass fiber, rock or slag wool), organic (paper, polystyrene, and polyurethane), and metallic. These types can be further subdivided into rigid, flexible, fill, and reflective insulation.

### Rigid Insulation

This is an insulating board that is manufactured primarily from wood or cane fibers. The fibers are reduced to a pulp and then assembled into large lightweight or low-density boards. The boards combine strength with heat- and sound-insulating properties. The boards are usually ½ to 1 inch, 4 feet wide, and 6 to 12 feet long.

### Flexible Insulation

Flexible insulation is manufactured in two basic classifications: blanket or quilt, and batt. Blanket insulation is usually manufactured in rolls or strips of various lengths and widths. The thickness of the insulation ranges from ½ to 3½ inches. The insulation is sometimes manufactured from loosely felted mats of wood fiber that are chemically treated for resistance to fire. This kind is equipped with nailing tabs on the sides. It also has a covering sheet of either Kraft paper (providing a vapor barrier on one side of the insulation) or aluminum foil (used for its heat-reflective value).

Batt insulation is similar to the blanket type, but it is usually thicker, and is made in widths to fit between framing members.

### Fill Insulation

This insulation is an organic or inorganic material that can be blown or poured into a given area. The density of the insulation will vary depending upon the application. One type of fill insulation is a loose, granular wool that is used primarily between open attic joists; another type is a free-flowing granular vermiculite that pours readily into cores or cavities of masonry. When vermiculite is used, it can fill cores or cavities up to heights of 20 feet in a single pour without bridging. This granular insulation is poured into a hopper placed on top of the wall. To prevent leakage of insulation around weep holes, glass fiber ropes are placed in any voids.

### Reflective Insulation

There are various forms of reflective insulation available, but in most cases it is used in combination with batt and blanket insulation and gypsum lath and wallboard. To be effective, the reflective surface must face a minimum air space of ¾ inch. If this material is in contact with another surface, it loses its reflective properties. The air space should not, however, be so great that convective air currents reduce the effect of the insulation.

## CONDENSATION

Air within a building contains a certain amount of moisture, and when it comes in contact with a cold surface the temperature of the air is lowered. When the temperature reaches a point at which the air can no longer contain its moisture, water

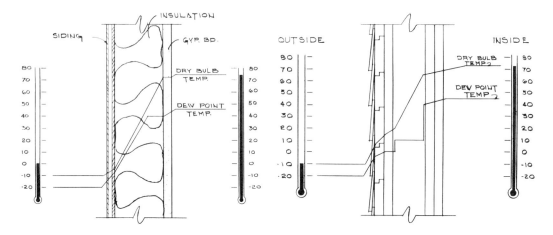

FIGURE 19-8.  Wall section without a vapor barrier                FIGURE 19-9.  Wall section with a vapor barrier

droplets are deposited on the cold surface. The relative temperature that allows for condensation to occur is called the "dewpoint."

Condensation on windows is a good example of the effect of lowering the atmospheric temperature below the dewpoint. When condensation occurs on the window, the temperature difference is great enough to lower the surface temperature of the glass below the dewpoint temperature of the atmosphere in the room.

Condensation occurring within walls is a common cause of exterior paint failure, as well as setting up favorable conditions for decay.

When insulation is placed in a wall cavity, the part of the wall on the cool side will be lower during cold weather than if no insulation were used. Therefore, there is likely to be more condensation on the wall material, unless a vapor barrier is used.

To illustrate this point, the wall section shown in Figure 19-8 does not have a vapor barrier; when the dewpoint temperature intersects with the room temperature of 73°, condensation of moisture vapor into water will occur and wet insulation will be the result. Now look at the wall section in Figure 19-9, in which a vapor barrier is used. The dewpoint temperature line drops when it strikes the vapor barrier and the two temperature gradient lines do not touch. Thus the insulation remains dry and performs at full efficiency.

## VAPOR BARRIERS

Because vapor passes relatively unchecked through most building materials, a vapor barrier is necessary to prevent this migration. For maximum effectiveness the vapor barrier should be placed at or near the warm face of the wall. Thus the temperature of the barrier will always be above the dewpoint temperature of the room.

Several different types of vapor barriers can be used:

- A vapor-resistant membrane installed on the inner face of the studs and under lath or finishing
- Vapor-resistant covering material on the inside face of the insulation

- A vapor-resistant membrane that is an integral part of the lath or finishing material
- Reflective insulation constructed of vapor-resistant materials.

## VENTILATION

To help prevent condensation, ceilings and attics should be properly vented so the air flow can carry the vapor away before it can condense. Vent openings should be designed to keep snow and rain from entering and should be screened to prevent the entrance of insects.

Some of the types of ventilators are:

- *Gable vents:* Placed at opposite gable ends for air flow, or used in conjunction with eave vents
- *Eave vents:* Placed under roof eaves; should be used with gable or roof vents for air flow.

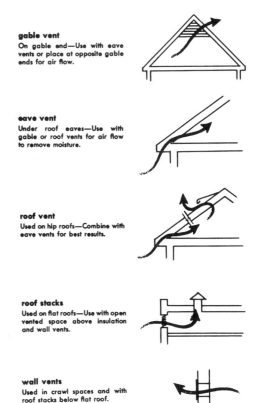

**gable vent**
On gable end—Use with eave vents or place at opposite gable ends for air flow.

**eave vent**
Under roof eaves—Use with gable or roof vents for air flow to remove moisture.

**roof vent**
Used on hip roofs—Combine with eave vents for best results.

**roof stacks**
Used on flat roofs—Use with open vented space above insulation and wall vents.

**wall vents**
Used in crawl spaces and with roof stacks below flat roof.

FIGURE 19-10. Five types of ventilator

- *Roof vents:* Usually used on hip roofs; should be combined with eave vents for best results
- *Roof stacks:* Usually used on flat roofs; should be used with open vented space above insulation and wall vents

**TABLE 19-7   MINIMUM RECOMMENDED VENTILATION REQUIREMENTS FOR ATTICS**

| Type of Screen of Louver Covering Vent Opening | Gross Area of Vent Required per 300 sq. ft. of floor area |
|---|---|
| $\frac{1}{8}$" mesh screen | $1\frac{1}{4}$ sq ft |
| $\frac{1}{16}$" mesh insect screen | 2 sq ft |
| Louvers and $\frac{1}{4}$" mesh hardware cloth | 2 sq ft |
| Louvers and $\frac{1}{8}$" mesh screen | $2\frac{1}{4}$ sq ft |
| Louvers and $\frac{1}{16}$" mesh insect screen | 3 sq ft |

Ventilation in excess of these minimums results in better performance by removing unforeseen excessive moisture in walls and reduces the solar temperature in the attic in summer.

Reprinted with permission of Owens-Corning Fiberglas Corporation.

■ *Wall vents:* Used both in crawl spaces and with roof stacks below flat roofs, (See Figure 19–10.)

Table 19–7 illustrates the determination of minimum recommended ventilation requirements for attics.

## INSTALLATION OF BATT INSULATION

For the most effective results, insulation materials must be properly installed. When 48-inch batt insulation is used, the first batt should be placed at the bottom of the stud space and pressed into place against the sheathing. The second batt is placed above the first and pressed into place. The batts may overlap or butt at joints and should fit snug at both plate lines. Some batts are designed without flanges and can be held in place by friction, while others have flanges that are stapled to the framing members.

Insulation should also be placed behind pipes to keep them from freezing

When batt insulation is placed in the ceiling, the batts should not extend past the exterior plate line to avoid covering the eave vents. The batts should extend only 1 or 2 inches beyond the inside edge, and the ends should be butted together to form a tight joint.

If a vapor barrier is desired or required, foil-backed gypsumboard is the simplest to apply; but if the ceiling is properly ventilated no vapor barrier is necessary.

# PLUMBING

AN ADEQUATE PLUMBING SYSTEM IS NECES-
sary to maintain minimum health standards and is a requirement in homes built
today. A typical plumbing system consists
of a water supply and a waste-disposal system.

Each system is a network of pipes,
connectors, and valves.

## MATERIALS

The materials used in the system vary, but
the most common are cast iron, vitrified
clay pipe, plastic, galvanized steel pipe, and
copper.

### Cast Iron

Pipes made from cast iron are usually used
for sanitary drainage and stacks in waste-
disposal systems. The pipe is usually sold
in 5-foot lengths and can have a large diameter hub at one end and a slight ridge
spigot at the other end. (See Figure 20–1.)
To join lengths of pipe, the spigot is placed
inside the hub, and oakum and lead are
packed down into the space between them.
The oakum is packed with a "yarning iron"
to within ¾" or 1" of the hub top. Molten
lead is then poured over the oakum and allowed to cool. As soon as it hardens it is
caulked by caulking irons tapped against
the lead surface. As the lead is caulked it
spreads against the surface of the hub.

Another means of connecting the kind of
cast iron pipe that has a bell and spigot is
to join the individual pieces of pipe with a
neoprene gasket. (See Figure 20–2.) The
gasket is inserted in the hub and mopped
with a lubricant; then the spigot end of the
connecting pipe is force-fitted into the gasket. This technique is faster than using
oakum and lead, and according to recent
research is an effective means of jointing 2
lengths of pipe.

Hubless cast iron soil pipe and fittings
are also being used in the plumbing industry. (See Figure 20–3.) This type of cast
iron pipe is jointed by means of a neoprene
sealing gasket designed to be compressed
around a butted pipe joint with a special
stainless steel retaining clamp. The piping
is available in 2-, 3-, and 4-inch diameters,
and in 5- and 10-foot lengths. The system is
easy to install, requires no heat or flame
during installation, and a standard 3-inch
pipe can fit into a 2- × 4-inch stud wall.

### Vitrified Clay Pipe

Vitrified clay pipe is frequently used for
public sewers, house sewers, and drains. It
is made of fired clay and water, and is
available in varying lengths. To join the
different lengths of pipe, the spigot is
placed inside the hub, and oakum is packed
between the inside of the hub and the
outside of the connecting pipe. A mixture

FIGURE 20–1. Cast-iron soil pipe

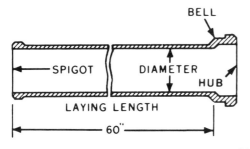

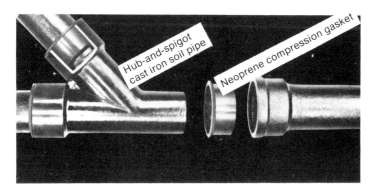

FIGURE 20-2  Connecting cast iron pipe with a Neoprene gasket. Fast, uniform installation; needs no heat or flame to install; proven performance, above and below ground; quiet—absorbs vibration; chemical resistant; lasting durability

of portland cement, sand, and water is then packed over the oakum. One part of portland cement is mixed with 2 parts of sand and just enough water to make an easily workable putty substance.

When the pipe is laid it should be placed either on stable soil or on stone or concrete foundations. If it is laid on an unstable foundation it might sag, which would cause the joint to crack.

Individual pipe lengths can also be joined by a mechanical compression joint made of polyvinyl chloride. This joint is first inserted into the hub, then mopped with a lubricant, and finally, the spigot end of the connecting pipe is force-fitted into the gasket.

## Plastic Pipe

Plastic pipe is presently being used in all areas of construction, but many local build-

FIGURE 20–3.  Hubless cast-iron soil pipe and fittings

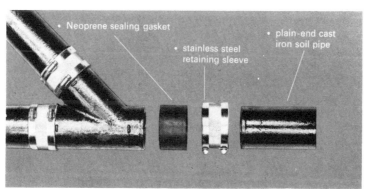

ing codes have still not accepted the relatively new material. It is readily cut with a sharp knife or any fine-toothed saw, and can be joined with either threaded fittings or a special plastic cement. To join this pipe with plastic cement, the end must first be coated with cement and then inserted into the fitting. As the pipe is being slid into position, it should be slightly twisted so that the cement will spread evenly. Within just a few minutes the cement will have set.

## Galvanized Steel Pipe

Galvanized steel pipe can be used to carry water or gas, but in most cases it has been replaced by other types of material. To join the pipe it is necessary only to dope the ends with pipe-joint compound and screw the fittings into place. But if necessary, the pipe can also be cut, reamed, and threaded with a ratchet die.

## Copper Tubing

Copper tubing is available in 3 types—K, L, and M. Each type represents a series of sizes with different wall thicknesses. All 3 types can be purchased in hard-tempered straight lengths, but only types K and L are also available in a soft or annealed temper, straight length or coil. Hard-tempered tubes can be joined by soldering, brazing, or by the use of capillary fittings; soft-tempered tubes can be joined by soldering, brazing, or by flare-type compression fittings.

To make a soldered joint, the tube must first be cut to the correct length with a hacksaw or a tube cutter. Then the small burrs should be removed, either by the

reamer on the cutter or by a half-round file. (See Figure 20–4.)

The surfaces to be joined should then be cleaned of any surface dirt or oil. One of the best ways to clean copper tubing is to lightly sand the tube ends with a strip of emery cloth. (See Figure 20–5.) Cleaning pads and special wire brushes, however, can also be used. Next the surfaces should be covered with a thin film of flux, applied with a brush. Flux removes residual traces of oxides and protects the surfaces to be soldered from further oxidation. (See Figure 20–6.) The most widely recommended flux is mildly corrosive. It contains zinc and ammonium chlorides in a petrolatum base.

Next the tube can be inserted into the fitting, adjacent to the stop. During this step, the tube should be slightly turned so that the flux will spread evenly over the adjoining surfaces. Then the excess flux is removed with a rag, and the joint is ready for soldering.

For proper soldering of the joint, an even distribution of heat should be played on the fitting, but the flame should not be pointed into the socket. When the surface has been heated, the solder should be touched to the joint. If the area has been correctly prepared, the solder will be drawn into the joint by the natural force of capillary action. (See Figure 20–7.) Under no circumstance should the flame be placed directly on the solder, nor should the fitting be overheated, because this may burn the flux from the joint or cause the fitting to crack. After the joint has been soldered, it should be allowed to cool at room temperature.

To make a flared joint, the copper tube should be cut to the correct length and the burrs removed. The tube is then clamped in

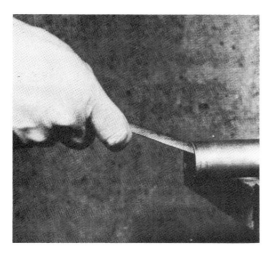

FIGURE 20–4 (top left) Removing burr from tubing
FIGURE 20–5 (center left) Sand the end of the tube with emery cloth
FIGURE 20–6 (bottom left) Applying flux to tubing
FIGURE 20–7 (above) Applying solder to the tubing joint

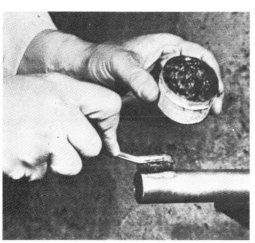

a flaring block so that the end projects about ⅛ inch past the face of the block. (See Figure 20–8.) The yoke of the flaring tools is then positioned so the beveled end of the compressor cone falls directly over the end of the tube. Pressure should be applied to the compressor screw, and a flare will form on the end of the tube. (See Figure 20–9.) Next the fitting is placed against the flare and the coupling nut is engaged with the fitting's threads. (See Figure 20–10.) Once the threads have engaged, two wrenches are used to tighten the joint.

If necessary, copper tubing can be bent. One technique is to use a lever-type hand bender. (See Figure 20–11.)

FIGURE 20–8. Clamping the tube in the flaring block

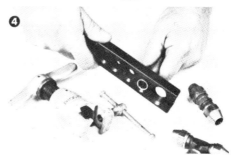

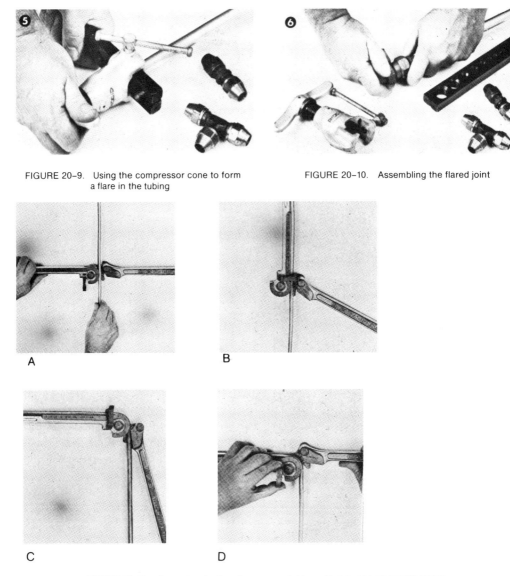

FIGURE 20–9.   Using the compressor cone to form          FIGURE 20–10.   Assembling the flared joint
a flare in the tubing

A                                    B

C                                    D

FIGURE 20–11.   Procedure for bending copper tubing with a lever-type hand bender

## BUILDING DRAIN AND BUILDING SEWER

The building drain and building sewer are the lowest horizontal portions of the drainage system, and are used to carry waste to the main sewer. (See Figure 20–12.) The *building drain* receives the discharge of all soil and waste stacks. It is a horizontal part of the system located under the building, and extending to a point 3 feet beyond the outside edge of the building. The *building sewer* receives the discharge from the building drain and carries this waste to the sewer.

To determine the size needed for the building drain and building sewer, the discharge of the expected or peak volume of water and waste must be calculated. First it will be necessary to ascertain the *total individual fixture unit values*. A fixture unit corresponds to 7½ gallons of water, which an ordinary lavatory with a normal 1¼-inch trap can discharge into a stack. Other fixtures have been tested for maximum flow, and fixture unit value has been assigned to each fixture. (See Table 20-1.) The number of fixture units must also be totaled. Table 20-2 illustrates how to determine the size of the drain and sewer.

*For example,* if a building has 2 water closets, 2 lavatories, a slop sink, a kitchen sink, a pair of laundry tubs, 2 floor drains, and a shower, there will be a total fixture value of 27. Using the 27 fixture units and assuming a fall of ¼ inch per foot, Table 20–2 indicates that a 3-inch building drain is needed. It should be noted that the building drain must have a graduated fall, usually of ½ inch per foot.

If it is necessary to change the direction of the house drain, this change should be made with fittings of long radius. Such fittings lessen the possibility of stoppage, which is a frequent problem. To correct such conditions, the house drain should be equipped with a cleanout. (See Figure 20–13.)

A cleanout should be placed at the base of all soil and waste stacks, as well as at 75-foot intervals.

FIGURE 20–12.  Schematic of the building drain and building sewer

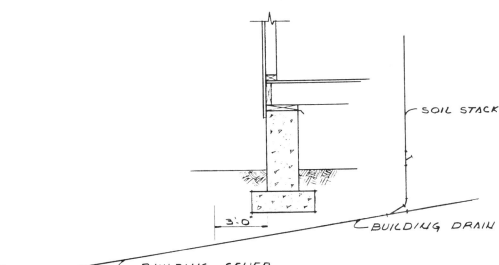

## TABLE 20–1 FIXTURE UNITS PER FIXTURE OR GROUP

| Fixture Type | Fixture unit value as load factors | Minimum size of trap (inches) | |
|---|---|---|---|
| 1 bathroom group consisting of water closet, lavatory, and bathtub or shower stall. | Tank water closet 6 Flush-valve water closet 8 | | |
| Bathtub (with or without overhead shower) | 2 | | $1\frac{1}{2}$ |
| Bathtub | 3 | | 2 |
| Bidet | 3 | Nominal | $1\frac{1}{2}$ |
| Combination sink and tray | 3 | | $1\frac{1}{2}$ |
| Combination sink and tray with food-disposal unit | 4 | Separate traps | $1\frac{1}{2}$ |
| Dental unit or cuspidor | $\frac{1}{2}$ | | $1\frac{1}{4}$ |
| Dental lavatory | 1 | | $1\frac{1}{4}$ |
| Drinking fountain | $\frac{1}{2}$ | | 1 |
| Dishwasher, domestic | 2 | | $1\frac{1}{2}$ |
| Floor drain | 1 | | 2 |
| Kitchen sink, domestic | 2 | | $1\frac{1}{2}$ |
| Kitchen sink, domestic, with food-disposal unit | 3 | | $1\frac{1}{2}$ |
| Lavatory | 1 | Small P.O. | $1\frac{1}{4}$ |
| Do | 2 | Large P.O. | $1\frac{1}{2}$ |
| Lavatory, barber, beauty parlor | 2 | | $1\frac{1}{2}$ |
| Lavatory, surgeon's | 2 | | $1\frac{1}{2}$ |
| Laundry tray (1 or 2 compartments) | 2 | | $1\frac{1}{2}$ |
| Shower stall, domestic | 2 | | 2 |
| Showers (group) per head | 3 | | |
| Sinks: | | | |
| Surgeon's | 3 | | $1\frac{1}{2}$ |
| Flushing rim (with valve) | 8 | | 3 |
| Service (Trap standard) | 3 | | 3 |
| Service (P trap) | 2 | | 2 |
| Pot, scullery, etc. | 4 | | $1\frac{1}{2}$ |
| Urinal, pedestal, syphon jet, blowout | 8 | Nominal | 3 |
| Urinal, wall lip | 4 | | $1\frac{1}{2}$ |
| Urinal stall, washout | 4 | | 2 |
| Urinal trough (each 2-foot section) | 2 | | $1\frac{1}{2}$ |
| Wash sink (circular or multiple), each set of faucets | 2 | Nominal | $1\frac{1}{2}$ |
| Water closet: | | | |
| Tank-operated | 4 | Nominal | 3 |
| Valve-operated | 6 | | 3 |

Reprinted with permission of Manas Public ations.

**TABLE 20–2  MAXIMUM LOADS FOR HORIZONTAL DRAINS**

| Diameter of Drain (in) | Horizontal Fixture Branch (dfu)[1] | Building Drain or Building Sewer Slope | | | |
|---|---|---|---|---|---|
| | | $\frac{1}{16}$ in/ft (dfu) | $\frac{1}{8}$ in/ft (dfu) | $\frac{1}{4}$ in/ft (dfu) | $\frac{1}{2}$ in/ft (dfu) |
| 1¼ | 1 | | | | |
| 1½ | 3 | | | | |
| 2 | 6 | | | | 26 |
| 2½ | 12 | | | 24 | 31 |
| 3 | 32[2] | | 36[3] | 42[2] | 50[2] |
| 4 | 160 | | 180 | 216 | 250 |
| 5 | 360 | | 390 | 480 | 575 |
| 6 | 620 | | 700 | 840 | 1000 |
| 8 | 1400 | 1400 | 1600 | 1920 | 2300 |
| 10 | 2500 | 2500 | 2900 | 3500 | 4200 |
| 12 | 3900 | 3900 | 4600 | 5600 | 6700 |
| 15 | 7000 | 7000 | 8300 | 10000 | 12000 |

[1] dfu—drainage fixture unit.

[2] Not more than two water closets or two bathroom groups.

[3] Less than 2 feet per second (fps).

Reprinted with permission of Manas Publications.

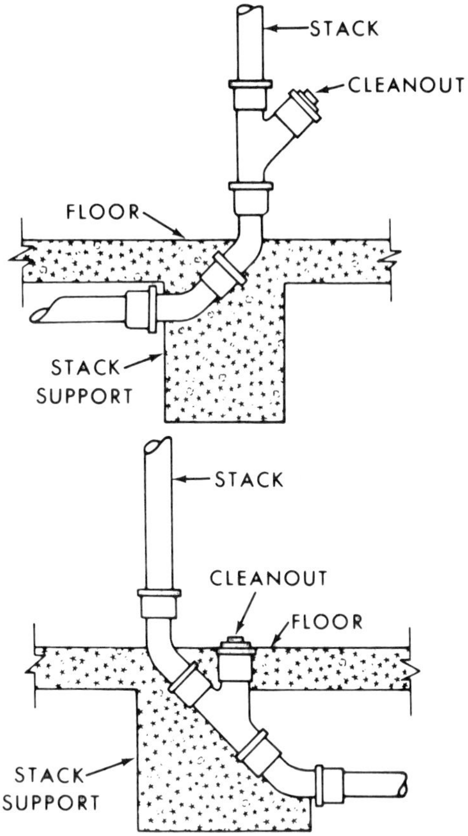

FIGURE 20–13 (right)  Two types of house drain with cleanout

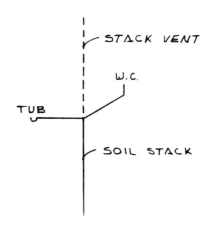

## SOIL AND WASTE PIPES

A soil pipe is a vertical portion of the plumbing system that receives the discharge of water closets, urinals, or fixtures having similar functions. (See Figure 20–14.) A waste pipe is similar to a soil pipe, except that it carries only liquid waste that is free of fecal matter.

The method of sizing a soil or waste stack is similar to that used in sizing a building drain. The maximum discharge in terms of fixture units is first calculated.

FIGURE 20–14 (right)  Schematic of a soil, or waste pipe

Then the pipe diameter is determined. (See Table 20–3.) The minimum size of a soil stack is 3 inches in diameter, but a pipe 4 inches in diameter is preferred.

**TABLE 20–3    MAXIMUM  LOADS FOR SOIL AND WASTE STACKS HAVING NOT MORE THAN THREE BRANCH INTERVALS**

| Diameter of Stack (inches) | Maximum Load | | Diameter of Stack (inches) | Maximum Load | |
| | On Any One Branch Interval (dfu)[1] | On Stack (dfu) | | On Any One Branch Interval (dfu)[1] | On Stack (dfu) |
|---|---|---|---|---|---|
| 1¼ | 1 | 2 | 4 | 100 | 240 |
| 1½ | 2 | 4 | 5 | 225 | 540 |
| 2 | 4 | 9 | 6 | 385 | 930 |
| 2½ | 8 | 18 | 8 | 875 | 2100 |
| 3 | 20[2] | 48[2] | | | |

[1] dfu—drainage fixture unit.

Reprinted with permission of Manas Publications.

[2] Not more than two water closets or bathroom groups within each branch interval nor more than six water closets or bathroom groups on the stack.

## SOIL BRANCH

A soil branch is a horizontal portion of the plumbing system that receives the direct discharge of water closets and sometimes of additional plumbing fixtures. (See Figure 20–15.) For the soil branch to be accessible it should be equipped with an adequate number of cleanouts. These should be installed wherever the soil branch changes direction, and should also be placed at the end of the branch farthest away from the soil stack. To assure the proper flow velocity, the soil branch should be placed on a ⅛- or ¼-inch grade per foot.

The soil branch should be adequately supported through its length so it won't sag. (See Figure 20–16.) In most cases it is recommended that cast iron pipe be supported at intervals of no more than 5 feet.

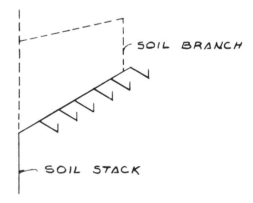

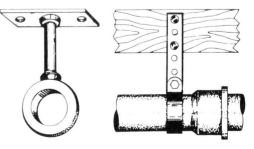

FIGURE 20–15 (upper right)   Schematic of a soil branch
FIGURE 20–16 (lower right)   Pipe supports

## TRAPS

Traps are used in the plumbing system to prevent the passage of sewer gases. (See Figure 20–17.) In its simplest form, a trap is a bend in the plumbing that is filled with water. The water completes a seal and makes it possible for gases to pass. Various types of mechanical devices have been tried as traps, but most have proved ineffective. The common seal trap has been used for years, is inexpensive, and has proved effective. The trap seal, however, can be lost by direct siphoning, evaporation, and capillary action.

Direct *siphoning* occurs when there is a pressure drop on the discharge side of a trap; when this happens, the atmospheric pressure on the fixture side pushes the water seal through the trap. This usually occurs when the trap is not vented.

When a water seal is lost because of

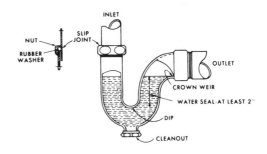

FIGURE 20–17. Trap assembly

*evaporation,* the trap has not been used for a considerable length of time. The rate of evaporation depends upon the room temperature and the humidity.

When a water seal is lost because of *capillary action,* some object is usually lodged in the trap; it acts as a wick and drains the water from the trap.

## VENTS

To prevent direct siphoning, there are several approved methods for venting a plumbing system. The particular method to use is determined by the layout and number of fixtures, as well as the design of the building. Some of the more common vents are branch, circuit, common, continuous, individual, loop, and stack vents and the vent stack.

A *branch vent* connects one or more individual vents with a vent stack or a stack vent. (See Figure 20–18A.)

A *common vent* serves the drains of two fixtures which are placed back to back or side by side. (See Figure 20–18B.)

A *circuit vent* serves 2 or more traps and extends from in front of the last fixture

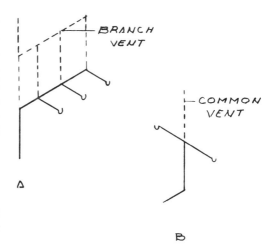

FIGURE 20–18. Methods of venting a plumbing system

connection of a horizontal branch to the vent stack. (See Figure 20–18C.)

A *continuous vent* is the continuation of a vertical pipe. (See Figure 20–18D.)

An *individual vent* is a vertical pipe that vents a particular fixture trap, and connects with the vent system above the fixture served. (See Figure 20–18E.)

A *loop vent* serves two or more traps and extends from in front of the last fixture

connection of a horizontal to the stack vent. (See Figure 20–18F.)

A *stack vent* is an extension of the soil or waste stack. (See Figure 20–18G.)

A *vent stack* is a vertical vent pipe that provides for the free circulation of air. (See Figure 20–18H.)

The sizes of continuous, circuit, and loop vents can be determined from Table 20–4; and the minimum diameters and maximum lengths of vent stacks and stack vents can be determined from Table 20–5.

FIGURE 20–18. (cont.)

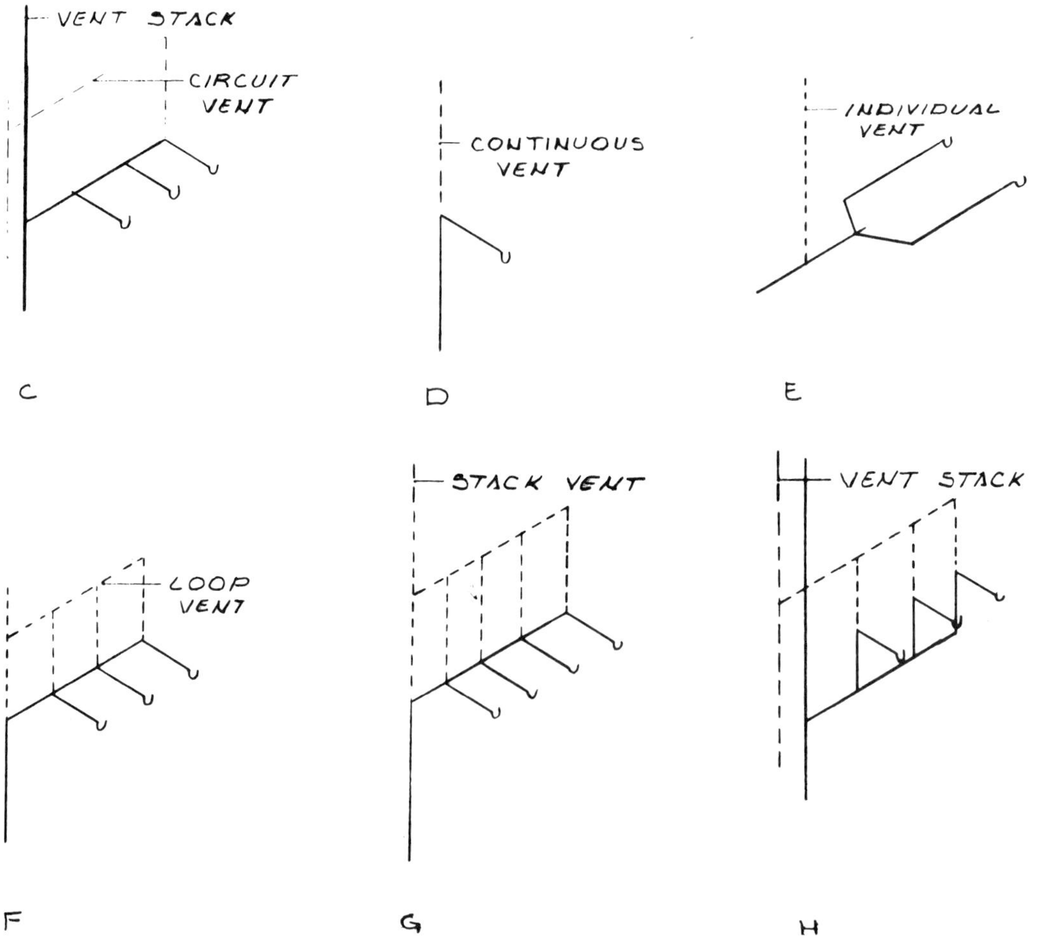

**TABLE 20–4  MINIMUM DIAMETERS AND MAXIMUM LENGTHS OF CONTINUOUS, CIRCUIT, AND LOOP VENTS FOR HORIZONTAL SOIL AND WASTE BRANCHES**

| Diameter of horizontal branch (inches) | Slope of horizontal branch (inches per foot) | Diameter of vent (inches) | | | | | | | | | |
|---|---|---|---|---|---|---|---|---|---|---|---|
| | | 1¼ | 1½ | 2 | 2½ | 3 | 4 | 5 | 6 | 8 | 10 |
| | | *(Maximum developed length of vent, in feet, given below)* | | | | | | | | | |
| 1¼ | ⅛ | NL¹ | | | | | | | | | |
| | ¼ | NL | | | | | | | | | |
| | ½ | NL | | | | | | | | | |
| 1½ | ⅛ | NL | NL | | | | | | | | |
| | ¼ | NL | NL | | | | | | | | |
| | ½ | NL | NL | | | | | | | | |
| 2 | ⅛ | NL | NL | NL | | | | | | | |
| | ¼ | 290 | NL | NL | | | | | | | |
| | ½ | 150 | 380 | NL | | | | | | | |
| 2½ | ⅛ | 180 | 450 | NL | NL | | | | | | |
| | ¼ | 96 | 240 | NL | NL | | | | | | |
| | ½ | 49 | 130 | NL | NL | | | | | | |
| 3 | ⅛ | | 190 | NL | NL | NL | | | | | |
| | ¼ | | 97 | 420 | NL | NL | | | | | |
| | ½ | | 50 | 220 | NL | NL | | | | | |
| 4 | ⅛ | | | 190 | NL | NL | NL | | | | |
| | ¼ | | | 98 | 310 | NL | NL | | | | |
| | ½ | | | 48 | 160 | 410 | NL | | | | |
| 5 | ⅛ | | | | 190 | 490 | NL | NL | | | |
| | ¼ | | | | 98 | 250 | NL | NL | | | |
| | ½ | | | | 46 | 130 | NL | NL | | | |
| 6 | ⅛ | | | | | 190 | NL | NL | NL | | |
| | ¼ | | | | | 96 | 440 | NL | NL | | |
| | ½ | | | | | 44 | 220 | NL | NL | | |
| 8 | ⅛ | | | | | | 190 | NL | NL | NL | |
| | ¼ | | | | | | 91 | 310 | NL | NL | |
| | ½ | | | | | | 38 | 150 | 410 | NL | |
| 10 | ⅛ | | | | | | | 190 | 500 | NL | NL |

¹ The abbreviation "NL" means "No Limit." Actual values in excess of 500 feet.

Reprinted with permission of Manas Publications.

**TABLE 20–5  MINIMUM DIAMETERS AND MAXIMUM LENGTHS OF VENT STACKS AND STACK VENTS¹**

| Diameter of Soil or Waste Stack (in.) | Total Fixture Units Connected to Stack (dfu) | 1¼ | 1½ | 2 | 2½ | 3 | 4 | 5 | 6 | 8 | 10 |
|---|---|---|---|---|---|---|---|---|---|---|---|
| | | *(Maximum developed length of vent, in feet, given below)* | | | | | | | | | |
| 1¼ | 2 | 30 | | | | | | | | | |
| 1½ | 8 | 50 | 150 | | | | | | | | |
| 1½ | 10 | 30 | 100 | | | | | | | | |
| 2 | 12 | 30 | 75 | 200 | | | | | | | |
| 2 | 20 | 26 | 50 | 150 | | | | | | | |

**TABLE 20-5 (cont.)**

| Diameter of Soil or Waste Stack (in.) | Total Fixture Units Connected to Stack (dfu) | 1¼ | 1½ | 2 | 2½ | 3 | 4 | 5 | 6 | 8 | 10 |
|---|---|---|---|---|---|---|---|---|---|---|---|
| | | (Maximum developed length of vent, in feet, given below) | | | | | | | | | |
| 2½ | 42 | 30 | 100 | 300 | | | | | | | |
| 3 | 7 | | 42 | 150 | 360 | 1040 | | | | | |
| 3 | 21 | | 32 | 110 | 270 | 810 | | | | | |
| 3 | 53 | | 27 | 94 | 230 | 680 | | | | | |
| 3 | 102 | | 25 | 86 | 210 | 620 | | | | | |
| 4 | 43 | | | 35 | 85 | 250 | 980 | | | | |
| 4 | 140 | | | 27 | 65 | 200 | 750 | | | | |
| 4 | 320 | | | 23 | 55 | 170 | 640 | | | | |
| 4 | 530 | | | 21 | 50 | 150 | 580 | | | | |
| 5 | 190 | | | | 28 | 82 | 320 | 990 | | | |
| 5 | 490 | | | | 21 | 63 | 250 | 760 | | | |
| 5 | 940 | | | | 18 | 53 | 210 | 640 | | | |
| 5 | 1400 | | | | 16 | 49 | 190 | 590 | | | |
| 6 | 500 | | | | | 33 | 130 | 400 | 1000 | | |
| 6 | 1100 | | | | | 26 | 100 | 310 | 780 | | |
| 6 | 2000 | | | | | 22 | 84 | 260 | 660 | | |
| 6 | 2900 | | | | | 20 | 77 | 240 | 600 | | |
| 8 | 1800 | | | | | | 31 | 95 | 240 | 940 | |
| 8 | 3400 | | | | | | 24 | 73 | 190 | 720 | |
| 8 | 5600 | | | | | | 20 | 62 | 160 | 610 | |
| 8 | 7600 | | | | | | 18 | 56 | 140 | 560 | |
| 10 | 4000 | | | | | | | 31 | 78 | 310 | 960 |
| 10 | 7200 | | | | | | | 24 | 60 | 240 | 740 |
| 10 | 11,000 | | | | | | | 20 | 51 | 200 | 630 |
| 10 | 15,000 | | | | | | | 18 | 46 | 180 | 570 |

¹ Does not apply to circuit, loop, or sump vents.

Reprinted with permission of Manas Publications.

## SEWAGE DISPOSAL FOR INDIVIDUAL HOMES

In some cases a home may not connect to a sewer, but must use a septic tank and a subsurface disposal field to dispose of waste and effluent. (See Figure 20–19.) A septic tank is a watertight container made usually of steel or concrete; it is the portion of an individual system in which the solid parts of sewage settle and are mostly changed into liquids or gases by bacteria. (See Figure 20–20.)

FIGURE 20-19.  Schematic of a septic tank and subsurface disposal field

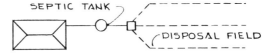

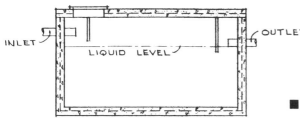

FIGURE 20–20 (left)   Side view of a septic tank

After the bacteria have worked on the solid waste, a heavy black semiliquid sludge remains and must be removed from the tank every few years. The size of a septic tank varies, but in most cases a one-bedroom home should have at least a 500-gallon tank. If the home has two bedrooms, a 750-gallon tank is recommended; for a three-bedroom home, a 900-gallon tank; for a four-bedroom home, a 1,150-gallon tank.

The septic tank effluent is disposed of in either soil absorption trenches or small oxidation ponds. Before the effluent can be placed in a subsurface irrigation field, these three general conditions must be met:

■ The soil percolation rate should be within the acceptable range

■ The maximum elevation of the ground water table should be below the bottom of the subsurface irrigation field

■ If an impervious stratum or clay formation is located under the irrigation field, it should be at a depth greater than 4 feet below the trenches

To determine the length of an absorption trench a percolation test must be made. To conduct this test, the following steps should be performed:

■ Locate and dig three separate test holes in various locations on the proposed absorption field. The holes should be 4 to

12 inches deep and have their sides carefully scratched with a knife blade to provide a natural absorption area.

■ Each hole should then be filled with clear water, which is allowed to stand overnight. Thus the soil is given ample opportunity to swell and to approach the operating condition it will have during the wet season of the year.

■ After the water has been allowed to stand overnight, the percolation rate measurement is made by adding water until the liquid depth is at least 6 inches, but not more than 12 inches. Then from a reference point, the water-level drop is measured over a 60-minute period.

■ The water-level drop in each of the three holes is measured and recorded. Their average is then used to determine the total length of the absorption trench. Table 20–6 shows how to determine the absorption length.

**TABLE 20–6   ABSORPTION TRENCH LENGTH REQUIREMENTS FOR INDIVIDUAL RESIDENCES**

| Average Water-Level Drop in 60 Minutes (in inches) | Length in Feet of Absorption Trenches Required per Bedroom[1] |
|---|---|
| more than 12″ | 77 |
| 12″ | 83 |
| 6″ | 110 |
| 4″ | 127 |
| 2″ | 166 |
| 1½″ | 200 |
| 1″ | 220 |
| less than 1″ | not acceptable for absorption field |

[1] In every case at least 160 linear feet should be provided.

For maximum efficiency the absorption trench system should comply with the following standards:

■ The minimum length of the field line is 160 linear feet. The trench should be 12 to 18 inches wide, and 24 to 36 inches deep. (See Figure 20–21.) An individual field line should not be over 100 feet in length, with the individual lines placed 6 feet apart.

■ The subsurface absorption trenches should be located at least 100 feet from any well and 10 feet from any dwelling or property line.

■ A minimum of two field lines must be used.

■ For even distribution of the effluent, the trench should not have any slope.

■ The field lines should be a minimum of 4

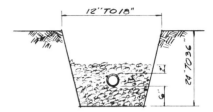

FIGURE 20–21.   Side view of a soil-absorption trench

inches in diameter. They should be laid on a slope of 2 to 3 inches per 100 feet, and should consist of either perforated nonmetallic pipe, agricultural drain tile, or vitrified clay bell and spigot sewer pipe laid with open joints.

■ The field line should be surrounded by wash gravel or crushed stone. This bed material should cover the top of the pipe to a depth of at least 2 inches, and extend to a depth of at least 6 inches below the bottom of the pipe.

## WATER-SUPPLY SYSTEM

Every occupied building must be provided with an ample water supply that is maintained in satisfactory working condition. If the building is classified as a dwelling unit, it is also required to have at least one kitchen sink equipped with hot and cold running water and a waste connection.

A water-supply system is composed of water lines, valves, and fixtures. The water lines, in most cases, are either copper or plastic and are connected by conventional techniques. The sizes of the lines vary, however, and can be computed by using Tables 20–7, 20–8, and 20–9.

In order to be able to correctly use the tables, each fixture must first be assigned a fixture value from Table 20–6. The fixture

units can then be totaled, using Table 20–7, and the size of the individual pipes determined. Then Table 20–8 is used to determine the minimum size of a fixture supply pipe.

### Valves

Valves in a water-supply system are used to control the flow of water. Of the many different types of valves the most common are gate, globe, and check valves. The *gate* valve has a retractable leaf machined to fit tightly against two sloping surfaces. A valve of this type is usually placed in a location where it is left open most of the time. (See Figure 20–22A.)

## TABLE 20–7  WATER-SUPPLY FIXTURES RATED IN FIXTURE UNITS

| Fixtures | Fixture Units |
|---|---|
| Water closet, flush valve | 10 |
| Water closet, tank supplied | 5 |
| Pedestal urinal, flush valve | 10 |
| Stall or wall urinal, flush valve | 5 |
| Stall or wall urinal, tank supplied | 5 |
| Lavatory, public use | 2 |
| Bathtub, public bath use | 4 |
| Shower stall, mixing valve head | 3 |
| Service sink, public | 3 |
| Kitchen sink, public | 4 |
| Drinking fountain, public | 1 |
| Kitchen sink, private | 2 |
| Lavatory, private | 1 |
| Bathtub, private | 2 |
| Shower stall, two valve, private | 2 |
| Fixture combinations, private | 3 |
| Laundry tray, private | 3 |
| Laundry washer, private | 3 |
| Sink with disposal | 3 |
| Dishwasher, private | 3 |

Reprinted with permission of Manas Publications.

## TABLE 20–9  SIZE OF FIXTURE SUPPLY

| Type of fixture or device: | Pipe size (inch) |
|---|---|
| Bathtubs | $\frac{1}{2}$ |
| Combination sink and tray | $\frac{1}{2}$ |
| Drinking fountain | $\frac{3}{8}$ |
| Dishwasher (domestic) | $\frac{1}{2}$ |
| Kitchen sink, residential | $\frac{1}{2}$ |
| Kitchen sink, commercial | $\frac{3}{4}$ |
| Lavatory | $\frac{3}{8}$ |
| Laundry tray, 1, 2 or 3 compartments | $\frac{1}{2}$ |
| Shower (single head) | $\frac{1}{2}$ |
| Sinks (service, slop) | $\frac{1}{2}$ |
| Sinks, flushing rim | $\frac{3}{4}$ |
| Urinal (flush tank) | $\frac{1}{2}$ |
| Urinal (direct flush valve) | $\frac{3}{4}$ |
| Water closet (tank type) | $\frac{3}{8}$ |
| Water closet (flush valve type) | 1 |
| Hose bibbs | $\frac{1}{2}$ |
| Wall hydrant | $\frac{1}{2}$ |

For fixtures not listed, the minimum supply branch may be made the same as for a comparable fixture.

Reprinted with permission of Manas Publications.

## TABLE 20–8  WATER-PIPE SIZING TABLE: Small Buildings

Sizes are computed to maintain a maximum velocity of 10 feet per second, based on water pressure drop of 5 psi per 100 feet.

| Line no. | Service Main Diameter | Inside Piping Diameter | Developed Length of Piping Feet | | Fixture unit Requirements Quantity | |
|---|---|---|---|---|---|---|
| 1 | $\frac{3}{4}''$ | $\frac{3}{4}''$ | Maximum | 50 ft | Maximum | 25 FU |
| 2 | $\frac{3}{4}''$ | $\frac{3}{4}''$ | '' | 100 ft | '' | 16 FU |
| 3 | $\frac{3}{4}''$ | $\frac{3}{4}''$ | '' | 150 ft | '' | 15 FU |
| 4 | $\frac{3}{4}''$ | $1''$ | '' | 50 ft | '' | 40 FU |
| 5 | $\frac{3}{4}''$ | $1''$ | '' | 100 ft | '' | 33 FU |
| 6 | $\frac{3}{4}''$ | $1''$ | '' | 150 ft | '' | 28 FU |
| 7 | $1''$ | $1''$ | '' | 50 ft | '' | 50 FU |
| 8 | $1''$ | $1''$ | '' | 100 ft | '' | 40 FU |
| 9 | $1''$ | $1''$ | '' | 150 ft | '' | 30 FU |
| 10 | $1''$ | $1\frac{1}{4}''$ | '' | 50 ft | '' | 96 FU |
| 11 | $1''$ | $1\frac{1}{4}''$ | '' | 100 ft | '' | 65 FU |
| 12 | $1''$ | $1\frac{1}{4}''$ | '' | 150 ft | '' | 55 FU |
| 13 | $1\frac{1}{4}''$ | $1\frac{1}{4}''$ | '' | 50 ft | '' | 150 FU |
| 14 | $1\frac{1}{4}''$ | $1\frac{1}{4}''$ | '' | 100 ft | '' | 100 FU |
| 15 | $1\frac{1}{4}''$ | $1\frac{1}{4}''$ | '' | 150 ft | '' | 65 FU |
| 16 | $1\frac{1}{4}''$ | $1\frac{1}{2}''$ | '' | 50 ft | '' | 250 FU |
| 17 | $1\frac{1}{4}''$ | $1\frac{1}{2}''$ | '' | 100 ft | '' | 160 FU |
| 18 | $1\frac{1}{4}''$ | $1\frac{1}{2}''$ | '' | 150 ft | '' | 130 FU |

Source: BMS 79, Water-Distributing Systems for Buildings.

Reprinted with permission of Manas Publications.

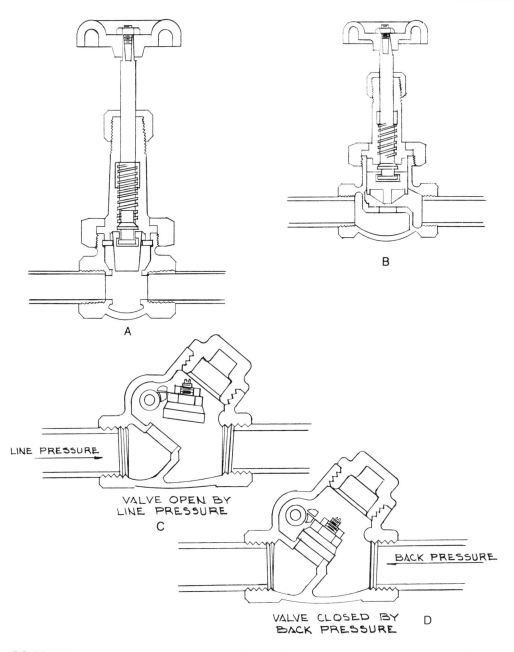

FIGURE 20–22.   Three types of valve in a water-supply system: A, a gate valve, full open, allows unrestricted flow with minimum pressure drop; B,... or regulates flow, depending upon how far it is opened; C, a swing valve is kept open by line pressure, or; D, when line pressure drops, gravity and back pressure close the valve, thus preventing back flow

A *globe* valve is similar to a gate valve, and is used for many of the same purposes. (See Figure 20–22B.) Faucets and hose connections are examples of globe valves.

A *check* valve is used to prevent flow in a direction opposite to that desired. (See Figure 20–22C.) The valve is designed with a hinged leaf that swings in the direction of flow but closes against attempted flow in the opposite direction (backflow).

## Pipe Supports

Because of the weight of the pipe and water, it is often necessary to support the pipe with individual supports. Without proper support the pipes might fracture, causing water leaks. If water-supply pipes are placed in a vertical position, copper pipes and plastic pipes 1½ inches and over should be supported at each story; 4-foot intervals are required for pipes smaller than 1½ inches. Screwed galvanized steel pipe should be supported at not less than every other story height.

If the piping is placed in a horizontal position, copper tubing and plastic pipe 1½ inches and smaller should be supported at 6-foot intervals, and piping 2 inches and larger should be supported at 10-foot intervals. If screwed pipe is used, it should be supported on 12-foot centers.

Regardless of the spacing used, the supports must be strong enough to sustain the weight imposed on them.

## Shock and Water Expansion

Water-supply systems are often noisy; when valves are abruptly shut off, pipes may rattle and cause excessive noise. One way to eliminate the "water hammer" is to

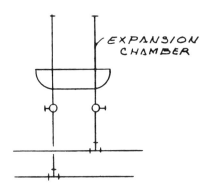

FIGURE 20–23. Extending a plumbing-fixture branch to eliminate rattling of pipes

extend the fixture branch approximately 18 inches. (See Figure 20–23.) The added pipe length traps air, which absorbs the impact of the water. There are commercial devices that can be used, but a capped air (expansion) chamber is usually sufficient.

## Hot-Water Systems

A hot-water system consists of a tank and a series of lines to feed the various fixtures. In some buildings the network of pipes is installed so that instant hot water is available, while others do not have continuous circulation of hot water.

The two systems used to provide continuous hot-water circulation are the upfeed and gravity return and the pump circuit systems.

The *upfeed gravity return system* operates on the principle that water expands and becomes lighter when it is heated above 39.2° F. In such a hot-water system the piping is arranged so that hot water is carried to the farthest fixture and is returned back to the heater with a continuation of the water line. (See Figure 20–24.)

Perhaps the most successful technique

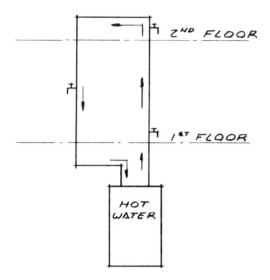

FIGURE 20–24. The upfeed gravity-return system
of hot-water circulation

## Fixtures

Of the many different types of plumbing
fixtures, some of the more common are lavatories, water closets, bathtubs, and sinks.

Lavatory bowls may be circular, square,
or oval-shaped and are usually made of vitreous china, but may be made of stainless
steel or fiberglass. The bowls usually hold 1
or 2 gallons of water, and are equipped with
a trap to prevent escape of sewer gases.

Five basic types of lavatory styles are
popular today. The *flush-mount type* is installed with a metal ring or frame to hold it
in place. (See Figure 20–26A.) This type of
lavatory is inexpensive, but presents a
small cleaning problem at the junction of
rim and countertop.

A *self-rimming lavatory* requires no
metal frame, but is designed so that the rim
of the lavoratory supports the entire assembly. (See Figure 20–26B.) When it is
properly installed and sealed, a lavatory of
this type does not present a maintenance
problem.

for circulating hot water in light construction is with the use of a circulating pump.
A centrifugal type of pump is usually used,
and is placed on the circulating main as
close to the heating unit as possible. (See
Figure 20–25.)

FIGURE 20–25.   The pump-circuit system of hot-water circulation

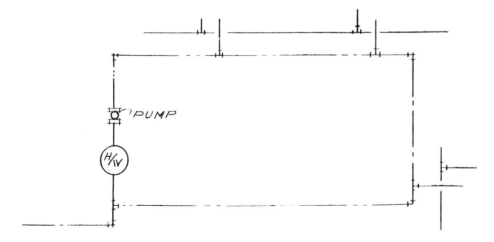

An *under-the-counter type of lavatory* is mounted below the countertop, which is usually marble or a synthetic material that resembles marble (not shown). The style of the lavatory is striking, but it does present a maintenance problem.

An *integral lavatory* and counter have become increasingly popular with the development of "synthetic marble" countertops. (See Figure 20–26C.) These lavatories are attractive and easy to clean.

A frequently installed lavatory is the conventional *wall-hung type,* often rectangular in shape. (See Figure 20–26D.) This type of lavatory is used where space is scarce, or where storage space is not required.

Bathtubs are usually manufactured from one of three materials: molded cast iron with a porcelain enamel surface, formed steel with a porcelain enamel surface, or molded gel-coated glass-fiber-reinforced polyester resin (fiberglass).

A *cast iron tub* with a porcelain enamel surface is available in 4-, 4½-, 5-, 5½- and 6-

C

FIGURE 20–26. Types of lavatory bowl

A      B      D

foot lengths. The widths range from 30 to 48 inches and the depths from 12 to 16 inches. A tub of this type can weigh as much as 500 pounds.

A *formed steel tub* with a porcelain finish is less expensive and weighs less than a cast-iron tub. This type is available in lengths of 4½ and 5 feet. Widths range from 30 to 31 inches and the depth is usually 15 to 15½ inches.

The *fiberglass bathtub* has only recently been accepted as a reliable plumbing fixture. Of the various styles available, most are manufactured in only 5 foot lengths.

Water closets may be either floor-mounted or wall-mounted, and are con-structed so that flushing will siphon out the contents. The water closet usually has either a water tank or a flush valve for flushing purposes. There are also several different types of water closet bowls.

The four basic kinds of residential water closets are washdown, reverse-trap, siphon-jet, and low-profile. Regardless of the type of bowl they are all installed in basically the same manner. The bowl is usually set with a wax-ring gasket and bolted to the floor or wall. As the bowl is being lowered into position, it should be twisted slightly. The twisting helps to settle the wax ring and bowl in the proper position. After the bowl has been placed, the closet water tank can be hung.

# CHAPTER 21

# INTERIOR WIRING, HEATING, AND AIR CONDITIONING SYSTEMS

## INTERIOR WIRING

Most residential wiring systems use alternating current (AC) as a power source. The power is usually supplied from a distribution transformer to the service entrance and from there to different points throughout the house.

### Wiring Plans

Wiring plans are an integral part of a set of building plans. They show wire and conduit size and the number of wires in the conduit, as well as explanatory symbols, notes, specifications, and a legend. (See Figure 21–1.) The legend gives full particulars of the important details of the wiring plans.

The wiring plan also shows the locations of the ceiling fixtures, convenience outlets, the distribution panel, and the individual circuits. The arrowhead on each circuit points toward the distribution panel, and the number next to the arrowhead shows the circuit breaker to which that circuit is connected.

FIGURE 21–1.   A typical wiring plan

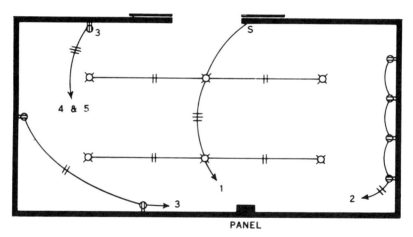

## FLOOR PLAN
### WIRING PLAN FOR SHOP

| LEGEND | |
|:---:|:---|
| ♢ | CEILING OUTLET, PORCELAIN PULL-CHAIN SOCKET UNLESS OTHERWISE NOTED |
| ●▬₃ | 50 AMP. 250V. THREE-POLE WALL RECEPTACLE AND PLUG |
| ⊖▬ | 15 AMP. 125V DUPLEX CONVENIENCE WALL RECEPTACLE |
| ▬ | ENCLOSED CIRCUIT BREAKER OR CIRCUIT-BREAKER PANEL |
| ───╫─── | THREE-WIRE CIRCUIT TO PANEL |

## Service Drop

A service drop is that portion of the electrical source that exists between the distribution transformer and the first point of attachment to the building. The wires for a service drop can be placed either overhead or underground, and are called service entrance conductors. The size of wire for the service drop will vary, but the minimum size should be a No. 8 AWG (American Wire Gauge).

To avoid electrical accidents, a service drop should have a 10-foot clearance over a sidewalk, 18 feet over an alley or public road, and a minimum of 12 feet over residential driveways.

The number of wires in the service usually dictates the type of distribution system and voltage. If the service has only two wires it is usually a single-phase, 120-volt system. When a system of this type is used, one wire is neutral or ground. If there are three wires entering the building, the service can be either straight three-phase or single-phase three wire. If there are four wires, the system is three-phase with a neutral wire.

The service entrance installation usually consists of a rigid conduit that has a weatherhead located on the outside end of the conduit and a fuse or main disconnect installed inside. (See Figure 21–2.)

## Panels and Boxes

The electrical service installation can normally be divided into two major procedures: roughing in and finishing. *Roughing in* is accomplished before plaster, gypsumboard, or paneling is placed, and consists of the installation of panels, outlet

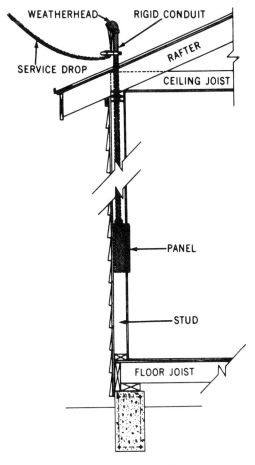

FIGURE 21–2.  Electrical service entrance

and junction boxes, cable, wire, and conduit. *Finishing,* one of the last construction steps, consists of the installation of switches, receptacles, covers, fixtures, and the completion of service connections.

## Main Disconnect

Every electric service that enters a building should have a service entrance switch that can disconnect all the circuits from the main power source. All commercial switches must also contain protective devices, such as fuses or circuit breakers. Each service switch should also be equipped with a handle mechanism that can open and close the switch.

## Panel Box

A panel box is used to group circuit switching and protective devices in a common place. The box has a set of copper buss bars, called mains, from which several circuits can be tapped. The number and location of the circuits are based on the number and length of each branch circuit. It is usually standard practice to leave one spare circuit for each five circuits installed.

## Fuse Panels

A fuse panel or submain is used to supply power to individual circuits. The panel should be in a convenient location that will eliminate long runs of conduit. The purpose of installing a fuse panel is to divide the main circuit into branch circuits. If an appliance has a power requirement over 10 amperes it should be on a single-branch circuit. Single branch circuits should also be provided for convenience outlets that service electric clothes dryers, automatic washers, air conditioning units, and food freezers.

## Grounds

There are two types of grounds: the system ground and the equipment ground. The *system* ground reduces the voltage of one of the wires of the system to zero potential above ground, removing the possibility of fire and shock.

The *equipment* ground is an additional ground that is attached to appliances and machinery. It provides added safety, because if a short-circuit occurs, the fuse protection will open the circuit.

In a system ground, the neutral wire is always grounded and has white insulation. But the equipment ground wire should have green insultation.

The ground connection, usually made to the cold-water pipe and as near the meter as possible, should be kept as short and direct as possible.

## Box Installation

Outlet and switch boxes are used to house the ends or splices in wires at the point where the insulation has been removed. The boxes are mounted directly to or attached by special brackets to the supporting framing members. When a box is correctly positioned, its forward edge or extension ring should be flush with the finished wall. A common method of attaching switch and outlet boxes is to fasten them to studs with 16d nails.

The bottoms and sides of the boxes have knockout sections, which can easily be removed to form openings for the wire to enter. It is possiblle to remove the knockouts by sharply tapping them with a hammer and screwdriver. Boxes are ordinarily made of steel or an insulating material, and are generally required to be 1½ inches deep.

A junction or a ceiling outlet box is either mounted directly to the ceiling joist or supported by cleats nailed to the joist.

## Conductors

Conductors carry electric current, and consist usually of aluminum or copper wire covered by an insulating material. The sizes of conductors vary from No. 14 to 40, and in larger sizes from 250 MCM to 2,000 MCM. Insulation on the conductors is usually made from rubber, thermoplastic, or asbestos, and is identified by the letters R, T, or A.

## Conduit Installation

In many cases electrical wiring is placed in conduits, tubing, or pipe that protects the wiring. There are two major classifications of conduits—rigid conduit and electrical metallic tubing (EMT), often called thin-wall. A conduit is much like a water pipe, but the inside is smoother, to prevent damage to the insulation on the conductors when they are pulled through it.

The size of conduit is determined by the inside diameter. The most common sizes are: ½, ¾, 1, 1¼, 1½, 2, and 2½ inches. Connection of rigid conduits is made by bushings and locknuts screwed onto their ends. Thin-wall conduit connections require special fittings. If rigid conduit is used, the end can be screwed into a standard outlet box or junction box. The junction or pullbox provides an intermediate point in a long run of conduit.

For proper conduit installation it is often necessary to make some field bends. Certain procedures must be followed to prevent the collapse of any portion of the conduit. Rigid conduit must be bent with a right bender or hickey, and three initial bends are necessary to shape the conduit to 90°. (See Figure 21–3.) The first bend is about 25° and is made about 2 inches in from the center line of the bend. The second bend brings the conduit to 45° and is bent from a point 2 inches past the first point, just to one side of the center line.

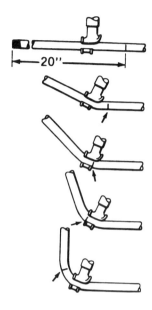

FIGURE 21–3. Conduit bending (each arrow points to the center line of the bend)

The third bend brings the conduit up to 90° and is made about 1 inch behind the second bend. Thin-wall conduit is bent in a similar fashion, but with one continuous movement.

Conduits must also be cut and threaded before installation. A hacksaw or standard pipe cutter can be used to cut rigid conduit, while thin-wall conduit can be cut with either a hacksaw or a tubing cutter. After cutting, any burrs should be removed with a ream or a file. Then the conduit can be threaded, usually with a special die that can be used only for conduits. The dies are usually nonadjustable ratchet types and range in sizes from ⅛ inch to 2 inches. (See Figure 21–4.)

When a conduit is placed in a structure it should be run as straight and directly as possible. It should also be supported by either strap or pipe hangers. If the hangers are placed over a wood frame, screws or

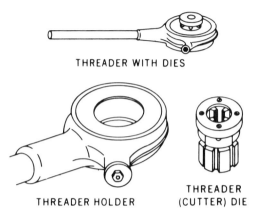

THREADER WITH DIES

THREADER HOLDER

THREADER
(CUTTER) DIE

FIGURE 21–4.  Rigid conduit threader and die

nails can be used to secure them; but if they are placed over a masonry surface, they should be fastened with machine screws attached to an anchor embedded in the masonry.

## Conductor Installation

Almost any type of wire can be used in conductor installation, but the most common used is TW. This type of wire has a thermoplastic, moisture-resistant insulation.

Most wire is pushed through the conduit to the different boxes, but it is sometimes necessary to use a fish tape to pull the wires through. A fish tape has a hook on one end and is pushed through the conduit. The conductor is then attached to the end of the fish tape and is pulled back through the conduit, usually from the distribution panel to the last box in the run. Six inches of free conductor should be left at each outlet and switch box, to make up joints for connection of fixtures.

Boxes can also be connected to the distribution panel by either armored cable or nonmetallic sheathed cable. *Armored cable,*

often called BX cable, consists of two or three rubber- or thermoplastic-covered wires encased in a flexible steel armor. Armored cable is manufactured as Type AC (without a lead sheath) and Type ACL (with a lead sheath). Armored cable has three conductors—red, white, and black.

*Nonmetallic* sheathed cable is available in two different types; NM and NMC. Type NM is designed for use in dry locations, while NMC can be used in moist, damp, and corrosive locations. Nonmetallic sheathed cable is manufactured in two or three wire combinations. For protection the individual wires are wrapped with plastic, thick spiral paper, or woven fabric braid.

Nonmetallic sheathed cable and armored cable should be secured to the framing members at 4-foot intervals and within 12 inches of the box. Staples are most often used to support the cables. But they can also be supported by holes drilled in the framing members or by mounting straps.

## Finishing Procedures

The splice and the tap are the two basic joining techniques used in electrical connections. A *splice* is the joining of the ends of two pieces of wire. A *tap* is the joining of the end of one wire to the middle of another wire. Taps and splices must be made in a box, and can be done by soldering, brazing, welding, or by an approved mechanical pressure device.

For a *solder tap,* the insulation must first be stripped from the wires. Then the intersecting conductor is wrapped tightly around the other conductor. (See Figure 21–5.) Heat is then applied to the joint until the conductor is hot enough to melt the solder. After the solder has been applied

FIGURE 21–5 (top).   Solder tap
FIGURE 21–6 (center).   Pigtail splice
FIGURE 21–7 (bottom).   Wirenut

and has cooled, it should be taped. A *pig-tail splice* is made in a similar fashion, but the splice is made at the ends of two conductors. (See Figure 21–6.) This splice is also made by stripping the conductors, twisting them together, soldering them, and then taping the splice.

The most common way to join two connectors is with a mechanical pressure device called a *wirenut*. (See Figure 21–7.) Usually made of plastic, rubber, or other insulating material, the wirenut has a threaded portion that screws onto the conductors.

## Installation of Switches and Outlets

Before switches and outlets are installed, it is important to understand which terminals to attach the "hot" (ungrounded) and the ground wires to. The terminal screw for a hot line is brass- or copper-colored, while that for the ground wire is light or nickel-colored.

Switches are used to open and close electrical circuits and can be operated either automatically or manually. A single-pole switch is used to control individual, low-amperage circuits and is installed in the hot wire of the circuit. To install a single-pole switch to a common lighting circuit, the hot wire (black) is first cut. Then both ends are attached to the two terminal screws.

A three-way switch has three terminals and is used to control a light from two different locations. A four-way switch is installed between two three-way switches and can control a light from more than two locations. A four-way switch has four terminals and is connected to the two wires running between two three-way switches. A diagram of how three-way and four-way switches work is provided in Figure 21–8.

FIGURE 21–8.   Three-way and four-way switch operation

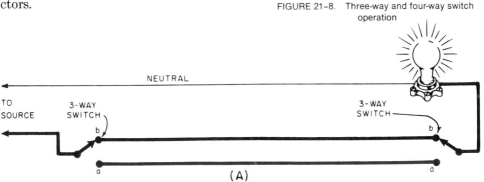

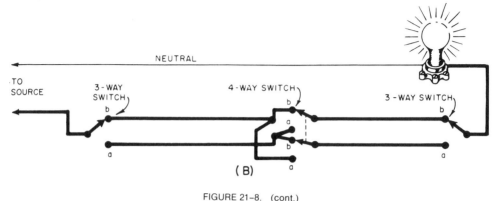

FIGURE 21–8.  (cont.)

Outlets are installed by connecting the hot (black) wire to the brass-colored termi-nal screw and the neutral or white wire to the nickel-colored screw.

## HEATING AND AIR CONDITIONING

Several different methods are used for heating and cooling buildings. Some of the more common systems are forced warm-air, electric, hydronic or hot water, space heat-ers, window units, and central systems.

### Forced Warm Air

In *forced warm-air* system air is heated in a furnace and forced through ducts by a blower. First, cool air is brought into the furnace through a return air grill and is heated. Then the furnace blower drives the warm air through a series of ducts. To dis-tribute the warm air, registers are placed at the ends of the ducts. The best arrange-ment is to place the registers in the floor below large areas of glass. The return air grills should be placed on interior walls and at high locations for maximum efficiency.

To achieve a proper balance in a forced-air system, the furnace should be located as near the center of the building as possible.

Typical components of a forced warm-air system include furnace, ducts, registers, and controls.

### The Furnace

The furnace is equipped with a fan (blower), motor, filters, heat-transfer sur-face, and a gas or oil burner. The burner in the furnace operates, in most cases, on fuel gas or fuel oil. If fuel gas is used as an en-ergy source it is fed to the burner under constant low pressure and is controlled by a pressure regulator. A room thermostat controls the operation of the burner, and a pilot light or an electric spark ignites the burner when the thermostat registers a pre-determined temperature. An oil burner op-erates in a similar manner, except that the fuel oil is pumped from a storage tank and a gun-type oil burner is used as a source of heat.

The furnace size is based on (1) the total heat loss calculated for a particular building and (2) the total amount of Btu's the unit can produce in an hour.

### Ducts

Ducts deliver the warm air and are constructed to a noncombustible material. At one time ducts were made exclusively from galvanized sheet metal, but in recent years asbestos and fiberboard have been used.

Ducts can be either round, square, or rectangular, but the round ducts are more efficient in terms of volume of air handled per perimeter distance.

To join sheet metal ducts, several different types of joints have been developed, among them drive cleats, double S slip, reinforced S cleat, and a button-punch snaplock. (See Figure 21–9.)

There are three basic types of duct systems: the perimeter system, the radial system, and the extended plenum system. The *perimeter system* has a duct that runs the perimeter of the building and is supplied by interconecting ducts. (See Figure 21–10A.) The *radial system* has ducts extending from the plenum. (See Figure 21–10B.) The extended plenum system (not shown) has a

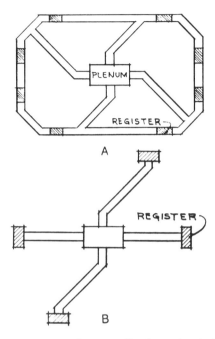

FIGURE 21–10. Two types of heating-system duct

long horizontal plenum with ducts branching off from it.

When a duct is placed in an unconditioned space, heat is often lost. To prevent this, the ducts should be insulated with an approved insulator. These three rules should be followed when duct insulation is being installed:

- If a supply duct is over 12 feet long, it should be insulated.

- Insulate all warm-air ducts that are placed in garages, attics, and unexcavated subfloor spaces.

- Insulate all cold-air ducts that are placed in attics, basements, and crawl spaces.

The size of a duct is usually determined by the total cubic feet of air per minute

FIGURE 21–9. Sheet-metal duct joints

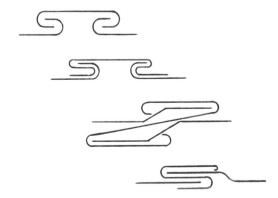

(CFM) necessary for heating or cooling a given area. The CFM that circulates through a system is usually recommended by the manufacturer of the heating or cooling system—typically between 300 and 420 CFM per ton. Once the CFM has been calculated, a table can be used to determine the duct size.

### Dampers

Dampers are used in a forced-air system to maintain an even distribution of air. The dampers can be located in the diffuser, grill, or duct. For maximum efficiency, they should be fitted snugly and operate with minimum leakage. The three damper types are multiple blade, butterfly, and split damper.

### Registers

Supply registers are used to disperse air in a given area. Each register should be equipped with a damper or vanes, and should be able to distribute the air in several directions.

## Electric Heating Systems

Electric heating systems are available in a variety of styles and voltages. Some common systems are baseboard, wall units, cable, ceiling units, central furnaces, infrared heaters, and heat pumps.

### Baseboard Units

*Baseboard* heating units are available in a variety of styles and models. Most models are compact in design and can easily be installed in areas of large heat loss. The units can be controlled by a centrally located thermostat or individually controlled.

### Wall Units

A *wall* unit may have resistance coils and fans or wires embedded in glass or ceramic. Some units are controlled by an individual thermostat, but most are manufactured without temperature controls.

FIGURE 21–11. Schematic of a cable system for electric heating

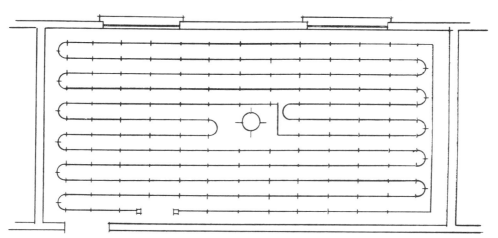

### Cable System

A *cable* system can also be constructed by embedding wires in the ceiling or floor. The system operates by reflecting heat rays from the different surfaces of the room. To control the temperature, a thermostat is placed in each room.

The cables should be installed at least 2½ inches apart. They should also be placed 6 inches from the wall and 8 inches from any ceiling outlet. (See Figure 21–11.) If the ceiling joists are placed 16 inches on center, only ten cables may be placed between the joists.

### Ceiling Units

*Ceiling* units, like wall units, can also be constructed with resistance coils and fans. Units of this type are usually found in bathrooms, but they can be used in other areas as well.

### Central Furnaces

A central electric furnace operates by heating air in a furnace and forcing it through ducts with a blower. Cool air is brought back into the furnace through a return air grill; the cool air is heated, and then the cycle is repeated.

## Heat Pump

A *heat pump* can also be listed as a form of electric heating. A heat pump operates by switching the conventional air conditioning refrigeration cycle. Once the cycle has been switched, the evaporation coils cool the air outside and the condenser coils warm the air inside. When a cooling unit is desired, a reversal of the evaporator and condenser switches the unit back to air conditioning.

## Hydronic Systems

A hydronic system in its simplest form is nothing more than a forced hot-water system. The water is first heated in a boiler and then circulated by means of a pump through pipes and into the convectors. Once the water has passed through the pipes, it is returned to the boiler and the process is repeated.

### The Series Loop

The series loop is one of the simplest hydronic systems. The supply lines are connected directly to the radiators. (See Figure 21–12.) Hot water is then forced through the supply lines and radiators.

The biggest disadvantage of this sytem is that the radiators are connected in series, and all of them, heat up and cool off at the same time. To regulate the radiators, only one thermostat is needed to turn the central heating unit on and off.

### The One-Pipe System

A one-pipe system is similar to a series loop, except that at each radiator location a branch pipe carries hot water to the radiator and another branch pipe carries the return water from the radiator back to the circulating pump. (See Figure 21–13.) With a system of this nature, individual room heat can be obtained.

## Space Heaters

Space heaters are sometimes used to heat individual rooms or buildings. Some building codes forbid their use on the grounds

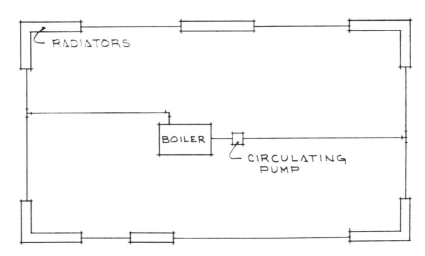

FIGURE 21-12.   Series-loop hydronic system

that unvented they can consume the oxygen and discharge combustion products in a room. Space heaters usually operate on electricity, oil, gas, or coal.

### Electric Space Heaters

An electric space heater does not consume the oxygen, nor does it discharge any combustion products, but in most cases it is not capable of heating a large area.

The most popular units have resistance coils and fans, or wires embedded in glass or ceramic. Most heaters of this type are limited in capacity to around 1250 watts and operate on 120 volts. Some permanent electric space heaters are of higher wattage, however, and require 240 volts.

To determine the needed watts, a rough rule of thumb is to have about 900 watts for each 10 feet of outside wall. This is only a rough estimate; there are other variables that can influence the heating efficiency.

### Oil and Gas Heaters

An oil or gas heater can be used to heat a building, but a flue or vent must be used in conjunction with the heater. (An oil heater uses a flue, and a gas heater requires a vent.) A typical oil or gas heater can produce around 50,000 Btu's per hour, although some may produce 90,000 or more. A full-sized house usually requires three or more heaters producing a minimum of 50,000 Btu's per hour.

FIGURE 21-13.   One-pipe hydronic system

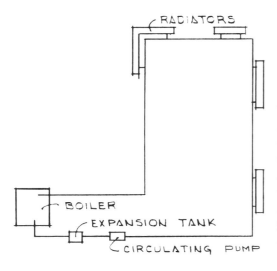

## Air Conditioning Systems

Air conditioning systems can be either central or unit systems. A *central system* can be an integral part of the heating system, using the same blower, filters, ducts, and registers, or it can be a separate system with its own distribution method.

*Unit systems* are usually built into the exterior wall, or may be placed in a window opening. Room-sized units can cool a building effectively, especially when the building has naturally defined zones.

### Electric Cooling Units

Electric cooling units have had wide acceptance. In most cases, they operate simply by changing a liquid to a gas. The gas most often used is freon. Liquid freon is first forced through coils. A blower then forces air around the coils. As the freon is being pumped through the coils, it changes into a gas. When a liquid is changed into a gas, it absorbs heat; thus, the air around the coils is cooled. At that point a blower forces the cool air through the ducts and into the house. After the freon has passed through the coils, it is pumped through an outside air unit called a condenser. In the condenser the freon is cooled and returns to its liquid state. Then it is forced back to the cooling coil, where the cycle is repeated.

### Cold-Water Air-Cooling System

A water chiller can be designed to work in combination with a forced hot-water heating system, or it can be a separate cooling system. The chiller usually consists of a compressor, condenser, and evaporator tank, and is used to chill water and circulate it to the same units that emitted heat from hot circulated water.

Another type of chilled-water system uses an absorptive-type water chiller. In such a system, the chiller is charged with lithium bromide and water, and refrigeration is produced by the absorptive principle.

# CHAPTER 22

# PAINTS AND PAINTING

MANY DIFFERENT TYPES OF PAINTS AND finishes are available to the construction industry, each designed for a specific purpose. A good-quality paint should be able to resist effectively moisture and mildew, stains from rust, and discoloration from industrial gases.

The two types of house paints commonly used today are linseed oil-based and water-based. An *oil-based* paint is usually a mixture of white lead, linseed oil, drier, and thinner, with colors added as desired. *Water-based* paints, extremely popular, use water for the thinner instead of turpentine or mineral spirits. These paints, usually called latex, are emulsions of polymers or copolymers in water. When water-based paint is applied, the emulsion coalesces, permitting the water to evaporate and leaving the vehicle and pigment.

Pigmented stains, often used to finish exterior and interior walls, are classified by the solvent in which they are dispersed. The most common solvents are water, mineral spirits, and oil. *Water stain* is often used to change the color of wood, but it raises the grain and needs to be sanded after the stain has dried. *Oil stains* give good penetration, but they dry more slowly than the other stains. When oil stain is applied, the strokes should be made along the grain. *Mineral-spirit* stains dry rapidly, and give good service.

Clear finishes such as varnish are popular in the construction industry. Polyurethane varnishes are available in a full range of glosses and give a hard, durable finish that is resistant to oil, water, alcohol, and heat. Shellac is not recommended as a wood finish because it gives a brittle finish that waterspots easily.

## PREPARATION OF THE SURFACES

Before any type of paint is applied, the surface to be painted should be properly prepared. For new surfaces, the following precautions should be taken:

■ The surface should be thoroughly dry before it is painted.

■ Siding and trim should be protected from the weather until they are installed.

■ Nonrusting nails should be used in the application of siding and trim.

■ Siding and trim should be primed immediately after installation.

■ Any structural defects that might permit the entrance of water into a cavity should be corrected before painting.

■ The surface to be painted should be free of dust and dirt.

■ Any open joints should be properly sealed with caulking compound.

■ If knots are present, they should be covered with a knot sealer.

If the surface to be painted is covered with old paint, the following precautionary steps should be followed:

■ The surface should be dusted before painting; if the surface is extremely dirty it should be washed with a mild synthetic detergent and rinsed with water.

■ The surface should be thoroughly dry before paint is applied.

■ Rust marks should be removed with sandpaper or steel wool.

■ All open joints should be sealed with caulking compound.

■ Loose, flaking, or blistering paint should be removed with a wire brush and scraper.

■ If the cracking or alligatoring of the old paint is extensive, the old film should be removed to the bare wood.

■ To obtain the best results, the paint should be applied in clear, dry weather with the temperature above 40° F.

## Application of Caulks, Sealants, Putties, and Glazing Compounds

Before any paint is applied, it is necessary to seal the structure against moisture and cold. The most commonly used sealing products are putty, glazing compounds, oil-based caulks, flexible sealants, water-based sealants, solvent-based acrylics, and elastomeric sealants.

*Putty* is a soft, doughlike substance that is manufactured by the blending of pigments and oil. It is applied with a knife and is most often used for face glazing. After the putty has set, it can be painted to extend its life.

*Glazing compounds* are similar to putty except that they are modified to provide a more plastic and resilient product for a longer period of time. Used for face glazing,

they provide a highly effective seal between glass and framing.

*Oil-based caulks* consist of oxidizing oils, pigments, and additives and are used to seal nonmoving cracks and joints. Oil-based caulks are usually applied with a caulking gun and should be allowed to skim over, or cure, before the surface is painted.

*Flexible sealants* are extremely stretchable and are used where hard-drying caulks are not practical. Most flexible sealants are butyls. A *butyl sealant* usually has a minimum life expectancy of ten years and is considered to be nonstaining. The biggest disadvantage of a butyl sealant is that it will normally shrink from 10 to 35 percent. Because of the excessive shrinkage, it is not recommended for joints larger than a quarter of an inch.

Acrylic latex is a *water-based sealant* that can be applied to damp surfaces, tooled with water, and painted almost immediately after application. The sealant is also flexible and does not become brittle with age. Polyvinyl acetate latex is usually recommended for interior use. It is quite brittle and nonflexible when it dries, but usually costs less than acrylic latex.

A *solvent-based acrylic* will adhere to almost any surface and has a life expectancy of twenty years. It is usually sold in cartridge form and must be heated before it is applied. It is an all-purpose sealant, but should not be used on traffic surfaces or where it would be submerged in water.

*Elastomeric sealants* have a rubberlike consistency and are available in two forms: one-part and two-part. One-part is used as it is supplied, but two-part requires the mixing of a base compound with an accelerator.

Before any type of sealant is applied, the

area to be caulked should be inspected for dirt and grease. Dirt can be removed with a mild solution of soap and water, while grease can be removed with a rag that has been soaked in mineral spirits. The joints to be caulked should be cleaned of any old caulking. All joints should be increased to ¼ inch in width. Wood and steel surfaces should be primed before applying sealant.

## Application of Caulking Compounds or Sealants

Gun-grade sealant is packaged in a cartridge which fits into a caulking gun. To operate the caulking gun, the nozzle must first be cut at an angle and the interseal punctured. With the gun held at an approximate angle of 45°, the compound is forced out in a uniform line when the gun is triggered. The gun should be slowly drawn along the crevice as the bead of sealant is flowing from the nozzle.

Some gun-grade sealants are also available in cans or collapsible tubes. When caulk is applied from a collapsible tube, the tube should be rolled up as the caulk is being applied.

Caulking compounds are also available in rope form, and are applied by simply being pushed into a crevice. This type of caulking compound is usually considered as a temporary seal and is used primarily around windows and screens.

## PAINTING TOOLS

The most common painting tools are paint brushes, rollers, and spray equipment. Paint brushes are usually available in three basic sizes: a 3½- or 4-inch brush is usually used for large areas; a 2½- or 3-inch brush is usually used for shutters, windows and door trim; and a 1- or 1½-inch oval brush is used for window sash and moldings.

Rollers are often used to paint large areas and are manufactured in a variety of textures and finishes. The 7-inch roller is the most popular one, but they can be purchased in lengths from 2 to 9 inches. A roller is ideal for painting large surfaces with latex, but should not be used for finishes that dry rapidly, such as shellac or lacquer.

Spray painting accounts for a large percentage of industrial and commercial finishes because it can be done very quickly. Most spray equipment includes a compressor, hose, paint container, and spray gun. To properly spray a surface, the gun should be held approximately 6 inches from the surface and moved in a line parallel to the surface. If the gun is kept perpendicular to the surface, a uniform deposit of paint will be placed; but if the gun is allowed to arch, an irregular deposit will result.

## INTERIOR PAINTING

Before an interior surface is painted, all cracks should be filled with spackling compound and sanded smooth. All hardware should be removed from the doors and windows, light fixtures should be loosened, and switch plates and convenience outlet covers should be removed or covered with masking tape and paper.

A dropcloth spread over the floors will catch paint drippings. To minimize drippage, the brush should not be dipped more than one-third of the bristle length into the paint. It is best to cover exposed areas of skin with a protective cream, which creates a film and can easily be removed when the painting has been completed.

The ceiling should be painted first, and worked across the width rather than lengthwise. If this procedure is followed, the paint can be lapped before the preceding course has dried. Laps, to be effective, should never be more than 2 inches wide.

Walls are usually started from the upper left-hand corner and the work proceeds down toward the floor. The woodwork is usually finished with a 1-inch brush for window sash and a 2- or 3-inch brush for the remainder of the window. If a roller is used to paint the walls, a 1-foot strip below the ceiling line should first be painted with a brush. The corners should also be painted from ceiling to floor with a brush before the roller is used. The roller cannot be used at intersections of wall and ceiling, or intersection of wall and wall. Once the corners have been painted, the roller is loaded and is started in an upward direction. After an area about 2 feet wide and 3 feet deep has been coated with up-and-down strokes, the roller should be moved back and forth. A cardboard guard should be used at the bottom of the freshly painted wall to keep the paint from touching the floor or woodwork.

## Painting New Plaster

Before plaster can be painted it should be allowed to dry thoroughly—usually for three to four weeks, but the time can vary according to temperature and relative humidity. If the plaster must be painted before it is dry, it is usually good practice to apply only one coat of a good latex.

After the walls have thoroughly dried, they should be inspected for uniformity of surface. If the surface is covered with an accumulation of "chalk" or a dry powdery material, it should be removed by vigorous brushing.

Plastered walls should be primed with a latex, alkyd, or oil-type primer-sealer. The finish coat can be a flat finish, semigloss, or a gloss paint.

## Painting Gypsum Wallboard

Latex emulsions are usually preferred for painting gypsum wallboard. Solvent-thinned primers are ordinarily not recommended because they raise the fibers of the paper, giving the wallboard a fuzzy appearance. Before gypsum wallboard is painted it should be properly taped, floated, and sanded.

## Painting Particleboard

Before particleboard is painted, its surface should be filled with a paste wood filler, or a sanding sealer should be brushed to the exposed surface. In some cases, particle boards have been pretreated at the factory and are ready for painting.

To prevent the migration of moisture, free-moving units should be painted on both sides, edges, top, and bottom. Tops and bottoms of doors are often overlooked and are not painted, but they should be well-sealed to reduce moisture intake.

An oil-based enamel undercoat is usually recommended as a first coat for particleboard. Latex paints can be used, but they do not produce a smooth surface and should be applied over a suitable primer coat.

## Painting Hardboard

Before hardboard is painted it should be clean, dry, and free of any grease. A good water-thinned or solvent-thinned primer-sealer can be painted on the hardboard. For best results all exposed edges should be painted and if an adequate vapor barrier is not used, the back side of hardboard siding should be primed.

## EXTERIOR PAINTING

### Painting Brick

Bricks can be sealed with either a latex primer or an enamel undercoat. There may be efflorescence—a fine white powder caused by salts in the brick and brought to the surface by the action of water. This should be removed by vigorous scrubbing and then treated with a clear resin sealer. Because of the alkali in fresh mortar, it is usually best to use paints designed for use on concrete.

### Painting Concrete and Concrete Masonry

Before concrete and concrete masonry can be painted, they should be cleaned of any bond-breaking materials. It is also necessary to use one or two coats of block filler for concrete blocks. It is not necessary to fill the voids, but doing so results in a smoother surface that resists dirt accumulation.

Concrete or concrete masonry should be primed with a latex, alkyd, or oil-type pri-

In some cases, it is not necessary to use a primer because the hardboard is primed at the factory.

After the hardboard has been properly sealed, the finish coat can be applied. For a smooth and better-looking job, the sealer should be lightly sanded before the top coat is applied. In most cases a good quality of water-thinned or solvent-thinned paint will perform satisfactorily on hardboard. But if hardboard is to be in a permanent horizontal position, a good grade of floor and deck enamel should be applied as the top coat.

mer-sealer. The finish coat can be flat finish, semigloss, or a gloss paint.

### Painting Wood

Either oil or latex paints can be used for painting wood. Oil paints have better coverage and hide defects better but latex paints are easier to use and my be applied over damp surfaces. Regardless of the type of paint used, the manufacturer's recommendations should be followed.

### Painting Plywood

Plywood will take almost any type of finish, but if it is painted, a top-quality acrylic latex house paint is usually recommended. It is necessary to seal the edges of plywood panels that are for exterior use. Sealing is easiest when the panels are stacked, but edges that are cut later should also be sealed.

An exterior panel should be primed and finished with a quality non-lead-base prime. A minimum of two coats should be applied, but three are better. If the plywood is rough-sawn or textured, a semitransparent or opaque stain can be used. Interior plywood panels can be finished with paint, enamel, or stains.

## COMMON PAINT PROBLEMS

Blistering and peeling are caused by excessive amounts of moisture in the object painted. In most cases the problem can be alleviated by installing a vapor barrier.

Wrinkling occurs when the paint has been applied too thickly, or if there is too much oil in the paint.

Mildew is caused by warm, damp conditions, but many paints now manufactured will minimize the mildew problem.

# REMODELING

# TECHNIQUES

DURING A REMODELING JOB MANY PROBLEMS will arise that are not found in new construction.

Thus there are standard construction practices and procedures that apply specifically to remodeling.

## FOUNDATIONS

Before any construction work is begun, the foundation should be inspected for structural stability. The foundation wall should be inspected for cracks and termite tubes. To check the soundness of sills and joists, a knife should be inserted. If it penetrates easily, the structural members are probably either decayed or infested with termites, or both. If it is necessary to replace supporting members such as beams, girders, or joists and they are within 12 inches of grade, the timbers should be treated with a wood preservative to ward off insect attacks and prevent decay.

If beams or girders are sagging, it will be necessary to elevate them. A screw jack can be used to raise a beam, but a permanent pier or column must be designed and built to support the existing load.

If the structure is built on piers or columns, they should also be inspected and replaced if they are deteriorating. If it is necessary to replace a pier, a screw jack should be placed next to the old pier and the structure raised until a slip of paper can pass between sill and pier. It is then possible to remove the old pier and add a new one.

If the structure has a leaking basement, it may be necessary to excavate around the perimeter of the foundation and then apply various dampproofing techniques. Some of the techniques include placement of a drain tile, parging the foundation wall, backfilling with wash gravel, and maintaining a proper grade around the building.

If the structure to be remodeled has a crawl space, the ground beneath should be covered with a suitable vapor barrier. in most cases a layer of polyethylene covered with gravel is sufficient. The vapor barrier prevents the rising water vapor and often eliminates cold and damp floors. It is also good practice to place insulation between the joists. Batt insulation is most often used, and can be held in place by wire mesh stapled to the edges of the joists or by pieces of heavy-gauge wire wedged between the joists. It is also a good idea to insulate the foundation wall by placing rigid insulation against the inside.

If a wing or a room is added to an existing structure, it is important to establish the proper elevation for the batter boards. To accomplish this, a builder's level is set up. Then the existing floor elevation is calculated by means of a reading made with the rod held in a doorway. This reading can then be transferred to establish the batter board heights.

## EXTERIOR AND INTERIOR FRAMED WALLS

Walls are classified as load-bearing and non–load-bearing. A *load-bearing* wall supports part of the weight of the structure, while a non–*load-bearing* wall has neither dead nor live loads imposed on it.

To determine whether or not a wall is

load-bearing, it is necessary to find out if any joists are resting on it, or if the wall is resting directly on another wall, pier, or column. Because load-bearing walls support part of the weight of a structure, it is best not to cut into them. But if this is necessary, a header should be placed in the opening.

When it is necessary to cut an opening in either an exterior or an interior wall, the following precautionary steps should be taken:

- Check to see if there are electrical wires in the wall. If such a wire is sawed or cut, it might cause extensive damage.

- Check to see if there is any plumbing in the wall. If there is, the wall may have to be cut elsewhere, or the plumbing may have to be relocated.

- When studs are sawed off in an existing wall, the cut should be made about 2 inches from the top or the bottom of the stud. If this procedure is followed, the nails in the stud should be missed.

- When breaking through an old wall, it is a good idea to cover the eyes and nose with a protective device. A pair of goggles will help protect eyes, and an inexpensive filtering device will protect the lungs from excessive dust.

Before a wall is cut, a line should be chalked to indicate the area to be cut. Materials such as brick or plaster can then be removed with a cold chisel and hammer or a power saw with a carbide-tip blade. It is necessary to use a special blade, because an ordinary one is not designed to cut plaster or masonry. If a cut through drop or plywood siding is necessary, a regular power-saw blade can be used; but any nails that are on the chalkline should first be removed.

It is also a good idea to use a large plastic dropcloth to protect the floors during the wall removal.

## SHINGLES

When roof shingles begin to curl, appear puffed up, and no longer provide adequate protection, it is necessary to replace the roof covering.

### Asphalt Shingles

If the roof is covered with asphalt shingles, the old shingles can be removed or new shingles can be placed over the existing roof. If an overlay is placed over the existing shingles, the deck should be free of any structural defects. Any loose or curled shingles must be nailed down or cut away. The stack flashing, hip and ridge units, and loose nails must be removed, and the roof swept clean.

The application of the overlay is started by cutting 2 inches off the bottom of a 7-inch starter shingle. The 5-inch starter shingle is then placed along the eave. The top of the starter butts the bottom of the old shingle. The nails used in an overlay should be long enough to penetrate ¾ inch into a solid wood deck or all the way through a plywood deck. Two inches should be cut off the width of the first course of shingles; this course is laid with a 1-inch rake overhang. The shingles in this course are butted to the bottom of the second and third courses of old shingles to prevent the roof from being uneven and irregular. The second and succeeding courses of shingles are laid with full-width shingles, and are butted to the bottom edges of the old shingles. The exposure will be automatic and will coincide with that of the old roof.

If the shingles have cutouts they must be in line with the cutouts on the old roof. To eliminate any possibility of misalignment, chalklines can be placed between the cutouts.

## Wood Shakes or Shingles

Placing *wood shakes* or *wood shingles* over an existing roof provides several advantages: it affords a double roof with added insulation and storm protection; and the interior is protected from sudden rains during the construction period.

Before any wood shakes or shingles are installed, the roof sheathing should be inspected for any structural defects. For a smooth application, any loose or curled shingles should be nailed down or cut away.

If a roof requires an overlay, the stack flashing usually must also be replaced: therefore, it is necessary to remove all the old stack flashing and install new flashing as the overlay is placed.

To assure a sound deck and a pleasing roofline, the old shingles should be cut back 6 inches from the rake edge and a 1 × 4 nailed on top of the sheathing adjacent to the rake edge. The old shingles are also cut back one course from the eave line. The ridge units should be removed and replaced with a thin strip of bevel siding, which should be placed bevel-side down. To keep the old valley flashing segretated from the new flashing, a 1 × 4 should be nailed to the old valley.

The new wood shakes or shingles are then applied, using 5d rust-resistant nails.

## EXTERIOR SIDEWALL COVERING

Settling foundations, freezing temperatures, and water seepage often cause fine cracks to develop in stucco. To repair such cracks, all the loose cement should be removed with a cold chisel. The crack should be opened to a width of about 1 inch and undercut to provide a key for the new stucco. The crack should then be wet thoroughly, and new stucco troweled over the crack. The mixture used for patching stucco is 1 part portland cement to 2½ parts fine sand.

If the exterior wall has loose or rotted boards, it is usually best to remove them and install new facing material over the existing siding. If the old siding is not removed, the surface should be made as true, level, and smooth as possible. Wood shakes and wood shingles make an excellent covering for old walls because they are relatively small and can readily conform to an irregularly shaped wall. Metal and plywood siding can also be used, and provide a durable and attractive wall.

Brick walls often become coated with a white powder called efflorescence. It is caused by salts such as sodium or magnesium in the brick or mortar, and has been brought to the surface by the action of water. Efflorescence can be cleaned and removed from a brick wall with a solution of diluted muriatic acid, followed by a clear water rinse.

Any holes in the mortar joints should be pointed and filled with additional fresh mortar.

## INTERIOR WALL FINISH

Before the interior wall finish can be installed the walls should be properly conditioned and prepared. If the wall finish material is to be placed over old walls that are irregular or over masonry walls, furring strips can be installed. These should be placed so that they run at right angles to the direction of panel application. These strips are usually nominal 1 × 2s, and are placed on 16-inch centers. If the furring strips are attached to masonry walls, masonry nails should be used.

Vertical furring blocks are used to provide a solid backing at the panel edges. The blocks are spaced 48 inches on center, and for ventilation purposes there should be a ½-inch space between the horizontal furring strips and the vertical furring blocks. If the furring strips are not level, shims can be used.

Once the walls have been straightened, the finish material can be placed in a conventional manner.

Small cracks in plaster and gypsumboard can be repaired with spackling compound. Before cracks are filled, they should be cleaned and undercut to provide a mechanical key. Then the compound can be worked into the crack with a putty knife. Once the compound has dried, it can be sanded and painted. For larger holes or deep cracks, it is usually necessary to apply the compound in two or three layers. The first application should fill about half of the depression, and be allowed to harden before the second application. When this procedure is followed shrinkage will be minimized.

Large holes in gypsumboard are often repaired by cutting out a section of the wallboard from the center of one stud to the center of the adjacent stud. A new piece of gypsum wallboard can then be cut to the appropriate size and placed in the opening. The cracks created by the addition of the new piece can then be properly taped and floated.

Grout in ceramic tile walls often turns dark, and in some cases the joints open up. Most of the dark stains can be removed by (a) scrubbing with an old toothbrush dipped into a solution of household laundry bleach. If the dark stains are extreme, or if the joints are open, it is best to regrout the joints. Many excellent grouts are available in handy tubes with a special applicator tip. The tip is cut from the applicator and a thin ribbon of caulk can be applied.

It is often necessary to hang or fasten objects to a wall covered with plaster or gypsum wallboard, but such walls will not support much weight. There are several kinds of wall fasteners, however, that can be used to hang heavy objects: metal expansion anchors, toggle bolts, and plastic anchors.

*Metal expansion anchors* have expansion sleeves that spread apart when a threaded bolt is placed in the center.

*Toggle bolts* have spring-actuated wings

that are hinged to a nut. A bolt is inserted into the nut, the wings folded back, and the assembly is pushed through an opening in the wall. Once through the opening, the

wings spread and the bolt can be tightened.

A *plastic anchor* is placed inside a drilled hole in the wallboard. When a screw is inserted, the anchor splits or expands.

## FLOORING

Before any type of resilient flooring is installed, it is usually necessary to cover the subfloor with an underlayment. The two basic types of underlayment are mastic and board.

A *mastic-type* underlayment is used to level worn or damaged areas, and can be troweled to a feather-edge. One of the most effective mastic-type underlayments contains a binder of latex, asphalt, or polyvinyl-acetate resins in the mix. A mastic that contains cement, gypsum, and sand can also be used, but it often breaks down under traffic.

*Board-type* underlayments are available in three basic types: hardboard, plywood, and particleboard. Hardboard, in most

cases, is used in remodeling because it minimizes the buildup of old floors.

Once the underlayment has been installed, flooring can be placed, with conventional techniques.

Wood floors often become loose, buckle, or squeak when pressure is applied to them. To eliminate squeaking, one of two corrective measures can be taken. Graphite can be squirted between the boards, or they can be renailed to the subfloor. In renailing wood flooring, finishing nails are usually used, driven at an angle to the flooring.

Wood flooring often buckles because of the presence of excessive moisture. This problem can often be solved by placing a vapor barrier beneath the crawl space.

## CEILING SYSTEMS

Ceiling members sometimes sag, causing an unsightly appearance; or a ceiling may be considered too high. In such cases the ceiling can be furred down, or, alternatively, a suspended acoustical ceiling can be installed.

If the ceiling is furred down, individual trusses will have to be constructed and placed; furring strips are then nailed to the

trusses, and finally the finished ceiling can be installed.

A furred-down ceiling is more costly than a suspended ceiling, and it requires more skill and craftsmanship to install.

A suspended ceiling consists of a simple metal grid framework suspended on wires from above. The ceiling panels are dropped into the framework.

## WINDOWS

Windows that are inoperative may only have broken sash cords or may need to be completely replaced. It is possible to install

windows that are prebuilt, preglazed, or prefinished.

When windows are merely sticking,

often because of wet weather, the stops can be repositioned, after which the window sashes should be sanded smooth and finished with an appropriate finish.

## DOORS

Doors are often rendered inoperative by sagging framing members or simply by the age of the hardware. After the framing members have been properly aligned, it may be necessary to remove the door and square it to the opening. After all necessary adjustments have been made, the door's edges should be properly sealed.

If the lock needs to be replaced, a new one can be repositioned on the door. A decorative escutcheon plate can be used to cover marks left by old lock hardware.

# LIGHT CONSTRUCTION REFERENCE TABLES

## TABLE A1.   FINISHES FOR WOOD SIDING

| General | Finish to Use | Instructions |
|---|---|---|
| Paints | Alkyd paints | Apply alkyd primer and 2 finish coats[1] |
| | Oil-base paints | Use a zinc-free primer plus 2 finish coats[1] |
| | Latex paints | Same as above[2] |
| Stains | Heavy-body, oil-base stain | 1 or 2 coats, Brush, dip, or spray.[4] |
| Solid, but somewhat soft color. Shows wood texture, but little grain. | Creosote stains | 1 or 2 brush applications[4] |
| | Semi-transparent oil base stain | 2 brush applications. May be sprayed and smoothed with brush.[5] |
| Light coloring, emphasis on wood grain show-through. | Semi-transparent resin stains | 2 brush applications[6] |
| Weathering agents | Commercial Bleaches | Brush 1 or 2 coats. Renew in 3 or 5 years if necessary.[7] |
| Repellents | Water repellent | 2 coats. Dip before installation, brush after.[8] |

[1] Alkyds are quick-drying, blister-resistant, and can be applied self-primed. Oil-base paints are not blister-resistant unless applied over a zinc-free primer. Seal back of siding with water repellent.

[2] Product development in this field rapid. Follow manufacturer's instructions.

[3] Particularly suited for rough and saw-textured products.

[4] A durable-type finish. Some brands suitable for subsequent painting after several years of weathering, if desired. Allows grain show-through.

[5] A natural for rough or saw-textured sidings. Gives transparent color which is durable and long-lasting.

[6] Fast-drying with good penetration and durability.

[7] Will give natural wood a weathered appearance.

[8] Excellent for retaining the natural wood look. Pigmented or dye stain may be added.

Reprinted with permission of **Western Wood Products Association.**

## TABLE A2.   STRESS-RATED BOARDS (standard sizes)

| Nominal | Surfaced Dry | Surface Unseasoned |
|---|---|---|
| | **THICKNESSES** | |
| 1″ | $\frac{3}{4}$″ | $\frac{25}{32}$″ |
| 1¼″ | 1″ | $1\frac{1}{32}$″ |
| 1½″ | 1¼″ | $1\frac{9}{32}$″ |
| | **WIDTHS** | |
| 2″ | 1½″ | $1\frac{9}{16}$″ |
| 3″ | 2½″ | $2\frac{9}{16}$″ |
| 4″ | 3½″ | $3\frac{9}{16}$″ |
| 5″ | 4½″ | $4\frac{5}{8}$″ |
| 6″ | 5½″ | $5\frac{5}{8}$″ |
| 7″ | 6½″ | $6\frac{5}{8}$″ |
| 8″ & Wider | $\frac{3}{4}$″ off nominal | $\frac{1}{2}$″ off nominal |

Reprinted with permission of Western Wood Products Association.

## TABLE A3.   PRODUCT CLASSIFICATION

Standard lengths of lumber generally are 6 feet and longer in multiples of 1′

| | Thickness In. | Width In. |
|---|---|---|
| board lumber | 1″ | 2″ or more |
| light framing | 2″ to 4″ | 2″ to 4″ |
| studs | 2″ to 4″ | 2″ to 4″ 10′ and shorter |
| structural light framing | 2″ to 4″ | 2″ to 4″ |
| joists and planks | 2″ to 4″ | 6″ and wider |
| beams and stringers | 5″ and thicker | more than 2″ greater than thickness |
| posts and timbers | 5″ × 5″ and larger | not more than 2″ greater than thickness |
| decking | 2″ to 4″ | 4″ to 12″ wide |
| siding | thickness expressed by dimension of butt edge | |
| mouldings | size of thickest and widest points | |

Reprinted with permission of Western Wood Products Association.

**TABLE A4.  WOOD DIMENSIONS/ALL SPECIES**

| | | |
|---|---|---|
| Light Framing (2″ to 4″ thick, 2″ to 4″ wide) | Construction Standard Utility Economy | This category for use where high strength values are not required; such as studs, plates, sills, cripples, blocking, etc. |
| Studs (2″ to 4″ thick, 2″ to 4″ wide) | Stud Economy stud | An optional all-purpose grade limited to 10 feet and shorter. Characteristics affecting strength and stiffness values are limited so that the "Stud" grade is suitable for all stud uses, including load-bearing walls. |
| Structural Light Framing (2″ to 4″ thick, 2″ to 4″ wide) | Select structural No. 1, No. 2, No. 3, Economy | These grades are designed to fit those engineering applications where higher bending strength ratios are needed in light framing sizes. Typical uses would be for trusses, concrete pier wall forms, etc. |
| Appearance Framing (2″ to 4″ thick, 2″ and wider) | Appearance | This category for use where good appearance and high strength values are required. Intended primarily for exposed uses. Strength values are the same as those assigned to No. 1 Structural Light Framing and No. 1 Structural Joists and Planks. |
| Structural Joists and Planks (2″ to 4″ thick, 6″ and wider) | Select structural, No. 1, No. 2, No. 3, Economy | These grades are designed especially to fit in engineering applications for lumber 6 inches and wider, such as joists, rafters, and general framing uses. |

Reprinted with permission of Western Wood Products Association.

**TABLE A5.  SIZES OF EXTERIOR SIDINGS**

| | Nominal Size | | Dressed Dimensions | |
|---|---|---|---|---|
| Product | Thickness In. | Width In. | Thickness In. | Width In. |
| BEVEL SIDING For WRC Sizes See footnote[1] | $\frac{1}{2}$ $\frac{3}{4}$ | 4 5 6 | $\frac{15}{32}$ butt, $\frac{3}{16}$ tip, $\frac{3}{4}$ butt, $\frac{3}{16}$ tip | $3\frac{1}{2}$ $4\frac{1}{2}$ $5\frac{1}{2}$ |
| WIDE BEVEL SIDING (Colonial or Bungalow) | $\frac{3}{4}$ | 8 10 12 | $\frac{3}{4}$ butt, $\frac{3}{16}$ tip | $7\frac{1}{4}$ $9\frac{1}{4}$ $11\frac{1}{4}$ |

| | | | | Face | Overall |
|---|---|---|---|---|---|
| RABBETED BEVEL SIDING (Dolly Varden) | $\frac{3}{4}$ 1 | 6 8 10 12 | $\frac{5}{8}$ by $\frac{5}{16}$ $\frac{13}{16}$ by $\frac{13}{32}$ | 5 $6\frac{3}{4}$ $8\frac{3}{4}$ $10\frac{3}{4}$ | $5\frac{1}{2}$ $7\frac{1}{4}$ $9\frac{1}{4}$ $11\frac{1}{4}$ |
| RUSTIC AND DROP SIDING (Dressed and Matched) | 1 | 6 8 10 12 | $\frac{23}{32}$ | $5\frac{1}{8}$ $6\frac{7}{8}$ $8\frac{7}{8}$ $10\frac{7}{8}$ | $5\frac{3}{8}$ $7\frac{1}{8}$ $9\frac{1}{8}$ $11\frac{1}{8}$ |

**TABLE A5 (cont.)**

| Product | Nominal Size | | Dressed Dimensions | | |
| --- | --- | --- | --- | --- | --- |
| | Thickness In. | Width In. | Thickness In. | Face | Overall |
| RUSTIC AND DROP SIDING (Shiplapped, $\frac{3}{8}$-in. lap) | 1 | 6 | $^{23}/_{32}$ | 5 | $5\frac{3}{8}$ |
| | | 8 | | $6\frac{3}{4}$ | $7\frac{1}{8}$ |
| | | 10 | | $8\frac{3}{4}$ | $9\frac{1}{8}$ |
| | | 12 | | $10\frac{3}{4}$ | $11\frac{1}{8}$ |
| RUSTIC AND DROP SIDING (Shiplapped, $\frac{1}{2}$-in lap) | 1 | 6 | $^{23}/_{32}$ | $4^{15}/_{16}$ | $5^{7}/_{16}$ |
| | | 8 | | $6\frac{5}{8}$ | $7\frac{1}{8}$ |
| | | 10 | | $8\frac{5}{8}$ | $9\frac{1}{8}$ |
| | | 12 | | $10\frac{5}{8}$ | $11\frac{1}{8}$ |
| LOG CABIN SIDING | $1\frac{1}{2}$ (6/4) | 6 | $1\frac{1}{2}''$ at | $4^{15}/_{16}$ | $5^{7}/_{16}$ |
| | | 8 | thickest | $6\frac{5}{8}$ | $7\frac{1}{8}$ |
| | | 10 | point | $8\frac{5}{8}$ | $9\frac{1}{8}$ |
| TONGUE & GROOVE (T&G) S2S AND CM | 1 (4/4) | 4 | $\frac{3}{4}$ | $3\frac{1}{8}$ | $3\frac{3}{8}$ |
| | | 6 | | $5\frac{1}{8}$ | $5\frac{3}{8}$ |
| | | 8 | | $6\frac{7}{8}$ | $7\frac{1}{8}$ |
| | | 10 | | $8\frac{7}{8}$ | $9\frac{1}{8}$ |
| | | 12 | | $10\frac{7}{8}$ | $11\frac{1}{8}$ |

[1] Western Red Cedar Bevel Siding available in $\frac{1}{2}''$, $\frac{5}{8}''$, $\frac{3}{4}''$ nominal thickness. Corresponding surfaced thick edge is $^{15}/_{32}''$, $\frac{9}{16}''$ and $\frac{3}{4}''$. Widths 8″ and wider $\frac{1}{2}''$ off.

Reprinted with permission of Western Wood Products Association.

**TABLE A6.   COVERAGE ESTIMATOR FOR EXTERIOR SIDINGS**

| Nominal Size | Width | | Area Factor[1] | Nominal Size | Width | | Area Factor[1] |
| --- | --- | --- | --- | --- | --- | --- | --- |
| | Dress | Face | | | Dress | Face | |
| **SHIPLAP** | | | | **PANELING PATTERNS** | | | |
| 1 × 6 | $5\frac{1}{2}$ | $5\frac{1}{8}$ | 1.17 | 1 × 6 | $5^{7}/_{16}$ | $5^{1}/_{16}$ | 1.19 |
| 1 × 8 | $7\frac{1}{4}$ | $6\frac{7}{8}$ | 1.16 | 1 × 8 | $7\frac{1}{8}$ | $6\frac{3}{4}$ | 1.19 |
| 1 × 10 | $9\frac{1}{4}$ | $8\frac{7}{8}$ | 1.13 | 1 × 10 | $9\frac{1}{8}$ | $8\frac{3}{4}$ | 1.14 |
| 1 × 12 | $11\frac{1}{4}$ | $10\frac{7}{8}$ | 1.10 | 1 × 12 | $11\frac{1}{8}$ | $10\frac{3}{4}$ | 1.12 |
| **TONGUE-AND-GROOVE** | | | | **BEVEL SIDING (1″ lap)** | | | |
| 1 × 4 | $3\frac{3}{8}$ | $3\frac{1}{8}$ | 1.28 | 1 × 4 | $3\frac{1}{2}$ | $3\frac{1}{2}$ | 1.60 |
| 1 × 6 | $5\frac{3}{8}$ | $5\frac{1}{8}$ | 1.17 | 1 × 6 | $5\frac{1}{2}$ | $5\frac{1}{2}$ | 1.33 |
| 1 × 8 | $7\frac{1}{8}$ | $6\frac{7}{8}$ | 1.16 | 1 × 8 | $7\frac{1}{4}$ | $7\frac{1}{4}$ | 1.28 |
| 1 × 10 | $9\frac{1}{8}$ | $8\frac{7}{8}$ | 1.13 | 1 × 10 | $9\frac{1}{4}$ | $9\frac{1}{4}$ | 1.21 |
| 1 × 12 | $11\frac{1}{8}$ | $10\frac{7}{8}$ | 1.10 | 1 × 12 | $11\frac{1}{4}$ | $11\frac{1}{4}$ | 1.17 |
| **S4S** | | | | | | | |
| 1 × 4 | $3\frac{1}{2}$ | $3\frac{1}{2}$ | 1.14 | | | | |
| 1 × 6 | $5\frac{1}{2}$ | $5\frac{1}{2}$ | 1.09 | | | | |
| 1 × 8 | $7\frac{1}{4}$ | $7\frac{1}{4}$ | 1.10 | | | | |
| 1 × 10 | $9\frac{1}{4}$ | $9\frac{1}{4}$ | 1.08 | | | | |
| 1 × 12 | $11\frac{1}{4}$ | $11\frac{1}{4}$ | 1.07 | | | | |

[1] Allowance for trim and waste should be added.

Reprinted with permission of Western Wood Products Association.

## TABLE A7. NAIL SIZE SPECIFICATION

| Size | Length (Inches) | | Siding Nails (Count per lb.) | | Approx, lbs Per 1,000 BF of Siding | |
|---|---|---|---|---|---|---|
| | 1 | 2 | 1 | 2 | 1 | 2 |
| 6d | 1⅞″ | 2″ | 566 | 194 | 2 | 6 |
| 7d | 2⅛″ | 2¼″ | 468 | 172 | 2½ | 6½ |
| 8d | 2⅜″ | 2½″ | 319 | 123 | 4 | 9 |
| 10d | 2⅞″ | 3″ | 215 | 103 | 5½ | 11 |

¹ Aluminum

² Hot Dipped Galv.

Reprinted with permission of Western Wood Products Association.

## TABLE A9. HOW TO ARRIVE AT THE AMOUNT OF HARDWOOD FLOORING REQUIRED

To determine the board feet of flooring needed to cover a given space, first find the area in square feet. Then add to it the percentage of that figure which applies to the size flooring to be used, as indicated below. The additions provide an allowance for side-matching plus an additional 5 percent for end-matching and normal waste.

The above figures are based on laying flooring straight across the room. Where there are bay windows or other projections, allowance should be made for additional flooring.

| | | | |
|---|---|---|---|
| 55% for | 25/32″ × 1½″ | 38⅓% for | ⅜″ × 1½″ |
| 42½% for | 25/32″ × 2″ | 30% for | ⅜″ × 2″ |
| 38⅓% for | 25/32″ × 2¼″ | 38⅓% for | ½″ × 1½″ |
| 29% for | 25/32″ × 3¼″ | 30% for | ½″ × 2″ |

Reprinted with permission of National Oak Flooring Manufacturers' Association.

## TABLE A8. STANDARD SIZES, COUNTS, AND WEIGHTS OF PECAN, BEECH, BIRCH, AND HARD MAPLE STRIP FLOORING

| Nominal | Actual | Counted | Wts. M. Ft.¹ |
|---|---|---|---|
| 25/32 × 3¼″ | ¾″ × 3¼″ | 1″ × 4″ | 2210 lbs. |
| 25/32 × 2¼″ | ¾″ × 2¼″ | 1″ × 3″ | 2020 lbs. |
| 25/32 × 2″ | ¾″ × 2″ | 1″ × 2¾″ | 1920 lbs. |
| 25/32 × 1½″ | ¾″ × 1½″ | 1″ × 2¼″ | 1820 lbs. |
| ⅜ × 2″ | 11/32 × 2 | 1″ × 2½″ | 1000 lbs. |
| ⅜ × 1½″ | 11/32 × 1½″ | 1″ × 2″ | 1000 lbs. |
| ½ × 2″ | 15/32 × 2″ | 1″ × 2½″ | 1350 lbs. |
| ½ × 1½″ | 15/32 × 1½″ | 1″ × 2″ | 1300 lbs. |
| **SPECIAL THICKNESSES²** | | | |
| 17/16 × 3¼″ | 33/32 × 3¼″ | 5/4 × 4″ | 2400 lbs. |
| 17/16 × 2¼″ | 33/32 × 2¼″ | 5/4 × 3″ | 2250 lbs. |
| 17/16 × 2″ | 33/32 × 2″ | 5/4 × 2¾″ | 2250 lbs. |
| **JOINTED FLOORING, I.E., SQUARE EDGE** | | | |
| 25/32 × 2½″ | ¾″ × 2½″ | 1″ × 3¼″ | 2160 lbs. |
| 25/32 × 3¼″ | ¾″ × 3¼″ | 1″ × 4″ | 2300 lbs. |
| 25/32 × 3½″ | ¾″ × 3½″ | 1″ × 4¼″ | 2400 lbs. |
| 17/16 × 2½″ | 33/32 × 2½″ | 5/4″ × 3¼″ | 2500 lbs. |
| 17/16 × 3½″ | 33/32 × 3½″ | 5/4″ × 4¼″ | 2600 lbs. |

¹Weight per thousand feet    ²Above tongued and grooved and end matched

Reprinted with permission of National Oak Flooring Manufacturers' Association.

## TABLE A10.   LUMBER DIMENSIONS

| Nominal Size[1] | Actual Size[2] | Nominal Size[1] | Actual Size[2] |
|---|---|---|---|
| 1 × 2 | ¾ × 1½ | 2 × 8 | 1½ × 7¼ |
| 1 × 3 | ¾ × 2½ | 2 × 10 | 1½ × 9¼ |
| 1 × 4 | ¾ × 3½ | 2 × 12 | 1½ × 11¼ |
| 1 × 5 | ¾ × 4½ | 4 × 4 | 3½ × 3½ |
| 1 × 6 | ¾ × 5½ | 4 × 6 | 3½ × 5½ |
| 1 × 8 | ¾ × 7¼ | 4 × 10 | 3½ × 9¼ |
| 1 × 10 | ¾ × 9¼ | 6 × 6 | 5½ × 5½ |
| 1 × 12 | ¾ × 11¼ | | |
| 2 × 2 | 1½ × 1½ | | |
| 2 × 3 | 1½ × 2½ | | |
| 2 × 4 | 1½ × 3½ | | |
| 2 × 6 | 1½ × 5½ | | |

[1] Original Cut Size In Inches
[2] Minimum Cut Size In Inches
Reprinted with permission of Lowe's Companies, Inc.

## TABLE A11.   NAIL AND SCREW SHANKS, HEADS, AND POINTS

| Type | Abbrev. | Remarks | Illustration |
|---|---|---|---|
| | | **SHANKS** | |
| Smooth | C | For normal holding power; temporary fastener. | 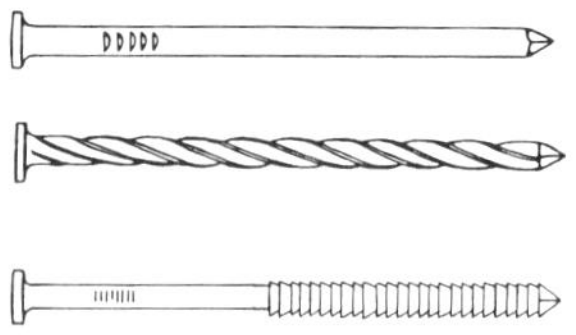 |
| Spiral | S | For greater holding power; permanent fastener. | |
| Ringed | R | For highest holding power; permanent fastener. | |
| Flat Countersunk | Cs | For nail concealment; light construction, flooring, and interior trim. | 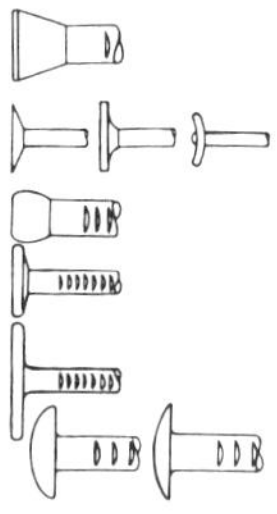 |
| Drywall | Dw | For gypsum wallboard. | |
| Finishing | Bd | For nail concealment; cabinetwork, furniture. | |
| Flat | F | For general construction. | |
| Large Flat | Lf | For tear resistance; roofing paper. | |
| Oval | O | For special effects; siding and clapboard. | |
| Diamond | D | For general use, 35° angle; length about 1.5 × diameter. | 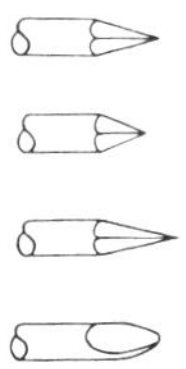 |
| Blunt Diamond | Bt | For harder wood species to reduce splitting, 45° angle. | |
| Long Diamond | N | For fast driving, 25° angle; may tend to split harder species. | |
| Duckbill | Db | For clinching small nails. | |

Reprinted with permission of Canadian Wood Council.

**TABLE A12.   NAIL AND SCREW MATERIALS, FINISHES, AND COATINGS**

| Type | Abbrev | Remarks |
|------|--------|---------|
| | | **MATERIALS** |
| Aluminum | A | For improved appearance and long life; increased stain and corrosion resistance. |
| Steel—mild | S | For general construction. |
| Steel—high-carbon hardened | Sc | For special driving conditions; improved impact resistance. |
| | | **FINISHES AND COATINGS** |
| Bright | B | For general construction; normal finish; not recommended for exposure to weather. |
| Blued | Bl | For increased holding power in hardwood; thin oxide finish produced by heat treatment, suitable for mouth holding. |
| Heat-treated | Ht | For increased holding power; black oxide finish, not suitable for mouth holding. |
| Phoscoated | Pt | For increased holding power; not corrosion resistant. |
| Electro-galvanized | Ge | For corrosion resistance; thin zinc plating; smooth surface; for interior use. |
| Hot-dip galvanized | Ghd | For improved corrosion resistance; thick zinc coating; rough surface; for exterior use. |

Reprinted with permission of Canadian Wood Council.

**TABLE A13.   GYPSUMBOARD NAILS AND SCREWS: SELECTOR GUIDE**

*Horizontal and Vertical Application:*

*Type S and G*—Metal studs or furring spaced 24″ on center. Screws should be spaced 12″ on center unless otherwise noted in the detailed assemblies shown in Georgia-Pacific literature.

*Type W*—Wood studs 16″ or 24″ on center. Screws should be spaced 12″ on center unless otherwise noted in the detailed assemblies shown in Georgia-Pacific literature.

Selection of the proper nail for each application is extremely important, particularly for fire-rated constructions. The nails recommended comply with performance standards adopted by the Gypsum Association.

| Illustration | Description | Applications |
|--------------|-------------|--------------|
| | $1\frac{1}{4}$″GWB-54 Annular Ring Nail $12\frac{1}{2}$ ga.; with a slight taper to a small fillet at shank. Bright finish; med. diamond pt. | $\frac{1}{2}$″, $\frac{3}{8}$″ and $\frac{1}{4}$″ Gypsumboard, $\frac{1}{2}$″ and $\frac{3}{8}$″ Gypsum Backer Board to wood frame |
| | $1\frac{3}{8}$″ Annular Ring Nail (Same as GWB-54 except for length) | $\frac{5}{8}$″ Gypsumboard to wood frame |
| | $2\frac{1}{2}$″ 7d Nail 13 ga., $\frac{1}{4}$″ diamond head | $\frac{5}{8}$″ Firestop® Gypsumboard face layers over $\frac{1}{2}$″ Sound Deadening Board or 2 layers $\frac{5}{8}$″ Firestop® to Wood Studs $\frac{3}{8}$″ |
| | $1\frac{7}{8}$″ 6d Nail 13 ga., $\frac{1}{4}$″ diamond head | and $\frac{1}{4}$″ Gypsumboard over existing surface, wood frame |
| | $1\frac{7}{8}$″ 6d Nail 13 ga., $\frac{1}{4}$″ diamond head | $\frac{5}{8}$″ Firestop® Gypsumboard to wood frame |
| | $1\frac{5}{8}$″ 5d Nail $13\frac{1}{2}$ ga., $\frac{15}{64}$″ diamond head | $\frac{1}{2}$″ Firestop® Gypsumboard to wood frame |

**TABLE A13 (cont.)**

| Illustration | Description | Applications |
|---|---|---|
| | 1⅛″ Matching Color Head Nail (Steel) | Predecorated Gypsumboard; to wood |
|  | 1⅞″ Matching Color Head Nail (Steel) | frame over existing surface, wood frame |
|  | 1⅜″ Matching Color Head Nail (Brass) | Predecorated Gypsumboard (colors) to wood frame |
| Gypsumboard to Metal Framing  | 1″ Gypsumboard Screw Type S | Single Layer Gypsumboard to 25 gauge Steel Studs. Gypsumboard to Resilient Channel. Gypsumboard to Metal Furring. |
|  | 1¼″ Gypsumboard Screw Type S<br>1⅜″ Gypsumboard Screw Type S | 1″ core units to L Runner in 2″ Solid and Semi-Solid Partition Systems. |
|  | 1⅝″ Gypsumboard Screw Type S | Double-Layer Gypsumboard to 25 gauge Steel Studs. Double-Layer Gypsumboard to Metal Furring. |
| 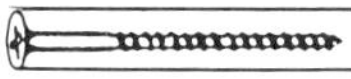 | 1⅞″ Gypsumboard Screw Type S | Double-Layer Gypsumboard. |
| 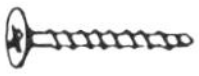 | 1″ Gypsumboard Screw Type S-12 Also available in ¾″, 1⅛″ & 1⁵⁄₁₆″ lengths | For single or multi-layer application of Gypsumboard to heavy gauge steel (up to 12 gauge). |
| Gypsumboard to Wood Framing  | 1¼″ Gypsumboard Screw Type W | Single Layer Gypsumboard to Wood Framing. |
| Gypsumboard to Gympsum Ribs 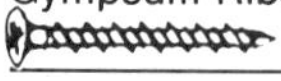 | 1½″ Gypsumboard Screw Type G<br>1⅝″ Gypsumboard Screw Type G | Gypsumboard to Gypsum Ribs in Semi-Solid Partition Systems. |
| Wood Trim to Metal Framing  | 1⅝″ Gypsumboard Screw Type S Tri Head | Wood trim over single layer of Gypsumboard on 25 gauge Steel Studs. |
| | 2¼″ Gypsumboard Screw Type S Trim Head | Wood trim over double layer Gypsumboard on 25 gauge Steel Studs. |
| Metal Studs to Door Frames and Runners 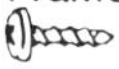 | ⅜″ Gypsumboard Screw S-12 Pan-Head | Metal Door frame to 12 gauge (max.) Steel Studs. Metal Studs to Metal Runners. |
| Metal Trim & Door Hinges to Metal Framing 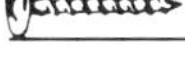 | ⅞″ Type S-18 Oval Head Also available in ¾″, 1⅛″ & 1 ⁵⁄₁₆″ lengths | Door hinges to door frame and aluminum components to metal. |

Reprinted with permission of Georgia-Pacific Corporation.

**TABLE A14.   NAILS FOR CONSTRUCTION**

| Type of Nail | Head | Shank | Point | Material | Finishes and Coatings | Size | |
|---|---|---|---|---|---|---|---|
| | | | | | | Diameter | Length |
| Common (Spike) | F | C,S | D | S | B | 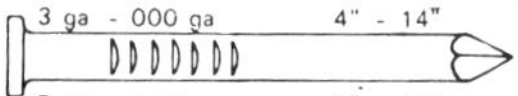 3 ga - 000 ga | 4″ - 14″ |
| Eavestrough (Spike) | Cs | C,S | N | S | B,Ghd | 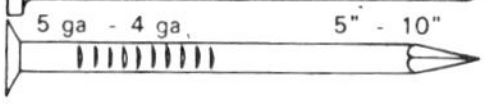 5 ga - 4 ga | 5″ - 10″ |

**TABLE A14 (cont.)**

| Type of Nail | Head | Shank | Point | Material | Finishes and Coatings | Size (Diameter) | Size (Length) |
|---|---|---|---|---|---|---|---|
| Standard or Common | F | C,R,S | D | A,S | B,Ge | 15 ga - 2 ga | 1" - 6" |
| Box | F,Lf | C,R,S | D | S | B,Pt | 17 ga - 8 ga | ¾" - 5" |
| Finishing | Bd | C,S | D | S | B,Bl | 17 ga - 9 ga | 1" - 4" |
| Flooring and Casing | Cs | C,S | Bt,D | S | B,Bl | 16 ga - 9 ga | 1⅛" - 3¼" |
| Concrete | Cs | S | Bt,D | Sc | Ht | 8½ ga - 5¾ ga | ½" - 3" |
| Siding and Clapboard | F,O | C,S | D | A,S | B,Ghd | 12 ga - 11 ga | 2" - 2½" |
| Clinch | F,Lf | C,S | Db | S | B | 15 ga - 11 ga | ¾" - 2½" |
| Gypsum Wallboard | Dw,F | C,R,S | D,N | S | Bl,Ge | 13 ga - 12½ ga | 1⅛" - 2" |
| Underlay and Underlay/Subfloor | F,Cs | C,R | D | S | B,Ht | 14 ga* - 10⅔ ga | ¾" - 2" |
| Roofing | Lf | C,R,S | D | A,S | B,Ghd | 13 ga - 9¾ ga | ¾" - 2" |
| Wood Shingle | F | C | D | S | B,Ghd | 14 ga - 12½ ga | 1¼" - 1¾" |
| Gypsum Lath | F | C,S | D,N | S | B,Bl,Ge | 13 ga | 1¼" |
| Wood Lath | F | C,S | D | S | Bl | 16 ga - 15 ga | 1" - 1⅛" |

Reprinted with permission of Canadian Wood Council.

**TABLE A15. GUIDE TO ENGINEERED GRADES OF PLYWOOD**

| Use these terms when you specify[1] | Description and Most Common Uses | Veneer Grade Face | Veneer Grade Back | Inner Plys | Most Common Thickness (inch)[2] |
|---|---|---|---|---|---|
| | INTERIOR TYPE | | | | |
| STANDARD C-D INT-DFPA[34] | Unsanded sheathing grade for wall and roof sheating, subflooring. Also available with intermediate glue or exterior glue. Specify intermediate glue where modern construction delays are expected; exterior glue where durability is required in long construction delays. For permanent exposure to weather or moisture, only exterior-type plywood is suitable. | C | D | D | ⁵⁄₁₆  ⅜  ½  ⅝  ¾  ⅞ |
| STRUCTURAL I C-D INT-DFPA STRUCTURAL II C-D INT-DFPA | Unsanded structural grades where plywood strength properti s are of maximum importance. Structural diaphragms, box beams, gusset plates, stressed-skin panels. Also for | C | D | D | ⁵⁄₁₆  ⅜  ½  ⅝  ¾  ⅞ |

**TABLE A15 (cont.)**

| Use these terms when you specify[1] | Description and Most Common Uses | Face | Back | Inner Plys | Most Common Thickness (inch)[2] | | | | |
|---|---|---|---|---|---|---|---|---|---|
| | | Veneer Grade | | | | | | | |
| | **INTERIOR TYPE** | | | | | | | | |
| | containers, pallets, bins. Made only with exterior glue. STRUCTURAL I limited to Group 1 species for face, back, and, inner plys. STRUCTUR-AL II permits Group 1, 2, or 3 species. | | | | | | | | |
| UNDERLAYMENT INT-DFPA[34] C-D PLUGGED INT-DFPA[34] | For underlayment or combination subfloor-underlayment under resilient floor coverings, carpeting. Used in homes, apartments, mobile homes, commercial buildings. Sanded or touch-sanded as specified. For utility built-ins, backing for wall and ceiling tile. Not a substitute for underlayment. Ply beneath face permits D grade veneers. Also for cable reels, walkways, separator boards. Unsanded or touch-sanded as specified. | C (plyd)  C | D  D | C[5] & D  D | ¼  ⁵⁄₁₆ | ³⁄₈  ³⁄₈ | ½  ½ | ⁵⁄₈  ⁵⁄₈ | ¾  ¾ |
| 2-4-1 INT-DFPA[36] | Combination subfloor-underlayment. Quality base for resilient floor coverings, carpeting, wood strip flooring. Use 2-4-1 with exterior glue in areas subject to excessive moisture. Unsanded or touch-sanded as specified. | C (plyd) | D | C & D | (available 1⅛″ or 1¼″) | | | | |
| | **EXTERIOR TYPE** | | | | | | | | |
| C-C EXT-DFPA[4] | Unsanded grade with waterproof bond for subflooring and roof decking, siding on service and farm buildings, crating, pallets, pallet bins, cable reels. | C | C | C | ⁵⁄₁₆ ³⁄₈ | ½ ⁵⁄₈ | ¾ ⁷⁄₈ | | |
| UNDERLAYMENT C-C PLUGGED EXT-DFPA[4] C-C PLUGGED EXT-DFPA[4] | For underlayment or combination subfloor-underlayment under resilient floor coverings where excessive moisture conditions may be present; for instance, bathrooms or utility rooms. Also use for tile backing where unusual moisture conditions exist. For refrigerated or controlled atmosphere rooms, for pallets, fruit pallet bins, reusable cargo containers, tanks and boxcar and truck floors and linings. Sanded or touch-sanded as specified. | C (plgd) | C | C[5] | ¼ | ³⁄₈ | ½ | ⁵⁄₈ | ¾ |

**TABLE A15 (cont.)**

| Use these terms when you specify[1] | Description and Most Common Uses | Veneer Grade | | | Most Common Thickness (inch)[2] |
|---|---|---|---|---|---|
| | | Face | Back | Inner Plys | |
| | **EXTERIOR TYPE** | | | | |
| STRUCTURAL I C-C EXT-DFPA | For engineered applications in construction and industry where full exterior-type panels made with all Group 1 wood are required. Unsanded | C | C | C | $\frac{5}{16}$  $\frac{3}{8}$  $\frac{1}{2}$  $\frac{5}{8}$  $\frac{3}{4}$ |
| B-B PLYFORM CLASS I & II EXT-DFPA | Concrete form grades with high reuse factor. Sanded both sides. Edge-sealed and mill-oiled unless otherwise specified. Special restrictions on species. Also available in HDO. | B | B | C | $\frac{5}{8}$  $\frac{3}{4}$ |

[1]All grades except Plyform available tongued-and-grooved in panels ½″ and thicker.

[2]Panels are standard 4 × 8-foot size. Other sizes available.     [3]Also available with exterior or intermediate glue.

[4]Available in Group 1, 2, 3, 4, or 5.     [5]Ply beneath face is a special C grade that limits knotholes to 1″.

[6]Available in Group 1, 3 or 3 only.

Reprinted with permission of American Plywood Association.

**TABLE A16.  GUIDE TO IDENTIFICATION INDEX ON ENGINEERED GRADES OF PLYWOOD[1]**

| Thickness (inch) | Standard (C-D) INT-DFPA[2] C-C EXT-DFPA | | | Structural 1[3] C-D INT-DFPA Str. 1 C-C EXT | Structural II[3] C-D INT-DFPA | |
|---|---|---|---|---|---|---|
| | Group 1 | Group 2 or 3[4] | Group 4[5] | Group 1 only | Group 1 | Group 2 or 3[4] |
| $\frac{5}{16}$ | 20/0 | 16/0 | 12/0 | 20/0 | 20/0 | 16/0 |
| $\frac{3}{8}$ | 24/0 | 20/0 | 16/0 | 24/0 | 24/0 | 20/0 |
| $\frac{1}{2}$ | 32/16 | 24/0 | 24/0 | 32/16 | 32/16 | 24/0 |
| $\frac{5}{8}$ | 42/20 | 32/16 | 30/12 | 42/20 | 42/20 | 32/16 |
| $\frac{3}{4}$ | 48/24 | 42/20 | 36/16 | 48/24 | 48/24 | 42/20 |
| $\frac{7}{8}$ | — | 48/24 | 42/20 | | | 48/24 |

[1] Identification Index numbers shown in the table appear in DFPA grade trademarks on Standard C-D, C-C, Structural 1 C-C and C-D and Structural II C-D grades. They refer to maximum recommended spacing of supports in inches when panels are used for roof decking and subflooring with face grain across supports. The left-hand number shows spacing for roof supports. The right hand number shows spacing for floor supports. Numbers are based on panel thickness and species makeup detailed in Product Standard PS 1-66. Under each grade, the table identifies the species classification of the veneer used for outer plys. Where face and back veneers are not from the same species group, the number is based on the weaker group.

[2]Also available with exterior or intermediate glue.     [3]Manufactured with exterior glue only.

[4] Panels made with Group 2 outer plys may carry the Identification Index numbers shown for Group 1 panels when they conform to special thickness and construction requirements detailed in PS 1-66.

[5] Panels made with Group 4 outer plys may carry the Identification index numbers shown for Group 3 panels when they conform to special thickness and construction requirements detailed in PS 1-66.

Reprinted with permission of American Plywood Association.

**TABLE A17.    CONVERSION DIAGRAM FOR RAFTERS**

To use the diagram select the known horizontal distance and follow the vertical line to its intersection with the radial line of the specified slope, then proceed along the arc to read the sloping distance. In some cases it may be desirable to interpolate between the 1-foot separations. The diagram also may be used to find the horizontal distance corresponding to a given sloping distance or to find the slope when the horizontal and sloping distances are known.

*Example:* With a roof slope of 8 in 12 and a horizontal distance of 20 feet the sloping distance may be read as 24 feet.

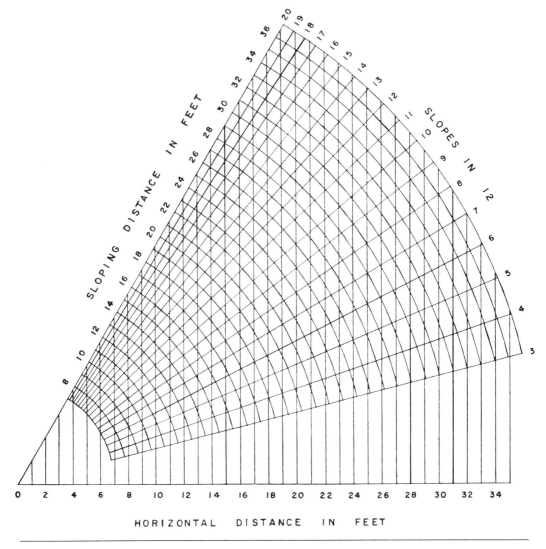

Reprinted with permission of National Forest Products Association.

**TABLE A18.   MAXIMUM CLEAR SPANS FOR APA GLUED FLOOR SYSTEM**

| Species-Grade | Joist Size | $\frac{5}{8}''$Plywood[1] Joists @ 16'' | $\frac{3}{4}''$Plywood[2,3] Joists @ 16'' | $\frac{3}{4}''$Plywood[3,4] Joists @ 24'' |
|---|---|---|---|---|
| Douglas Fir - Larch - No. 1 | 2 × 6 | 11'-1'' | 11'-6'' | 9'-5'' |
| | 2 × 8 | 14-4 | 14-9 | 12-5 |
| | 2 × 10 | 18-1 | 18-6 | 15-10 |
| | 2 × 12 | 21-9 | 22-2 | 19-3 |
| Douglas Fir - Larch - No. 2 | 2 × 6 | 10-6 | 10-6 | 8-7 |
| | 2 × 8 | 13-10 | 13-10 | 11-3 |
| | 2 × 10 | 17-7 | 17-7 | 14-5 |
| | 2 × 12 | 21-5 | 21-5 | 17-6 |
| Douglas Fir - Larch - No. 3 | 2 × 6 | 8-0 | 8-0 | 6-7 |
| | 2 × 8 | 10-7 | 10-7 | 8-8 |
| | 2 × 10 | 13-6 | 13-6 | 11-0 |
| | 2 × 12 | 16-5 | 16-5 | 13-5 |
| Douglas Fir - South - No. 1 | 2 × 6 | 10-5 | 10-9 | 9-1 |
| | 2 × 8 | 13-5 | 13-10 | 12-0 |
| | 2 × 10 | 16-10 | 17-3 | 15-4 |
| | 2 × 12 | 20-3 | 20-8 | 18-8 |
| Douglas Fir - South - No. 2 | 2 × 6 | 10-1 | 10-1 | 8-3 |
| | 2 × 8 | 13-2 | 13-4 | 10-10 |
| | 2 × 10 | 16-6 | 16-11 | 13-10 |
| | 2 × 12 | 19-10 | 20-3 | 16-10 |
| Hemlock - Fir - No. 1 | 2 × 6 | 10-3 | 10-3 | 8-5 |
| | 2 × 8 | 13-7 | 13-7 | 11-1 |
| | 2 × 10 | 17-2 | 17-4 | 14-2 |
| | 2 × 12 | 20-7 | 21-1 | 17-2 |
| Hemlock - Fir - No. 2 | 2 × 6 | 9-4 | 9-4 | 7-7 |
| | 2 × 8 | 12-4 | 12-4 | 10-0 |
| | 2 × 10 | 15-8 | 15-8 | 12-10 |
| | 2 × 12 | 19-1 | 19-1 | 15-7 |
| Southern Pine KD 15% - No. 1 | 2 × 6 | 11-3 | 11-8 | 9-9 |
| | 2 × 8 | 14-7 | 15-0 | 12-11 |
| | 2 × 10 | 18-4 | 18-9 | 16-6 |
| | 2 × 12 | 22-1 | 22-6 | 20-0 |
| Southern Pine - No. 1 | 2 × 6 | 11-1 | 11-6 | 9-5 |
| | 2 × 8 | 14-4 | 14-9 | 12-5 |
| | 2 × 10 | 18-1 | 18-6 | 15-10 |
| | 2 × 12 | 21-9 | 22-2 | 19-3 |
| Southern Pine KD 15% - No. 2 | 2 × 6 | 9-11 | 9-11 | 8-1 |
| | 2 × 8 | 13-1 | 13-1 | 10-8 |
| | 2 × 10 | 16-8 | 16-8 | 13-7 |
| | 2 × 12 | 20-3 | 20-3 | 16-7 |
| Southern Pine - No. 2 | 2 × 6 | 9-6 | 9-6 | 7-9 |
| | 2 × 8 | 12-7 | 12-7 | 10-3 |
| | 2 × 10 | 16-0 | 16-0 | 13-1 |
| | 2 × 12 | 19-6 | 19-6 | 15-11 |
| Southern Pine KD 15% - No. 2 Med. Grain | 2 × 6 | 10-10 | 10-10 | 8-10 |
| | 2 × 8 | 14-2 | 14-3 | 11-8 |
| | 2 × 10 | 17-9 | 18-2 | 14-10 |
| | 2 × 12 | 21-5 | 21-10 | 18-1 |

## TABLE A18 (cont.)

| Species-Grade | Joist Size | 5/8" Plywood[1] Joists @ 16" | 3/4" Plywood[2,3] Joists @ 16" | 3/4" Plywood[3,4] Joists @ 24" |
|---|---|---|---|---|
| Southern Pine - No. 2 Med. Grain | 2 × 6 | 10-6 | 10-6 | 8-7 |
| | 2 × 8 | 13-10 | 13-10 | 11-3 |
| | 2 × 10 | 17-5 | 17-7 | 14-5 |
| | 2 × 12 | 21-0 | 21-5 | 17-6 |
| Southern Pine KD 15% - No. 3 | 2 × 6 | 8-4 | 8-4 | 6-10 |
| | 2 × 8 | 11-0 | 11-0 | 9-0 |
| | 2 × 10 | 14-1 | 14-1 | 11-6 |
| | 2 × 12 | 17-1 | 17-1 | 14-0 |
| Southern Pine - No. 3 | 2 × 6 | 7-11 | 7-11 | 6-5 |
| | 2 × 8 | 10-5 | 10-5 | 8-6 |
| | 2 × 10 | 13-3 | 13-3 | 10-10 |
| | 2 × 12 | 16-2 | 16-2 | 13-2 |

[1] *Underlayment INT-DFPA* Group 1, 2, or 3 may be used for combined subfloor underlayment. Underlayment grade may be 19-32". If separate underlayment or a structural finish floor is installed, Standard DFPA 1/2", 32/16 or 42/20 may be used.

[2] *Underlayment INT-DFPA* Group 1, 2, 3, or 4 may be used for combined subfloor underlayment. Underlayment grade may be 23/32". If separate underlayment or structural finish floor is installed, Standard DFPA 3/4", 36/16, 42/20, or 48/24 may be used.

[3] 7/8" or 1" *Underlayment* of any Group, or Standard with appropriate identification Index numbers, may be substituted for lesser thicknesses if desired.

[4] *Underlayment INT-DFPA* Group 1 may be used for combined subfloor underlayment. Underlayment grade may be 23/32". If separate underlayment or structural finish floor is installed, Standard INT-DFPA 3/4", 48/24 may be used.

Reprinted with permission of American Plywood Association.

## TABLE A19.　PLYWOOD UNDERLAYMENT

For application of tile, carpeting, linoleum, or other nonstructural flooring.

*Installation:* Apply UNDERLAYMENT just prior to laying finish floor or protect against water or physical damage. Stagger panel end joints with respect to each other and offset all joints with respect to the joints in the subfloor. Space panel ends and edges about 1/32". For maximum stiffness, place face grain of panel across supports and end joints over framing. Unless subfloor and joists are thoroughly seasoned and dry, countersink nails 1/16" just prior to laying finish floor to avoid nail popping. Countersink staples 1/32". Do not fill nail holes. Lightly sand any rough areas, particularly around joints or nail holes.

| Plywood Grades and Species Group | Application | Minimum Plywood Thickness | Fastener size (approx.) and Type (set nails 1/16") | Fastener Spacing (Inches) | |
|---|---|---|---|---|---|
| | | | | Panel Edges | Inter- mediate |
| Groups 1, 2, 3, 4 UNDERLAYMENT INT-DFPA (with interior, intermediate or exterior glue) UNDERLAYMENT EXT-DFPA C-C Plugged EXT-DFPA | over plywood subfloor | 1/4" | 18 Ga. staples or 3d ring-shank nails[1,2] | 3 | 6 each way |
| | over lumber subfloor or other uneven surfaces | 3/8" | 16 Ga. staples[1] | 3 | 6 each way |
| | | | 3d ring-shank nails[2] | 6 | 8 each way |

**TABLE A19 (cont.)**

| Plywood Grades and Species Group | Application | Minimum Plywood Thickness | Fastener size (approx.) and Type (set nails $\frac{1}{16}''$) | Fastener Spacing (Inches) | |
|---|---|---|---|---|---|
| | | | | Panel Edges | Inter-mediate |
| Same Grades as above, but Group 1 only | over lumber floor up to 4″ wide. Face grain must be perpendicular to boards | $\frac{1}{4}''$ | 18 Ga. staples or 3rd ring-shank nails[1,2] | 3 | 6 each way |

[1] Crown width $\frac{3}{8}''$ for 16 ga., $\frac{3}{16}''$ for 18 ga. staples; length sufficient to penetrate completely through, or at least $\frac{5}{8}''$ into, subflooring.

[2] Use 3d ring-shank nail also for $\frac{1}{2}''$ plywood and 4d ring-shank nail for $\frac{5}{8}''$ or $\frac{3}{4}''$ plywood.

Reprinted with permission of American Plywood Association.

**TABLE A20. PLYWOOD WALL SHEATHING[1]**

Plywood continuous over 2 or more spans.

| Panel Identification Index | Panel Thickness (inch) | Maximum Stud Spacing (inches) Exterior Covering Nailed to: | | Nail Size[2] | Nail Spacing (inches) | |
|---|---|---|---|---|---|---|
| | | Stud | Sheathing | | Panel Edges (when over framing) | Intermediate (each stud) |
| 12/0, 16/0, 20/0 | $\frac{5}{16}$ | 16 | 16[3] | 6d | 6 | 12 |
| 16/0, 20/0, 24/0 | $\frac{3}{8}$ | 24 | { 16 / 24[3] | 6d | 6 | 12 |
| 24/0, 30/12, 32/16 | $\frac{1}{2}, \frac{5}{8}$ | 24 | 24 | 6d | 6 | 12 |

[1] When plywood sheathing is used, building paper and diagonal wall bracing can be omitted.

[2] Common smooth, annular, spiral-thread, or galvanized box, or T-nails of the same diameter as common nails (0.113″ dia. for 6d) may be used. Staples also permitted at reduced spacing.

[3] When sidings such as shingles are nailed only to the plywood sheathing, apply plywood with face grain across studs.

Reprinted with permission of American Plywood Association.

**TABLE A21. EXTERIOR PLYWOOD OVER SHEATHING**

Recommendations apply to all species groups.

| Plywood Siding | | Max. Stud Spacing (in) | | Nail Size (Use nonstaining box, siding, or casing nails) | Nail Spacing (in) | |
|---|---|---|---|---|---|---|
| Description | Nominal Thickness (in) | Face Grain Vertical | Face Grain Horizontal | | Panel Edges | Interme-diate |
| | | **PANEL SIDING** | | 6d for panels $\frac{1}{2}''$ thick or less | | |
| MDO EXT-APA | $\frac{5}{16}$ | 16[1] | 24 | | 6 | 12 |

**TABLE A21** (cont.)

| Plywood Siding | | Max. Stud Spacing (in) | | Nail Size (Use nonstaining box, siding, or casing nails) | Nail Spacing (in) | |
|---|---|---|---|---|---|---|
| Description | Nominal Thickness (in) | Face Grain Vertical | Face Grain Horizontal | | Panel Edges | Interme-diate |
| A-C EXT-APA B-C EXT-APA | $^3/_8$ | 16[1] | 24 | | | |
| C-C Plugged EXT-APA MDO EXT-APA | $^1/_2$ & thicker | 24 | 24 | 8d for thicker panels [2] | | |
| 303-16 o.c. Siding EXT-APA | 5/16 & thicker | 16[1] | 24 | | | |
| 303-24 o.c. Siding EXT-APA | 7/16 & thicker | 24 | 24 | | | |
| **LAP SIDING** | | | | | | |
| MDO EXT-APA | 5/16 | — | 16[3] | 6d for siding $^3/_8''$ thick or less; | | |
| A-C EXT-APA B-C EXT-APA | $^3/_8$ | — | 16[3] | | 4'' @ vertical butt joints; one nail per stud along bottom edge | 8'' @ each stud, if siding wider than 12'' |
| C-C Plugged EXT-APA MDO EXT-APA | $^1/_2$ & thicker | — | 24 | | | |
| 303-16 o.c. Siding EXT-APA | 5/16 or $^3/_8$ | — | 16[3] | | | |
| 303-16 o.c. Siding EXT-APA 303-24 o.c. Siding EXT-APA | 7/16 & thicker | — | 24 | 8d for thicker siding. | | |

[1] May be 24″ with $^1/_2$″ plywood, or lumber sheathing, if panel is also nailed 12″ o.c. between studs, provided panel joints fall over studs.

[2] Use next regular nail size when sheathing (other than plywood or lumber) is thicker than $^1/_2$″.

[3] May be 24″ with plywood or lumber sheathing.

[4] Use next larger nail size when sheathing is other than plywood or lumber, and nail only into framing.

Reprinted with permission of American Plywood Association.

**TABLE A22.  CLASSIFICATION OF SPECIES PLYWOOD VENEER GRADES**

| Group 1 | Group 2 | Group 3 | Group 4 | Group 5 |
|---|---|---|---|---|
| Birch | Cedar, Port Orford | Alder, Red | Aspen | Fir, Balsam |
| Sweet | Douglas Fir[2] | Cedar, Alaska | Bigtooth | Poplar, Balsam |
| Yellow | Fir | Pine | Quaking | |
| Douglas Fir 1[1] | California Red | Jack | Birch, Paper | |
| Larch, Western | Grand | Lodgepole | Cedar | |
| Maple, Sugar | Noble | Ponderosa | Incense | |
| Pine, Caribbean | Pacific Silver | Spruce | Western Red | |
| Pine, Southern | White | Redwood | Fir Subalpine | |

**TABLE A22 (cont.)**

| Group 1 | Group 2 | Group 3 | Group 4 | Group 5 |
|---------|---------|---------|---------|---------|
| Loblolly | Hemlock Western | Spruce | Hemlock, Eastern | |
| Longleaf | | Black | Pine | |
| Shortleaf | Lauan | Red | Eastern, White | |
| Slash | Almon | White | Sugar | |
| Tancak | Bagtikan | | Poplar, Western[3] | |
| | Red Lauan | | Spruce, | |
| | Tangile | | Engelmann | |
| | White Lauan | | | |
| | Maple, Black | | | |
| | Mengkulang | | | |
| | Meranti | | | |
| | Pine | | | |
| | Pond | | | |
| | Red | | | |
| | Western White | | | |
| | Spruce, Sitka | | | |
| | Sweetgum | | | |
| | Tamarack | | | |

**N** Special order "natural finish" veneer. Select all heartwood or all sapwood. Free of open defects. Allows some repairs.

**A** Smooth and paintable. Neatly made repairs permissible. Also used for natural finish in less demanding applications.

**B** Solid surface veneer. Circular repair plugs and tight knots permitted.

**C** Knotholes to 1″. Occasional knotholes ½″ larger permitted providing total width of all knots and knotholes within a specified section does not exceed certain limits. Limited splits permitted. Minimum veneer permitted in exterior-type plywood.

**C** Improved C veneer with splits limited to ⅛″ in width and knotholes and borer holes limited to ¼″ by ½″.

**D** Permits knots and knotholes to 2½″ in width and ½″ larger under certain specified limits. Limited splits permitted.

[1]*Douglas Fir 1:* Washington, Oregon, California, Idaho, Montana, Wyoming, British Columbia, Alberta.

[2]*Douglas Fir 2:* Nevada, Utah, Colorado, Arizona, New Mexico.

[3]*Black Cottonwood*

Reprinted with permission of American Plywood Association.

**TABLE A23.   U.S. TO METRIC CONVERSIONS**

| Meters | Yards | Inches | Centimeters | Inches | Feet | Kilometers | Miles |
|--------|-------|--------|-------------|--------|------|------------|-------|
| 1.000 | 1.093 | 39.37 | 1.00 | .394 | .0328 | 1.000 | .621 |
| .914 | 1.000 | 36.00 | 2.54 | 1.000 | 1/12 | 1.609 | 1.000 |
| | | | 30.48 | 12.000 | 1.000 | | |

**TABLE A23 (cont.)**

| Grams | Ounces | Pounds | Kilograms | Ounces | Pounds | Liters | Pints | Quarts | Gallons |
|--------:|--------:|--------:|--------:|--------:|--------:|--------:|--------:|--------:|--------:|
| 1.00 | .035 | .002 | 1.000 | 35.274 | 2.205 | 1.000 | 2.113 | 1.057 | .264 |
| 28.35 | 1.000 | $\frac{1}{16}$ | .028 | 1.000 | $\frac{1}{16}$ | .473 | 1.000 | $\frac{1}{2}$ | $\frac{1}{8}$ |
| 453.59 | 16.000 | 1.000 | .454 | 16.000 | 1.000 | .946 | 2.000 | 1.000 | $\frac{1}{4}$ |
| 1,000.00 | 35.274 | 2.205 | | | | 3.785 | 8.000 | 4.000 | 1.000 |

| Length | | Capacity | | Weight | |
|---|---|---|---|---|---|
| meter (m) = 100 cm | = 1,000 mm | 1 liter (1) = 100cl | = 1,000 ml | 1 gram (g) = 100 eg | = 1,000 mg |
| millimeter (mm) | = .001 m | 1 milliter (ml) | = .001 l | 1 milligram (mg) | = .001 g |
| centimeter (cm) | = .01 m | 1 centiliter (cl) | = .01 l | 1 centigram (cg) | = .01 g |
| decimeter (dm) | = .1 m | 1 deciliter (dl) | = .1 l | 1 decigram (dg) | = .1 g |
| decameter (dkm) | = 10 m | 1 decaliter (dkl) | = 10 l | 1 decagram (dkg) | = 10 g |
| hectometer (hm) | = 100 m | 1 hectoliter (hl) | = 100 l | 1 hectogram (hg) | = 100 g |
| kilometer (km) | = 1,000 m | 1 kiloliter (kl) | = 1,000 l | 1 kilogram (kg) | = 1,000 g |

# Glossary

## A

ACCESS: A passageway; a corridor between rooms.

ACCORDION DOORS: Doors that fold and are supported by rollers inserted in a track.

ACOUSTIC: Pertaining to the act or sense of hearing, heard sound, or the science of sound.

ACOUSTICAL BOARD: A board, such as insulating board, that is used to control sound.

ACRYLIC RESIN: A type of transparent thermosetting plastic, or resins made from acrylic acid.

ADHESIVE: A natural or synthetic material that is used to fasten or glue boards together, lay tile, secure plastic laminates, etc.

AGGREGATE: A material such as sand, stone, perlite, gravel, or clay that is mixed with cementing material to form concrete or pilaster.

AIR ENTRAINING: The ability of a material or process to develop tiny air bubbles in concrete.

AIR SPACE: A space between walls; or the space between the brick veneer and the wall frame.

AKALINE SOIL: A type of soil that contains an unusual amount of soluble mineral salts.

ALLIGATORING: A defect in a painted surface.

ANCHOR BOLT: A bolt that is used to secure the frame of a building against wind and vibration forces.

ANNULAR RING NAIL: A nail that has a shank with a series of protruding rings.

APRON: A piece of window trim that is placed under the stool to cover the joint with the wall.

APRON FLASHING: Waterproof material that is used with vertical wall flashing to cover the joint between the roof and the chimney.

ARCHITECT: One whose profession is to design and draw up the plans for buildings, etc., and supervise their construction.

AREA DRAIN: A drain used to carry water away from a given area.

AREAWAY: An open space around a basement door or window.

ARMORED CABLE: Insulated wire covered with a flexible steel covering.

ASHLAR: (1) Dressed stones that are used for facing a masonry wall; (2) small vertical pieces of wood that extend from the attic floor to the rafters.

ASPHALT: A residue that is left from the refining of petroleum; it is a solid, brownish black, combustible mixture of bituminous hydrocarbons.

ASPHALT-SATURATED FELT: A roofing and waterproof membrane composed of asphalt-saturated rag felt.

ASPHALT SHINGLES: Composition roof shingles made from asphalt-impregnated felt covered with mineral granules.

ASHPIT: An area at the bottom of a chimney used to store ashes.

ATTIC: The space or area that is located directly below the roof.

AWNING-TYPE WINDOW: A window that has lights (pieces of glass) that open outward.

# B

BACKER STRIP: A strip of wood, metal, or felt used to help elevate shingles or siding.

BACKFILL: Earth or other material placed in an area that has been excavated.

BACKHOE: A machine used for excavation that has its bucket attached to the end of a boom.

BACKING PAPER: A building material placed between the exterior wall covering and the frame wall.

BACKSTOP: A small strip of wood placed inside a window jamb.

BALLOON FRAME: A type of framing system in which the vertical framing members are in one continuous piece from sill to roof plate.

BALUSTER: A thin column that is used to support a railing.

BASE: A board that is placed in a horizontal position along the bottom of a wall next to the floor.

BASEBOARD: A horizontal piece of trim that is placed at the intersection of wall and floor.

BASE COURSE: A layer of gravel or crushed stone placed beneath a concrete slab.

BASE FLASHING: Sheet metal that is used to cover

the intersection of a sloping roof and a vertical wall.

BASEMENT: A full-story space that is located below grade.

BASE SHOE: A piece of interior trim that is placed adjacent to the baseboard.

BATT: A type of insulation that is placed between framing members.

BATTEN: A thin strip of wood that is used to cover the intersection of two panels.

BATTER BOARD: A horizontal board nailed to two stakes that marks the elevation of a slab or foundation wall; used also to support the building line.

BEAM: A structural member used to support dead and live loads.

BEARING PARTITION: A partition that supports its own weight plus any other weight imposed on it.

BEARING PLATE: A structural member placed under a loaded truss beam, girder, or column; and used to distribute the load.

BED: A layer of mortar in which bricks or stones are placed.

BED JOINT: A horizontal mortar joint upon which masonry units rest.

BEVEL: Any inclination of two surfaces other than 90°.

BEVEL SIDING: A type of siding placed on the exterior of a frame wall, and usually manufactured in a wedge-shaped piece.

BID: An offer to build a structure at a specified price.

BIRDSMOUTH: A portion of a structural member that has been cut out so the member may fit over a cross-timber.

BITUMINOUS: A building material that contains bitumen materials such as asphalt, macadam, tar macadam, or asphalt cement.

BLIND DADO: A dado or groove that does not span the entire width of a board.

BLISTERING: Raised spots on a finished surface.

BLOCK FLOORING: Small squares of flooring that are cut from narrow, short strips of regular flooring.

BOARD AND BATTEN: A type of siding that

consists of wide boards and narrow strips called battens.

BOND: A joining or adhering to.

BOND BEAM: A horizontal reinforced concrete or concrete masonry beam.

BOND FAILURE: The breaking of a masonry unit from its mortar bed.

BORDER TILE: Floor or ceiling tile that is placed adjacent to the wall.

BOSTON HIP ROOF: A construction technique used to cover the joint, or hip, of a hip roof.

BOTTOM PLATE: A horizontal member placed at the bottom of a framed wall.

BRACE: A diagonal piece of framing used to stiffen a wall or floor frame.

BRANCH VENT: A vent that connects one or more individual vents with a vent stack or a stack vent.

BRAZING: To join the surface of two metals by partial fusion with a layer of soldering alloy.

BRICK MOLDING: Molding that is used to trim out exterior window and door frames. It closes the space between the brick and the door or window.

BRICK VENEER: An outer covering of brick laid adjacent to and fastened to a wall frame.

BRIDGING: Framing members placed between the floor joist to brace the joist and to spread the dead and live loads.

BROWN COAT: In plastering, the first coat of plaster that is applied over the lath.

BTU: Amount of heat neccessary to raise 1 pound of water 1 degree F.

BUCKET: A piece of the excavation equipment that digs, lifts, and carries dirt.

BUCKLING: A term describing a structural member that is bent or warped.

BUILDING CODE: Legal requirements of governing agencies to establish minimum construction requirements.

BUILDING DRAIN: The second lowest horizontal portion of a building drainage system that receives the discharge from soil and waste pipes inside the walls of a building, and discharges it to the building sewer. (See Figure 20-12.)

BUILDING LINE: A stretched string used to mark the boundaries of a building, and to mark a desired elevation.

BUILDING PAPER: Paper that is placed between sheathing and outside wall covering.

BUILDING SEWER: The lowest portion of a building drainage system that receives the discharge from the building drain and carries the waste to the main sewer.

BUILT-UP ROOF: A roof composed of several layers of rag felt, saturated with tar or asphalt.

BUTTERING: The process of spreading mortar on the edge of a brick.

BUTT HINGE: A hinge fastened to the edge of a door and the face of a door jamb.

# C

CANTILEVERED: A beam that projects over an area and is supported on only one end.

CANT STRIP: A strip of wood placed under shingles or siding.

CAPILLARY ACTION: The ability of water to move in an area above the horizontal plane of the supply of existing water.

CAPILLARY STOP: An area between two surfaces that is large and porous enough to stop capillary action.

CAP MOLDING: A horizontal piece of molding placed on top of work, such as a window or door.

CAP PLATE: A horizontal structural member that is nailed to the top plate.

CAP UNIT: The last ridge unit placed in the laying of an asphalt-shingled roof.

CARBONATION: The operation of precipitating lime by mixing with carbon dioxide.

CARPORT: A partially enclosed shelter for an automobile.

CARRIAGE: A wooden support for stair treads and risers.

CASED OPENING: An opening that is finished with jambs and casing, but has no door.

CASING: Trim placed around window and door openings.

CASING NAIL: A wire nail with a flaring head.

CAST-IN-PLACE: Concrete that is placed in forms, and will not be moved from its original location.

CAULKING: The filling of joints with a compound to prevent leaking.

CAULKING IRON: A hand tool that is used to put oakum into a pipe joint.

CEILING JOIST: A timber or beam that is used to support ceiling and is placed parallel to the floor.

CELLULOSE: The substance in the framework of wood cells.

CENTER-CRIMPED: Flashing that is bent down the center.

CERAMIC TILE: A decorative piece of thin fired clay that can be attached to walls, floors, or countertops.

CHAIR: A device used to elevate reinforcement.

CHAIR RAIL: A piece of molding placed along the wall of a room to keep a chair from marring the wall.

CHALKLINE: A line that has been coated with chalk and is used to provide a straight line on a surface.

CHECK: A lengthwise crack or separation of the wood, usually extending across the rings of annual growth.

CHECKCRACKING: Small shallow cracks that develop on a surface.

CHECKOVER: A technique used in the alignment of asphalt shingles.

CHECK VALVE: A valve used in plumbing to prevent the backflow of water.

CHIMNEY: An area of a house that contains the flues for drawing off smoke.

CIRCUIT: The path taken by an electrical current.

CIRCUIT BREAKER: A device used to break an electric circuit.

CIRCUIT VENT: A vent that serves two or more traps and extends from in front of the last fixture connection of a horizontal branch to the vent stack.

CLAY: A cohesive soil that has the ability to expand when wet and to contract when dry.

CLEANOUT: A pipe that has a removable plug affording access into plumbing lines.

CLEATS: A strip of wood fastened to another piece to strengthen it, or to furnish a bearing surface.

CLOSED VALLEY: On the roof, a valley in which the different courses of shingles are alternately overlapped.

CLOSURE BLOCK: A masonry unit that is used to close a course.

COAL-TAR PITCH: A dark bituminous substance used as a binder in roofing.

COAT: A layer of substance that covers another.

COBBLE: A rock that has a diameter of 2½ to 10 inches.

COHESIVE: The ability of parts of a substance to stick together.

COLLAR BEAM: A beam used to connect two rafters; called a wind beam or rafter tie.

COLUMN: A vertical compression member whose height must exceed four times its least lateral dimension.

COMMON BOND: Sometimes called a running bond; a bond pattern where the head joints are in alignment on alternate courses.

COMMON NAIL: A large-headed nail used in rough carpentry.

COMMON RAFTER: A rafter that spans from a cap plate to the ridge.

COMMON VENT: A vent that serves two fixture drains that are placed back to back or side by side.

COMPOSITE PILE: A pile or post constructed of two materials, usually wood and concrete.

COMPRESSION: A force that presses or squeezes a structural member and tends to make it contract or shorten.

CONCAVE: A curved recess.

CONCRETE: A mixture of sand, gravel, or stone, and cement.

CONDUIT (ELECTRICAL): A metal pipe in which electrical wiring is placed.

CONGLOMERATE: A type of clay that contains different-sized rocks.

CONTACT ADHESIVE: An adhesive that bonds two surfaces together on contact.

CONTINUOUS VENT: The extension of a vertical waste pipe.

CONTRACT(n): An agreement between buyer and seller.

CONTROL JOINT: A joint in concrete used to regulate the location and amount of cracking.

CONVECTOR: An electric hot-water or steam-heating room unit.

CONVENIENCE OUTLET: An electrical outlet into which may be plugged electrical appliances and equipment.

COPED JOINT: The intersection between two pieces of molding in which a portion of one piece is cut away to receive the molded part of the other piece.

CORD: The outermost framing member of a truss.

CORNER: Studs spiked together to fix and secure converging sides in a frame structure.

CORNER BOARD: Board that is used as trim for the external corners and serves as a termination point for the siding.

CORNER BRACING: Diagonal braces, or sheets of plywood, placed at the corners of a frame structure to strengthen the walls.

CORNER LEAD: A portion of a masonry wall that is rocked back on successive courses and is used to hold the mason's line.

CORNER POLE: A vertical pole used in masonry construction to hold the mason's line.

CORNERITE: Metal lath that is cut in strips and bent at a right angle; used to prevent plaster from cracking.

CORNICE: A part of the exterior trim that is located where the roof and sidewalls meet.

CORRIDOR: A gallery or passageway, usually having rooms opening upon it.

CORRUGATED FASTENER: A metal device used to secure two wooden members.

COUNTER FLASHING: Flashing that is inserted into a vertical masonry wall such as a chimney, and extends over the roof flashing; used to prevent moisture entry.

COUNTERSINK: A tool used to recess or form a depression in a board so it will receive the head of a nail, bolt, or screw.

COURSE: In roofing, a horizontal row of shingles; in masonry, a horizontal row of units that extend the length and thickness of the wall.

COVE MOLDING: A piece of concave molding.

CRAWL SPACE: The space between the first floor and the surface ground.

CRAZING: The development of small hairline cracks caused by shrinkage.

CREOSOTE: A distillate of coal-tar that is used as a wood preservative.

CRICKET: Two sloping surfaces placed behind a vertical wall to divert water.

CRIPPLE: A stud that is cut less than full length.

CROSS-TEE: A T-shaped member in the grid-work of a suspended ceiling.

CROWN: The bow in a horizontal timber when it is placed on its edge.

CURING: The process that allows satisfactory hydration in concrete.

CURING COMPOUND: A substance that can be placed on top of a concrete slab to assure satisfactory hydration.

CUSHION: A soft pad placed beneath a carpet.

CUSHION BACK-CUTTER: A device used to cut carpet.

CUT: An excavation made to lower an existing grade.

CUTOUTS: An opening cut in a piece of building material.

# D

DADO: A rectangular groove cut across the grain of a board.

DAMPER: A device used to regulate the draft in a flue.

DAMPPROOFING: The preparation of a wall to prevent water penetration.

DARBY: Wooden or aluminum concrete finishing tool 3 to 4 inches in width and about 4 to 8 feet in length.

DEAD LOAD: A stationary load imposed on a structure; a constant weight.

DEADWOOD: Wood that serves as a nailing base, but serves no structural purpose.

DECAY: Disintegration of wood through the action of fungi or bacteria.

DECKING: A term that is sometimes used for sheating.

DEFLECTION: The bending of a structural unit.

DEFORMED BAR: A reinforcing bar that has lug-like ridges around it to provide a right bond with surrounding concrete.

DEGREE: One-360th part of the circumference of a circle.

DELAMINATION: The separation of two surfaces.

DETAIL DRAWING: A drawing that shows individual parts of a structure.

DEWPOINT: The point at which the temperature becomes oversaturated with moisture and the moisture condenses.

DIAMETER: A chord that passes through the center of a circle.

DIRECT LIGHTING: A lighting system that directs all of its light downward.

DOOR BUCK: A part of a framed wall that consists of studs, cripples, and headers; a rough door frame.

DOOR CASING: A piece of trim used to cover the intersection of the jamb and the wall.

DOOR JAMB: The vertical and horizontal pieces of wood that are placed in rough door openings.

DOORSTOP: A thin strip of wood nailed to the door jamb to keep the door from swinging through.

DOUBLE COURSE: A course of shingles placed directly over an existing course.

DOUBLE JOIST: The spiking of two joists together.

DOVETAIL JOINT: An interlocking joint made by cutting two boards to fit each other.

DOWEL: A small wooden pin used to strengthen a joint.

DOWNSPOUT: A vertical pipe used to carry rainwater from the roof to ground level.

DRAIN TILE: A pipe used to carry ground water to a storm sewer or dry stream bed.

DRAWER GUIDE: A guide that directs the movement of a drawer.

DRIP EDGE: A strip of metal that is placed along the rake edge and eave line.

DRIP NOTCH: A small groove on the underside of a window sill which prevents water from running back under the sill.

DROPCLOTH: A large cloth spread over an area to keep debris or paint from falling on it.

DROP SIDING: Exterior sidewall covering usually ¾ inch thick and 6 inches wide, with tongued-and-grooved or shiplapped joints.

DRYWALL: A type of interior wall finish that is placed without the use of water.

DUCT: A round or rectangular pipe that is used for distributing warm or cool air.

DUPLEX OUTLET: An electrical wall outlet that has two plug receptacles.

# E

EASEMENT: An acquired right for someone other than the land-owner to have access to a piece of land.

EAVES: A part of the roof that projects over the sidewalls.

EAVE FLASHING: Flashing that extends from the eave line to a point inside the wall frame.

EAVE VENT: A vent placed in the soffit.

EDGER: A concrete finishing tool used to produce a smooth, rounded corner.

EFFLORESCENCE: A white powder that forms on a brick or stone wall.

EFFLUENT: A liquid sludge that is discharged from a septic tank or oxidation pond.

ELBOW: A plumbing fitting used to make a 90° turn.

ELEVATION: A modified orthographic drawing that shows one side of an object.

EMBOSSING: An ornamental design that is raised above a surface.

EMULSION: A mixture of liquids which are insoluble in one another.

ENAMEL: A type of paint in which the vehicle is a drying oil or combination of drying oil and resin.

ENGLISH BOND: A masonry bond in which one course is composed entirely of headers and the next course entirely of stretches.

EPOXY: A type of adhesive with excellent adhesion properties.

EXCAVATE: To remove earth below grade level.

EXPANDED METAL LATH: Sheets of metal with small openings that are used to lock plaster in place.

EXPANSION JOINT: A space between members that allows for possible expansion and contraction.

EXTERNAL FLASHING: Visible flashing placed on the exterior of a building.

EXTRUSION: The forming of shapes by pushing hot or cold metal thorugh a shaped die.

# F

FABRICATE: To build or construct.

FACE-NAILING: Nailing that is perpendicular to a surface.

FACE PLY: The outer piece of plywood.

FASCIA: A board placed at the lower end of the rafter tails, and used by itself or combined with moldings.

FEATHERING: Blending of the edges of a new surface into an old surface.

FELT: A mat of organic fibers: a term that is sometimes applied to underlayment.

FIBERBOARD: A type of building material made from wood or plant fibers.

FIELD CUT: A cut that is made at the job site.

FILL: Soil or loose rock that is used to raise a grade.

FILL COAT: The first coat of paint, varnish, or similar material.

FILLER STRIP: A small strip of wood, rubber, plastic, or metal that is placed between two materials.

FILLET: Rounded interior corner.

FINGER JOINT: A type of wood joint whose intersecting pieces have interlocking fingers.

FINISH COAT: The last coat of plaster, stucco, or paint.

FINISH NAIL: A small-headed type of nail used for millwork.

FIREBOX: The combustion chamber of the fireplace.

FIRE-RATED: A building material designed to resist standard fire tests.

FIRE RESISTANCE: The ability of a material to withstand fire or give protection from it.

FISHMOUTH: A wrinkle in building paper.

FISH TAPE: A flexible wire that is placed in small openings to pull wire or remove objects.

FISSURE: A narrow opening.

FIXTURE UNIT: A mathematical factor that is assigned to each plumbing fixture for determining pipe size.

FLANGE: A projection from an object that is used for attachments.

FLARING BLOCK: A device used to flare the ends of copper pipe.

FLASHING: A special type of material that is used to cover open joints to make them waterproof.

FLITCH BEAM: A built-up beam constructed by placing a metal insert between two wooden members.

FLOAT COAT: In plastering, the second of three coats of plaster.

FLOATING ANGLE JOINT: A joint formed at an inside corner by two intersecting pieces of gypsumboard, one board being nailed, the other not.

FLOOR FRAME: The composite parts that make up a floor frame.

FLOOR JOIST: A part of the floor frame; a common joist.

FLOOR SEAL: A type of finish used to fill the pores in a hardwood floor.

FLUE: The open area in the chimney that allows for the passage of air and gases.

FLUSH: Even with adjacent surfaces.

FLUSH DOOR: A door that has two flat surfaces.

FLUX: In soldering, a substance which coats a joint and prevents oxidation.

FOOTING: An enlargement of the lower portion of the foundation that spreads and transmits the load of the superstructure to a bearing surface.

FOUNDATION WALL: A load-transmitting element placed beneath the floor frame.

FRIEZE: A horizontal band that is placed at or near the top of a wall.

FROST HEAVE: An upthrust of ground or pavement caused by freezing of moist soil.

FROSTLINE: The depth of frost penetration.

FURRDOWN: A drop in ceiling height, usually found over kitchen cabinets.

FURRING: Thin strips of wood that are used to level a surface.

FURROW: A narrow channel made in a soft substance.

FUSE: An electrical safety device used to interrupt a circuit when the current is excessive.

# G

GABLE END STUDS: Vertical framing members placed in the gable end of a building.

GAUGE BLOCK: A small block used to align a particular construction member.

GALVANIZING: The dipping of iron or steel into molten zinc; this process prevents rusting.

GASKET: A flat washer-type object used to prevent water leakage.

GENERAL CONTRACTOR: The principal contractor.

GIRDER: A main supporting element of the floor frame.

GLAZING: The placing of glass in an opening.

GLAZING COMPOUND: A substance used for setting window panes.

GLUE BLOCK: A small block that is glued in place to reinforce a right-angle butt joint.

GLUE LINE: The intersection of two surfaces that have been bonded together.

GRADE: The slope, elevation, or face of the ground.

GRADE BEAM: A horizontal foundation member.

GRADING: The alteration of the ground by a series of cuts and fills.

GRAIN: The direction, size, quality, arrangement, or spacing of wood fibers.

GRAVEL: A granular material that can be retained in a No. 4 sieve and is a result of the disintegration of rock.

GRAVEL STOP: A strip of metal that is used to keep the gravel around the edge of a built-up roof.

GRID: The network of cross-tees and main runner in a suspended ceiling.

GROUND: (1) A thin strip of wood used in a plastered wall to keep it straight; the strip serves as a nailing base for interior trim; (2) a conducting connection between an electrical circuit or equipment and the earth, or something that takes the place of earth.

GROUNDING WIRE: A conducting connection between an electrical circuit and the earth.

GROUND WATER: Water in the soil that is found below the standing water level.

GROUT: A fluid mixture of portland cement, sand, and water that is used to fill joints of masonry and tile.

GROUT LINE: The space between two adjacent pieces of ceramic tile.

GUTTER: A trough that is used to collect rainwater from the eave line.

GYPSUM LATH: A plaster base in sheet form made from gypsum faced and backed with paper.

GYPSUM WALLBOARD: A type of wallboard constructed from gypsum faced with paper.

GYPSUM WALLBOARD: A type of wallboard constructed from gypsum faced with paper.

# H

HAND-TAMP: A process in which the earth is compacted.

HANGER WIRE: Wire used to support various building materials.

HANGING STRIP: A strip of wood that serves as a nailing base.

HARDBOARD: A board that is composed of wood fibers joined together under pressure.

HAWKBILL KNIFE: A knife with a curved plate that is used for cutting flooring and gypsum wallboard.

HEADER: In carpentry, a wood lintel or beam that has been placed at a right angle to floor or ceiling joist.

HEADER COURSE: A course of masonry units placed with the long dimension perpendicular to the wall frame.

HEADER JOIST: A wooden structural member that is placed at right angles to the joist.

HEAD JAMB: The horizontal top member of a door or window frame.

HEADLAP: A portion of a shingle that is covered by another shingle.

HEAD JOINT: The vertical joint between two masonry units.

HEARTH: The floor of a fireplace or that portion of the fireplace directly in front of the firebox.

HEARTWOOD: Wood extending from the pith to the sapwood.

HEATER: A heating unit.

HEAT SEAMING: The process in which a hot iron is passed over seaming tape so that two pieces of carpet may be joined together.

HIP: The angle formed by the intersection of two sloping roofs.

HIP JACK RAFTER: A short rafter that extends from the plate to the hip rafter.

HIP ROOF: A roof inclined from all four sides.

HIP UNIT: An individual shingle used to cover the hip line on an asphalt-shingled roof.

HOLD-DOWN CLIP: A small clip that fits over a cross-tee and is used to hold down acoustical ceiling panels.

HONEYCOMBS: Voids in concrete.

HOPPER: A funnel device used to discharge a substance through a chute in the bottom.

HOUSE DRAIN: The main lower horizontal pipe that receives the discharge of soil and waste stacks.

HOUSE SEWER: The lowest part of a plumbing system that extends from the building to the main sewer.

HUB: A small stake used to locate a corner of a building on the building site.

## I

I-BEAM: A steel beam whose cross-section resembles a letter I.

IMPREGNATION: The placing of preservatives in the cell structure of timbers.

INSULATION: A material used for the reduction of heat gain and heat loss.

INSULATING GLASS: Glass manufactured in two thicknesses with a dead air space placed between them.

INTERLAYMENT: 30-pound asphalt-saturated felt that is placed between courses of wood shakes.

INTERNAL FLASHING: Flashing placed inside a wall, and used to divert water to the exterior.

ISOMETRIC: A drawing that shows the dimensions of height, width, and depth.

## J

JACK RAFTER: A short rafter that spans from the top plate to a hip rafter, or from a valley rafter to a ridge.

JAMB: The exposed interior lining of an opening.

JAMB BLOCK: A concrete block that forms a door jamb.

JITTERBUG: A tool used in the finishing of concrete to place large aggregate slightly below the finished surface.

JOINT COMPOUND: A puttylike substance used in the finishing of gypsum wallboard.

JOINTER: A concrete finishing tool used to score freshly placed concrete.

JOIST: A structural member used to support floor and ceiling loads.

JOIST HANGER: A metal device used to support and secure a joist to a girder.

JUTE: A type of fiber used in making rope.

## K

KERF: A groove in a board, made by a saw blade.

KEY: (1) A protruding piece of metal, wood, or concrete that is used to hold one or both parts of a joint together; (2) an area that has been cut so that plaster or mortar can be inserted into the joint.

KILN-DRIED LUMBER: Lumber that has been placed in a kiln to extract excessive moisture.

KNEE KICKER: An instrument used to stretch carpet.

KNIFING: The process of making relief cuts in the placement of resilient flooring.

KNOB: A round projecting handle.

KNOT: An irregular lump formed at the intersection of a limb and the trunk of a tree.

KRAFT PAPER: A type of strong brown paper.

# L

LADDER: A wood frame constructed to resemble a ladder, and used to support a soffit.

LALLY COLUMN: A cylindrically shaped steel member used as a girder or beam and sometimes filled with concrete.

LAMINATION: The joining of several layers of paper, textile, or veneer with a synthetic resin.

LAP BOND: A type of masonry bond in which the individual masonry units overlap each other on each succeeding course.

LAP JOINT: The overlapping of two pieces of wood.

LATH: A base for plastering, usually expanded metal or wood.

LATERAL BRACING: Bracing used on the side of a structure.

LATEX: A type of paint with a water base.

LEAD: A masonry wall rocked back on successive courses for holding a mason's line.

LEADER: A vertical pipe connected to a gutter and used to carry rainwater from it to the ground or drain.

LEDGER: A horizontal strip of wood that supports a horizontal structural member.

LEGEND: A statement of the meaning of symbols used on a drawing.

LEVEL: A precision instrument used for locating points in a horizontal plane.

LEVELING COAT: A thin coat of plaster used to level a given surface.

LEVELING ROD: A rod used in surveying that is marked in feet and fractions of feet.

LIGHT CONSTRUCTION: Building that is primarily residential in nature, but can include small commercial structures.

LINE WIRE: Wire placed over a wall frame and used to support building paper in the application of stucco.

LINOLEUM: A floor covering made of linseed oil and ground cork, being oxidized upon a fabric base.

LINTEL: A horizontal structural member placed over an opening; a header.

LIVE LOAD: A variable load that may be placed upon a structure.

LOADER: A piece of equipment that loads building materials onto or into something.

LOAM: An organic topsoil that is capable of growing grass and plants.

LOCK BLOCK: A block of wood placed between the veneers of a hollow-core door, and to which the lock is fitted.

LOCKNUT: A second nut that is used to prevent the first nut from becoming unscrewed.

LOOKOUT: The supporting agency for the soffit; nailed to the rafter tail and frame wall.

LOOM: A type of machine on which thread or yarn is woven into fabric.

LOOP VENT: A pipe that connects to the stack vent and is used to prevent back siphonage.

LOOSE-LAY: The application of finish flooring without securing it to the subfloor or to tackless strips.

LOUVER: An opening used for ventilating closed areas.

# M

MAIN RUNNER: A piece of the grid work in a suspended ceiling.

MALLET: A small hammer made usually of wood.

MANDREL: A piece of wood or metal (1) placed inside a cylindrical object so that it may be driven, or (2) used to enlarge or remove distortions.

MASONRY: A type of construction made of shaped or molded units.

MASON'S LINE: A stretched string that serves as a guide in placing masonry units.

MASTIC: A natural or synthetic material used to lay tile.

MECHANICAL KEYING: The physical attachment of plaster to a base or lath.

MILL FILE: A tool of hardened steel with cutting ridges; used to shape metal or smooth the edge of plastic laminate.

MILLWORK: Building materials that are made of finished wood and are manufactured or fabricated in a shop.

MITER: The joint formed by the intersection of two abutting pieces.

MOLDING: Thin strips of wood that are used for esthetic purposes, or to cover various joints.

MONOLITHIC CONCRETE: Concrete cast with no joints.

MORTAR: A mixture of sand, water, and cement; it is used as a bonding agent for masonry units.

MORTAR BED: A horizontal layer of mortar placed on a masonry unit.

MORTAR BOND: The adhesion of a masonry unit to the mortar.

MORTGAGE: A legal document used to hold property as security for a debt.

MORTISE: A cutout in one framing member that will receive the tenon of another framing member.

MULLION: The structural member between two windows.

# N

NAIL POP: The protrusion of a nail from a finished surface.

NAILED OFF: A term used to describe material that has been properly nailed to a frame.

NAILING STRIP: A wooden strip that serves as a nailing base for intersecting members.

NEOPRENE: Oil-resistant synthetic rubber.

NEWEL: The supporting post that is found at the foot of a staircase.

NOMINAL SIZE: The size of lumber before it is planed.

NOSING: The projecting part of a window sill or stair tread.

NOTCH: A small cut.

NOTCHED TROWEL: A trowel with a series of notches along its back edge.

# O

OAKUM: Hemp fiber used for caulking joints and seams.

ON CENTER (O.C.): The measurement from the center of one structural member to the center of another.

OPEN VALLEY: A valley-flashing technique in which the shingles of the intersecting roof slopes leave an open space covered by metal flashing.

OUTLET: A distribution source for electrical current.

OVERHANG: The projection of a roof beyond a vertical wall.

OVERTURNING: The radical tipping of a complete structure, caused by forces destroying stable equilibrium.

OVOLO: A piece of convex molding.

# P

PANEL BOX: A metal box in which electrical fuses and switches are located.

PAN SCREWS: Small screws used to fasten sheet metal together.

PARGE: To coat a surface with plastic.

PARTICLEBOARD: A composition board made from wood chips bonded together with an adhesive.

PARTITION: A wall that separates spaces within any one story of a building.

PATIO: A surfaced area that is an extension of the residence, though partially or entirely enclosed by it.

PATTERN BOND: A pattern formed by masonry units and the mortar joints.

PENNY: A measure of nail length, abbreviated by the letter "d."

PERCOLATION: The ability of water to move through the void spaces of earth.

PERFORATED: Pierced with holes.

PERIMETER: The distance around a particular object; the boundary.

PERLITE: An aggregate made from volcanic minerals.

PERSPECTIVE: A drawing used to illustrate design principles.

PHILLIPS HEAD: A screw that has a head with a cross-shaped indentation.

PIER: A support for beams and girders, and usually constructed of concrete or masonry.

PIGMENT: A substance that gives color to paint.

PILASTER: A column projection that is part of the foundation wall.

PILE: A structural member or post driven into the earth to help support a structure.

PILE CAP: A structural member placed over a pile and used to transmit loads to it.

PITCH: The degree of inclination of a roof.

PLANCIER: The underside of an eave, sometimes called the soffit.

PLANK FLOORING: Flooring constructed from a broad board, and usually more than 1 inch thick.

PLASTER: A mixture of lime, cement, and sand, which is used as both an exterior and interior wall finish.

PLASTICIZER: A material that, when added to cement paste, mortar, or a concrete mix, will increase its plasticity.

PLASTIC LAMINATE: A type of material usually used for topping cabinets.

PLATE: A flat horizontal member that is connected to both the top and bottom of studs.

PLATFORM FRAME CONSTRUCTION: A framing technique in which the floor joist of each story rests on the top plate of the story below.

PLENUM: A chamber or area that forms part of an air conditioning system.

PLUMB: To adjust in exact vertical alignment, using a plumbline.

PLUMB BOB: A weight pointed on one end hung from a string, and used to check vertical alignment.

PLUMB CUT: A cut made on a structural member that is perpendicular to the horizon or floor.

PLY-CLIP: A small metal clip placed between rafters when plywood is used for roof sheathing.

PLYWOOD: A wood product fabricated from several layers of veneer that have been bonded together.

POLYETHYLENE: An organic compound fabricated into large protective sheets; used for vapor barriers, and for covering concrete during the curing period.

POLYSTYRENE RESIN: A synthetic resin.

PONDING: A method of curing concrete by flooding freshly placed concrete with water.

PORCH: A roofed structure placed at an entrance to a building.

PORES: Wood cells with open ends; small cavities in soils.

PORTLAND CEMENT: A hydraulic cement used in construction that consists of silica, lime, and aluminum thoroughly mixed.

POST: A vertical member used to support loads.

POWER STRETCHER: A device used to stretch carpet.

PRECAST: Concrete shapes that are formed before they are placed in a structure.

PREFINISHED: A building material purchased with the finish already applied.

PREGROUTED: A term referring to ceramic tile that has its joints grouted at the factory.

PRIMER: The first coat of paint in a two-coat application.

PRIMING: The first coat of paint applied to a surface for preserving wood.

PURLIN: A horizontal member that is used to brace rafters.

PURLIN STUD: A roof-bracing member that spans from rafter to load-bearing wall.

PUTTY: A doughlike substance used for setting window panes and filling imperfections in wood.

# Q

QUARTER-ROUND: A small piece of molding that has a quarter-circle profile.

# R

RABBET: A rectangular groove cut in the edge of a board.

RACKED BACK: The placing of masonry units so that each successive course is shorter than the previously placed one.

RAFTER: A structural member used to support roof loads.

RAFTER TAIL: The portion of a rafter that extends past the frame wall.

RAIL: The horizontal member of the framework of a sash, door, or any paneled assembly.

RAKE EDGE: The inclined edge of a gable roof.

RAKED JOINT: A mortar joint that has been shaped with a square-edged tool.

RASP: A slender steel instrument with rows of teeth that is used to smooth the edges of a board.

RATCHET: A tool, such as a socket wrench, with a set of teeth so slated as to allow movement in only one direction.

REAM: To enlarge or smooth the inside of a pipe.

REGISTER: The end of a duct, usually covered with grillwork.

REINFORCEMENT BARS: Metal bars that are embedded in concrete.

REINFORCING TAPE: A paper tape placed over the intersection of two gypsumboard panels.

RESAWING: The process of cutting or ripping stock after the initial cut.

RESILIENCE: The ability of a material to return to its original shape after temporary deformation.

RESIN: A solid or semisolid organic substance excreted from various plants and trees.

RESORCINOL: A form of synthetic resin.

RETURN AIR PASSAGE: An area that allows the return of cold air back to the heating unit.

RIBBON: A narrow board that is let into the studding to help support the joist; or a thin strip of mastic.

RIDGE: The horizontal junction of two sloping rods.

RIDGE UNIT: An individual shingle unit used to finish a ridge line.

RIFFLE: A groove.

RIGID INSULATION: Sheet-form insulation that is manufactured from cellulose fibers, and is nailed over the frame wall.

RING-THREAD SHANK: A type of nail with a series of grooves formed around the shank.

RISE: The upward slope of a roof.

RISER: The vertical board of a stair that is placed under the nosing.

ROLL ROOFING: A type of asphalt roofing, manufactured in 36-inch-wide rolls.

ROLL VALLEY: A type of valley flashing made from galvanized sheet metal that has not been center-crimped.

ROOF PITCH: The degree of inclination of a roof.

ROOF SHEATHING: Material that is fastened to the rafter and on which the roof covering is laid.

ROOF VENT: A space used for ventilation purposes that protrudes above the roofing membrane.

ROSIN-SIZED SHEATING PAPER: Part of a built-up roof; the first piece of a built-up roof placed over a wood board deck.

ROTATION: The process in which a building is rotated on its foundation due to strong wind forces.

ROUGHING-IN: Preliminary work not considered to be finish work.

ROUGH OPENING: An unfinished opening.

ROUGH-SAWN: Lumber that has been ripped, but not surfaced.

ROUND: An arch of semicircular nature.

ROUTER: A hand-held machine used to finish the edges of wood and plastic laminate, to cut various joints.

RUBBLE: Rough stones of irregular shape and size.

RUN: In roof construction, the horizontal distance between the ridge and the face of a wall; in stair construction, the horizontal distance covered by the stairs.

RUNNING BOND: A masonry bond with the head joint of every other course in alignement.

# S

SADDLE: Two sloping surfaces placed directly behind a chimney or other vertical surface.

SAND: A fine aggregate used in concrete mix; a granular material that can pass through a No. 4 sieve, but is retained in a No. 200 sieve.

SANDING SEALER: A hard finish used to fill the pores of wood without hiding its grain.

SASH: The framework that holds glass in a window.

SCAB: A small board nailed to two other boards for reinforcement.

SCARIFYING: The making of small scratches or cuts in a surface.

SCHEDULE: A collection of organized notes.

SCORE: A thin cut or scratch made in a surface.

SCRATCH COAT: The first coat of stucco or plaster, usually cross-raked to assure a firm bond.

SCREED: An auxiliary aid that is used to aid in striking concrete to grade.

SCREED CHAIR: Metal devices used to support a screed.

SCRIBE: To cut a member irregularly so it will fit properly.

SEALANT: A sealing compound that prevents liquid from entering a finished surface.

SEALER: A clear finishing material applied to a wood surface to keep subsequent coats of varnish from seeping into the wood.

SEALING STRIP: A strip of adhesive used to waterproof the intersection of two surfaces.

SEPTIC TANK: A part of an individual sewage system, used to collect waste until decomposition can occur.

SERVICE DROP: A line bringing electricity into a building.

SETBACK: An imaginary line established by law or deed restriction, fixing the distance from the property line to the face of a building.

SETTING BLOCK: A small block of wood or rubber placed in the glass groove of the bottom rail of an insulating glass sash.

SHAKE: Hand-split shingle.

SHEAR: The failure of a structural member where the lines of force are perpendicular to the member.

SHEATING: Exterior wall or roof covering, usually plywood or nominal 1 x 6 inch boards.

SHEATHING PAPER: A material placed on a wall frame, roof sheathing, or subfloor to provent air infiltration.

SHELLAC: A clear finish made by dissolving lac in alcohol.

SHIM: A thin strip of wood that is used to level framing members.

SHINGLE STRIP: A small piece of trim that is nailed to the fascia.

SHIPLAPPED LUMBER: A type of lumber that has rabbeted edges.

SIDE JAMB: The exposed lining on the side of a window or door opening.

SIDING: The structural exterior covering of a frame wall.

SILL: A horizontal structural member positioned on top of a foundation wall. In a door frame, the wall is a horizontal board placed at the bottom of the frame.

SINGLE COURSE: An individual course of shingles or masonry units.

SIPHON: A tubular device that utilizes atmospheric pressure to effect or control the flow of water.

SITE: A geographical location of a project that is usually defined by legal boundaries.

SLAB: A flat horizontal section of molded concrete.

SLEEPER: A wood strip that is placed over concrete to receive a finished wood floor.

SLIDING DOORS: Doors that operate on an overhead track and that can be slid from side to side.

SLIP SHEET: A sheet of paper placed between a substrate and a piece of plastic laminate.

SLOP SINK: A low, deep sink.

SLOPE: The inclination of a surface from the horizontal.

SLUMP TEST: A test used to measure the consistency of concrete.

SOFFIT: A board attached to the underside of the rafter tail.

SOIL BRANCH: A horizontal portion of the plumbing system that receives the direct discharge of water closets, with or without additional plumbing fixtures.

SOIL STACK: A vertical portion of the plumbing system that receives the discharge of water closets, urinals, or fixtures having similar functions.

SOLDERING: A technique used to join two pieces of pipe by application of a molten alloy.

SOLVENT: The distance between two structural supports.

SPACED SHEATHING: Sheathing placed at regular intervals, usually 10 inches on center.

SPACE HEATER: A small heating unit usually operated on natural gas, butane, or electricity.

SPACER BLOCK: Small blocks used to effect a given distance between two or more objects.

SPACKLING COMPOUND: A puttylike substance used for filling small holes and cracks in gypsum wallboard.

SPADE: A flat hand tool with a long handle; a shovel.

SPAN: The horizontal distance between structural supports.

SPECIFICATION: A written document that stipulates the kind, quality, and quantity of materials and workmanship required for a construction project.

SPIGOT: The end of a piece of pipe that fits into the large end of another pipe; a faucet.

SPLICE: The joining together of two similar materials.

SPLIT: A small crack.

SPOT: The placing of joint compound over the nailheads in gypsum wallboard construction.

SPREADER: A metal or wooden device used to spread wall forms to the proper width.

SPRINKLING: A concrete curing technique in which a fine water spray is directed over the concrete surface.

SQUARE: A unit of measure —namely, 100 square feet.

STACK: A vertical arrangement of piping.

STACK BOND: A masonry bond in which all head joints are in alignment.

STACK VENT: A vertical extension of a soil or waste pipe.

STAIN: Any liquid that can penetrate the pores of wood and change its color.

STAPLE: A metal fastener in the shape of a U.

STARTER STRIP: A strip of composition roofing that is placed along the eave line of a roof.

STEP FLASHING: Flashing placed against a wall or chimney where the masonry joint entered by the flashing must be changed at regular intervals.

STEPPED FOOTINGS: Footings placed at different depths.

STICKING: The shaping of molding in a mill.

STIFFBACK: Two framing members spiked together in the shape of an L, and used to level and strengthen the ceiling joist.

STILE: A vertical framing member in a panel door.

STIRRUP: A U-shaped piece of reinforcement that is placed perpendicular to a longitudinal reinforcement.

STOOL: A narrow piece of trim used to trim out a window, being placed at its bottom.

STORM-SASH: An extra window placed on the outside of an existing window.

STRAIGHT-NAILING: A technique whereby a nail is driven in a position perpendicular to the framing member; face-nailing.

STRAPPING: A series of horizontal boards placed over an existing wall; furring strips.

STRATA: A layer of homogeneous rock.

STRETCHER BLOCK: A masonry unit that has its longest dimension placed parallel to the wall frame.

STRETCHER COURSE: A row of masonry units positioned with the long side parallel to the wall frame.

STRIATED: Fine linear markings on a surface.

STRIKEOFF ROD: A tool used to smooth off excess plaster or concrete.

STRINGER: The sloping member of a stair that stretches from one level to another, on which the treads rests.

STRIP FLOORING: A type of wood flooring that is manufactured in long narrow strips.

STRIKING: The removing of excess mortar, plaster, or concrete from a given area.

STRUCTURAL BOND: The technique used to lock or tie individual masonry units together.

STUCCO: A mixture of portland cement or masonry cement, lime, sand, and water.

STUCCO NETTING: Metal reinforcement used in stucco work.

STUDS: Vertical framing members in a wall.

SUBSTRATE: The base to which plastic laminate is bonded.

SUCTION BOND: A bond created when plaster is parged over a concrete or concrete-block wall.

SUPERSTRUCTURE: The part of a building that is supported by the foundation system.

SUSPENDED CEILING: A ceiling system that is supported by hanging wires attached to overhead structural framing.

SWITCH: A mechanical device used to break the flow of electric current.

SWITCH BOX: A box where electrical circuits are brought to be connected or disconnected.

SYMBOL: A mark, character, or figure that represents the name of something.

# T

TACKLESS STRIP: A thin strip of wood fitted with slanting pins; used to secure carpet.

TAP: The place where the end of a wire is joined to the middle of another wire.

TAPER: A uniform decrease in diameter of a round or rectangular object.

TARPAULIN: A waterproofed canvas.

TEE: A structural framing member fabricated from three full-length studs and three spacer blocks; a T-shaped pipe fitting.

TEMPERED: Having the elements of a product properly treated, as thoroughly mixed mortar or concrete; or case-hardened metal.

TEMPLATE: A pattern used to make layouts; usually wood, paper, cardboard, or metal.

TENON: A piece of wood that is cut with a projection to fit into a mortise.

TENSILE STRAIN: A stretching or pulling in a longitudinal direction.

TERMITE BARRIER: A barrier, visible or invisible, used to prevent termites from entering a structure.

TERMITE SHIELD: A metal shield of noncorrodible metal that is used to prevent the passage of termites.

TERRAZZO: A flooring that is constructed by placing marble chips in a concrete slab.

THERMOPLASTIC: An adhesive that becomes soft when heated.

THERMOSTAT: An automatic device used to regulate the temperature of a room.

THIN-SET: A skim coat of grout used to bond ceramic tile.

THRESHOLD: A member used to close the space beneath the bottom of a door.

TIE ROD: A steel rod that is used to hold structural parts together.

TILE: A unit of cured clay used for roofing, or for wall or floor covering.

TOENAILING: Driving a nail at an angle so that it penetrates two intersecting boards.

TOE SPACE: The recessed space at the bottom of a kitchen cabinet or other built-in unit.

TONGUED-AND-GROOVED: A type of sheathing in which one edge has a projecting tongue.

TOOLED JOINT: A masonry joint formed by compressing the mortar into the joint.

TOP LAP: The distance one shingle overlaps another.

TOP PLATE: A horizontal member that is nailed to the top of the studding.

TRACTOR: A piece of heavy equipment used for towing or pushing heavy loads.

TRANSIT: An instrument used by surveyors; has four basic parts: telescope, spirit level, vernier, and tripod.

TRANSLATION: Process by which a building is moved.

TRAP: A device used to prevent sewer gases from escaping through a plumbing fixture.

TREAD: The horizontal board in a stair; the step.

TRENCH: An excavation made for the laying of pipe and electrical lines; a ditch.

TRIG: A supporting agent for a mason's line.

TRIM: Finish materials in a building, such as window casings, door casings, ceiling molding, and base.

TRIM CUT: A sloping cut made on the end of a ceiling joist; allows rod sheathing to be placed flush with the rafters.

TRIMMER: A part of the floor frame that is placed around an opening in the floor.

TRIPOD: A three-leg stand used to support a level or transit.

TROWEL: A flat steel tool that is used to spread mortar and plaster.

TROWELED JOINT: A mortar joint cut flush with the masonry units.

TRUSS: An assemblage of structural members fastened together to form a support for the roof sheathing.

T-SQUARE: An instrument used in the drawing of horizontal lines.

TUCKED: The placing of carpet between the tackless strip and the wall.

# U

UNDERLAYMENT: Asphalt-saturated felt that is place over roof sheathing.

UREA: A colorless crystalline compound.

# V

VALLEY: The intersection of two sloping roofs.

VALLEY FLASHING: Pieces of sheet metal or roll roofing that are placed in a valley to prevent water penetration.

VALLEY RAFTER: The diagonal rafter placed at the intersection of two sloping roofs.

VAPOR BARRIER: A material used to prevent the passage of vapor or moisture into walls and floors.

VARNISH: A clear finish made from drying oil and resin.

VEHICLE: The liquid portiom of a finishing material, consisting of binders and thinners.

VENEERED CONSTRUCTION: A method of construction by which brick or other facing material is applied to a frame wall.

VENT: A pipe or opening that is used to allow trapped gases to escape.

VENT STACK: A vertical pipe that carries foul gases from a building and prevents back siphonage.

VERMICULITE: A type of bulk insulation or an aggregate in acoustical plaster.

VITREOUS CHINA: A porcelain enamel fused on metal, used for most plumbing fixtures.

VITRIFIED CLAY: A substance used in the making of pipes.

VOIDS: Spaces between grains of sand or gravel.

# W

WALER: A horizontal member that is placed across the studding in concrete form construction.

WALL ANGLE MOLDING: L-shaped metal molding placed around the perimeter of a room to support the gridwork of a suspended ceiling.

WALL FRAME: The various parts of a wall fitted together into a skeleton form.

WALL TIE: A small piece of metal used to bind brick veneer to a wall frame.

WARPAGE: The twisting or distortion of a board from its true shape.

WASTE PIPE: Vertical portion of a plumbing system that carries only liquid waste that is free of fecal matter.

WATER CLOSET: A flushable toilet.

WATER HAMMER: A rattling of plumbing pipes caused by the sudden checking of water.

WATER SEAL: Water contained in a trap to prevent the passage of gases from one side of the trap to another.

WEATHER EXPOSURE: The amount of exposed shingle.

WEATHERSTRIP: Narrow strip of material that is installed around doors and windows.

WEB: The vertical wall that connects the face shells of a hollow concrete masonry unit; the interior framing members of a truss.

WEDGE BLOCK: A piece of wood or metal that is tapered to a thin edge; used to adjust elevation.

WEEP HOLE: Small openings left in masonry walls to permit drainage and reduce pressures.

WIND BEAM: A horizontal roof framing member that is used to resist wind force.

WINDOW BUCK: A part of a framed wall consisting of studs, cripples, headers, and rough sill.

WIRE MESH: A series of longitudinal and transverse wires welded at right angles to each other; used to reinforce concrete.

WIRENUT: A fastener used to connect electrical wires.

WOOD FILLER: A heavily pigmented substance that is used to fill pores in open-grain woods.

WOOD SHAKE: A roof or wall covering that is manufactured from split-wood units.

WOOD SHINGLE: A roof or wall covering that is manufactured from sawn-wood units.

WOVEN VALLEY: A valley in which the different courses of shingles are alternately overlapped.

WYTHE: A single width of masonry.

# Y

YARD: A common term for cubic yard.

# Z

ZONING PERMIT: A permit issued by appropriate governmental agencies so a specific plot of land can be used for a specific purpose.

# Index

403